AIRLIFT TANKER

History of U.S. Airlift and Tanker Forces

TURNER PUBLISHING COMPANY
Paducah, Kentucky

TURNER PUBLISHING COMPANY
Publishers of America's History
412 Broadway, P.O. Box 3101
Paducah, KY 42002-3101
Phone: (502) 443-0121

"The Airlift/Tanker history book could not have been possible without the help of Air Force historians, especially John Leland at HQS Air Mobility Command who provided most of the USAF history documents and photos used in compiling this book. Airlift/Tanker folks have historically been poor at recording their exploits unlike the fighter/recce bunch who spent an hour or so in the cockpit each day and the rest of the day talking about it. "Trash Haulers and Texaco" have little of the glory but as long as wars are fought somebody will have to haul the bullets, beans, and blood. Flying into battle also beats walking and you can't do much without gas to get you there and back home."

Col. Melvin N. Ledbetter,
USAF (Ret.)

Airlift/Tanker Association Book Committee:
Col. Melvin N. Ledbetter, USAF (Ret.), Advisor
Collin Bakse, Author of Airlift/Tanker histories

Turner Publishing Company Staff:
Douglas W. Sikes, Publishing Consultant
Pamela G. Wood, Project Coordinator
Herbert C. Banks II, Designer

Library of Congress Catalog Card No. 95-70461
ISBN: 1-56311-125-X
Limited Edition.
Printed in the U.S.A.

Additional copies may be purchased directly from Turner Publishing Company.

All photos unless otherwise noted are courtesy of the United States Air Force.

On the Cover: The YC-141B, with capabilities for in-flight refueling. Jan. 1977.

This Page: The C-130 shown in the foreground, is a model that has undergone much change. The C-141 is shown in the background.

Table of Contents

Publisher's Message

Dave Turner, President

Turner®

(502)443-0121
(502)443-0128

P.O. Box 3101
Paducah, KY 42002-3101

It is truly an awesome sight to see planes refueling in mid-air or a massive airlifter transporting troops and supplies to those areas of the world in need. The photographs alone capture the true spirit of America and her people's willingness to do what it takes to get the job done right. But we have combined photographs and historical text to bring to you one of the most interesting stories of the military – the story of the Airlift/Tanker Force and her crews. Whether they are refueling planes, transporting food to hungry refugees or airlifting troops to a secret mission to free hostages, the full story is told in the words of the men and women who flew these massive planes. This outstanding book would not have been possible without the help of those who "lived it" and those who "were there," and we must acknowledge those individuals who were so instrumental in this project's success.

A special thanks goes to John Leland, Ph.D and the staff of the USAF Archives at Scott AFB, IL for support in providing us with many outstanding photographs and articles. We also thank the many Airlift/Tanker veterans who willingly sent us their personal photographs and stories for publication.

We are indebted to Mr. Collin Bakse for writing the excellent historical overiew and providing promotional support and are especially appreciative of Mr. Hal Lansing for his introduction of the project to the TA. Gen. Alfred Hansen, USAF (Ret.) deserves recognition for his direction, foresight, and initiative to commence on such a project.

Our deepest appreciation goes to Col. Melvin N. Ledbetter, USAF (Ret.) whose diligence, persistence, and untiring devotion to the Airlift/Tanker Association ensured that the book you now hold was the best that it could be by telling the story of the greatest Airlift/Tanker Force in the world.

Dave Turner,
President

Publishers of America's History

Introduction

THE AIRLIFT/TANKER ASSOCIATION
P.O. Box 27918 Washington, D.C. 20038-7786

25 August 1995

Throughout the history of America Mobility Forces have played an important role in meeting our national objective. Both during military operations and natural disasters Mobility forces have stood ready to accomplish any mission required of them. Offered the unique opportunity for a military force to be both destructive and constructive, Mobility members are known worldwide for their professionalism and "can do" attitude. They have been responsible for saving countless lives both on and off the battlefield.

This book captures the accomplishments and people who have either made or contributed to the making of mobility history. It is the deeds of these people who have built our heritage and carved the exploits of mobility into the history of our great nation.

My personal thanks to all who made it possible and especially Mel Ledbetter, Colonel USAF, (Retired) for his dedication.

Al Hansen

Gen. Alfred G. Hansen,
USAF (Retired)

Gen. (Ret.) Alfred G. Hansen

"We have learned and must not forget that from now on, air transport is an essential element of air power, in fact, of national power."

General H.H. "Hap" Arnold -1945

Douglas C-47 "Skytrains" 12th Air Force Troop Carrier Wing, loaded with paratroopers on their way for the invasion of southern France, 15 August 1944.

Airlift History

Early Military Air Transport

Supplying troops across vast distances has been critical to the success or failure of military campaigns throughout history. Mountains, rivers, and opposing armies have always hampered resupply efforts, but the appearance of manned hot air balloon flights in the late eighteenth century gave rise to thoughts of employing aerial transportation to overcome those difficulties. Napoleon, blocked by the British Royal Navy, dreamed of using balloons to transport his armies to England. During the siege of Paris in the Franco-Prussian War, Parisians flew balloons over the surrounding Prussian Army to maintain vital communications. Yet, flying remained somewhat unpredictable until the advent of the airplane in the twentieth century.

Aerial transportation has revolutionized modern warfare. Through airlift, it has become possible to move troops and supplies directly and rapidly into the battle zone. Nevertheless, the potential advantages of airlift were neither readily apparent to many military leaders nor initially feasible due to the technical limitations of early flying machines. As a result, the development of the United States' military airlift system followed an evolutionary course.

American Pre-World War II Airlift

Aviation officers in the Army Signal Corps did not fail to note the Airplane's commercial and military uses. They were well aware of its wartime application as demonstrated by the Europeans for reconnaissance, aerial combat, target acquisition, delivery of explosives and incendiaries, and the rapid transportation of messages and high-ranking officers. During Brigadier General John J. Pershing's Punitive Expedition into Mexico in 1916, the military primarily used airplanes for observation, aerial photography, and for transporting mail and dispatches. The Army Signal Corp's planes lacked the power and stability to carry more useful loads.

After the United States entered the First World War, the Army Signal Corps began to employ airplanes to transport personnel, cargo, and wounded pilots.

The first recorded American demonstration of transporting troops by air occurred on September 7, 1918, when several airplanes carried 18 enlisted men from Chanute Field to nearby Champaign, Illinois. Overseas in France, American pilots occasionally transported senior officers and couriers and later dropped Allied propaganda leaflets over the German lines.

The first effort to resupply combat troops from the air took place during the American Expeditionary Forces' Argonne Forest offensive in October 1918. Airmen from the 50th Aero Squadron dropped supplies and aided in the rescue of 550 American soldiers, later referred to as the "Lost Battalion," after they had become surrounded by German troops.

There were others in World War I who dreamed of larger airlift operations. Although Brigadier General Billy Mitchell is remembered primarily for his advocacy of an independent air force, he conceived an ambitious plan which included airdropping the First American Infantry Division behind German lines.

After the war, Army fliers became preoccupied with creating a separate air force. Military air transportation received little consideration from commanders of either the Army Air Service or its successor, the Army Air Corps, created in 1926. Military airlift continued to evolve without sustained

Built in 1918, the Handley Page was one of America's first bombers. It could also carry several hundred pounds of cargo.

During World War I, airplanes played a major role in warfare for the first time. On occasion, airplanes transported high-ranking officers.

Curtiss C-46 Commando over the "First Ridge" of the Hump en route to China, along the routes of the India-China Division, ATC, India.

Search and Rescue crew members in front of the plane in which Capt. John Porter was shot down and killed five days later.

One of the most outstanding photographs taken in the China-Burma-India theatre is this striking aerial view of the Taj Mahal with a Curtiss C-46 "Commando" ATC flying overhead. 14 Oct. 1944.

attention, however. The Army employed a variety of aircraft for its transport needs, sometimes purchasing only one model of a specific airplane. By the late 1920s the Air Corps had started purchasing cargo airplanes, assigning them to the supply depots and Army air fields. When necessary, bombers still served as transports. Limited funding forced the Army aviators to concentrate on their wartime mission, which was attacking the enemy. Thus, while civilian air transport matured during this period and proved its viability, the Army's senior aviators remained mostly concerned with creating a separate air force and perfecting combat tactics. Accordingly, they spent much of their time demonstrating that airplanes could sink battleships, bomb targets with great accuracy, and fly distances of several hundred miles. The latter indirectly aided the air transport cause.

An organized military air transport service first appeared in June 1922, when the Army Air Service began scheduled passenger and cargo flights over the Model Airways, a nationwide air system sponsored by the government. A nationwide network of air routes and airfields was deemed essential to the nation's defense and the advancement of aviation, both commercial and military. Through Model Airways the government saved several thousand dollars in express railway freight charges; savings for government travelers using the airways versus commercial transportation were far greater.

Air Service officials pronounced the airways program a success and, in 1925, appointed a board of officers to prepare regulations governing the continued air transport service. Before this could be done, however, Congress reorganized the Army Air Service into the Army Air Corps in 1926. Along with the Air Mail and the Air Commerce Acts of 1925 and 1926, respectively, the Air Corps Act took government agencies out of aviation activities that private enterprise could provide. In the process, Congress disbanded the Model Airways, and regularly scheduled military air transport service disappeared.

Further development of America's military air transportation system proceeded slowly over the next several years. In part, this pace reflected the isolationist and antiwar atmosphere of the 1920s. These popular sentiments forced Army Air Corps fliers to emphasize the defensive capabilities of military aircraft. Small defense budgets forced the military establishment to reduce its planned airplane purchases and concentrate on low-cost measures. The lack of funds prevented Air Corps officials from prepositioning men and supplies at every flying field airplanes would use to defend the United States. Therefore, the Army Air Corps tried to prove that their limited number of combat airplanes, which were dispersed around the nation's borders, could rapidly concentrate on either coast to repel any invader. To respond quickly to a threatened area, the Army Air Corps needed aerial transportation to move the supplies and personnel supporting these planes at their new locations. The combat units could not afford to wait for their support elements to arrive by rail or road. The Air Service/Air Corps maneuvers of 1925 and 1927 essentially confirmed these views.

Following the 1927 maneuvers, Air Corps officials agreed with the recommendation of Lieutenant Colonel Clarence C. Culver, Commandant of the Air Corps Tactical School, that the next maneuver should demonstrate the feasibility of supplying deployed Air Corps troops by air. To this end, during the 1928 aerial operations exhibitions, 14 bombers airlifted 73, 721 pounds of equipment and personnel. This success led to the recommendation that transport planes be employed to facilitate supply and transportation requirements. The next three maneuvers increasingly employed air transports to move supplies and personnel, including the injured. During the 1930 maneuvers, Major Henry H. "Hap" Arnold, the provisional wing supply officer, used one Douglas C-1, three Fokker C-2As, and one Keystone LB-7 aircraft to move 36,548 pounds of cargo. That requirement was massive for its time; the cargo took 36 missions. During the 1931 maneuvers, air transports, a collection of some 48 planes of 9 different models, supplied food, medical, and personal equipment items for 1,400 troops as they covered a dozen states over a 15-day period. The maneuvers affirmed that Air Corps units could be sustained by air.

Accordingly, at the 1931 Army Engineering Supply Conference, Major Hugh J. Knerr, Chief of the Field Service Section of the Materiel Division, proposed the creation of a new air cargo system. Knerr believed the division and its four geographical air depots should each receive two cargo planes. The Army Air Corps subsequently purchased four Bellanca YIC-14s that had been adapted for cargo hauling. The Air Corps assigned one plane and one enlisted pilot to each of its four air depots, and early in 1932, Major Knerr inaugurated a local Air Transport Supply Service at each depot. At the Army Engineering Supply Conference that fall, Major Knerr and Lieutenant Colonel Albert L. Sneed, commander of the Fairfield Air Depot, pushed the concept of airlift further. Knerr explained the purpose behind the Air Transport Supply Service." If an Air Force is tied to rail heads and its services of supply dependent upon motor transportation, its mobility is that of the flat car and truck. The ideal situation is one wherein the Air Force is maintained and accomplishes all of its transportation by air."

C-46 line-up on Alert Feb. 9, 1945.

The site of a C-46 accident, Dec. 28, 1944, India.

World War II and Aid to Allies

With the German invasion of Poland in September 1939, the start of World War II brought great changes in American military air transportation. Rapid German victories and the defeat of France in 1940 left only Great Britain to face the Nazi menace in Western Europe. Protected by the English Channel and the valiant effort of the Royal Air Force (RAF) during the Battle of Britain, the British withstood German air attacks. However, the only means available to the British for striking directly at Nazi Germany was aerial bombardment. Desperately needing bombers to carry out their air assault on Germany, the British looked to the United States' aircraft industry as a source of supply, the administration of President Franklin D. Roosevelt supported Britain's efforts to obtain bombers.

Neutrality legislation passed by Congress during the 1930s prevented the United States from supplying weapons to belligerents, but the Roosevelt Administration secured congressional support to allow the British to purchase American-made munitions if they transported them in their own ships on a "cash and carry" basis. Soon, however, the British ran short of money to pay for the weapons, and President Roosevelt responded by obtaining the passage of the Lend-Lease Act in March 1941. This new legislation allowed the United States to supply Great Britain with the weapons needed to continue the fight against Germany.

Flying the airplanes to Great Britain offered the quickest method of delivering them across the North Atlantic; airplanes that lacked the range to make the long flight went by ship. Before the Lend-Lease Act, American officials used an awkward procedure to transport the planes to Europe. Under the cash-and-carry legislation, the factories hired civilian pilots to fly the planes from the West Coast, where most of the factories were located, to Montreal, Canada. From there, civilian pilots employed by the British Atlantic Ferrying Organization (ATFERO) flew the planes to Newfoundland and then the final 2,100 miles across the North Atlantic to Scotland. This procedure reduced delivery times from three months to only ten days, but the British had difficulty finding pilots who were willing to ferry the planes. The lack of civilian pilots forced them to withdraw RAF pilots from combat.

In April 1941, after the passage of the Lend-Lease legislation, General Arnold, now in command of the Air Corps, suggested using Air Corps pilots to ferry the planes. The British readily accepted his proposal. On May 28, 1941, President Roosevelt directed the Secretary of War, Henry L. Stimson, to take responsibility for deliveries to Newfoundland; however, American military pilots lacked the authorization to fly the aircraft across the Atlantic. The following day, Colonel Robert Olds of the Plans Division, Office of the Chief of the Air Corps, received verbal orders to organize the ferrying service, and the Air Corps Ferrying Command (ACFC) was officially constituted as of May 29, 1941. Its mission included the movement of aircraft and the maintenance of "such special air ferry service (or air transport service) as may be required to meet specific situations." This broad charter allowed the Air Corps Ferrying Command to extend its operations around the world and authorized the creation of a regular military air transport system between the United States and Great Britain. From June 6 until December 7, 1941, the command ferried approximately 1,350 aircraft to American Allies. Over 90 percent of the trips to the British in Canada were from factories on the West Coast or on the Atlantic seaboard.

Growing cooperation between Great Britain and the United States also necessitated an American air transport service for diplomatic mail and personnel. Using the second part of its charter, the Air Corps Ferrying Command inaugurated the trans-Atlantic military air transport service on July 1, 1941 when Lieutenant Colonel Caleb V. Haynes flew a modified B-24 from Washington, DC, to Scotland, via Montreal and Newfoundland. As the number of flights between Great Britain and the United States increased, the British respectfully referred to this service, as the "Arnold Line." By the end of summer, ACFC was routinely operating six round trips over the North Atlantic every month. This effort continued unabated until the harsh winter weather set in.

To allow American military aircrews to ferry Lend-Lease aircraft outside the Western Hemisphere, President Roosevelt authorized the Air Corps Ferrying Command on November 24, 1941 to deliver planes "to such other places and in such manner as may be necessary to carry out the lendlease program." Although implicit before this time, President Roosevelt, with this decision, formally gave the ACFC a global mission. On December 30, 1941, the Office of the Chief of the Air Corps reorganized the command to reflect its expanded role Two subordinate units were created under its Headquarters: the Domestic Division and the Foreign Division.

Air Corps Ferrying Command and Contract Airlift

Although the Air Corps Ferrying Command had the authority to operate a world wide air transportation system by the end of 1941, it lacked the necessary resources to meet the Allied demands for aircraft. The leaders of ACFC were convinced that the command was stretched to the breaking point in operating the North Atlantic route and, therefore, proposed using civilian contractors to fly the South Atlantic ferrying missions. Colonel Olds agreed to contract with Atlantic Airways, Ltd., a Pan American Airways subsidiary, to fly ferrying missions to Africa along the South Atlantic route.

By the first part of August 1941, Pan American had created three subsidiaries to handle ferrying operations. The most important of these was Pan American Air Ferries, Inc., which actually flew the missions; the other two were essentially support organization for the ferrying company. It should be added, however, that it took several months for Pan American to begin operating smoothly. By the time the Japanese attacked Pearl Harbor, Pan American Air Ferries had delivered only 12 aircraft, all transports for use in North Africa. Deliveries increased every month thereafter, and by the end of 1942, Pan American aircrews had delivered about 460 aircraft to the Middle or Far East over the South Atlantic route.

Inextricably linking ACFC and civilian airlines was the quest of more modern and efficient aircraft, especially those with a sufficiently long range to travel intercontinental distances. For example, in early December 1941, the command had only 11 four-engine planes. The airlines possessed just over 400 transports. To meet the critical shortage of military transport aircraft after the attack on Pearl Harbor, Roosevelt signed an executive order on December 13, directing the Secretary of War to take possession of any portion of any civilian airlines needed for the war effort. Accordingly, twin-engined transports such as the DC-3 (C-47 and C-53 variants), which made up the bulk of the civilian airline fleet, were incorporated into ACFC's operations. They were serviceable aircraft and became, along with the C-46s, C-54s and C-87s, the military's air transport workhorses.

Pan American Airways, Transcontinental, and Western Air quickly signed contracts to provide aircraft ferrying and air transport services over numerous worldwide routes, primarily in the first year using civilian versions of the C-47 aircraft. Eventually, every major airline provided some type of contract service. At first, the civilian airlines handled most of the military's air transportation needs because they had the airplanes and pilots. Although many of the airlines' 2,600 pilots were Army reservists, the Army could not call them up for active duty without disrupting the vital air transportation system. Initially in 1942, the airlines provided almost 88 percent of the air transportation, but by 1945 the military planes and crews were providing 81 percent.

Instructors provided intense training which was necessary in the CBI Theater at Pilots Ground School. Jan. 16, 1945.

Weather normal - always bad.

Reorganizing to Meet the Wartime Challenge

Army aviation underwent a major reorganization in June 1941. The General Staff established the Army Air Forces (AAF) and named General Arnold its chief. Within the AAF, the General Staff created two subordinate commands: the Air Force Combat Command (AFCC), those units formerly comprising the GHQ Air Force; and the Air Corps, which controlled all other AAF functions. Not considered a combat command, the Air Corps Ferrying Command remained part of the Office of the Chief of the Air Corps until March 1942. On March 9 to effect a better division of responsibilities between the Army Air Forces and the other War Department agencies, the War Department established three Army branches - Army Ground Forces, Army Air Forces and Services of Supply. Included in this realignment was the elimination of the Office of the Chief of the Air Corps. This left the ACFC, now termed the Ferrying Command, directly subordinate to General Arnold, the commanding general of the Army Air Forces. Nevertheless, the Ferrying Command continued to grow and overlap with other units. In particular, the Ferrying Command duplicated the duties of the Air Service Command, which had been created to meet the increasing requirements for the supply and maintenance of Army aircraft. To solve the problem, General Arnold assigned the responsibility for transporting aviation technical supplies to bases or units within the Western Hemisphere to the Air Service Command in late March 1942. That command had the additional duty of creating transport squadrons capable of delivering airborne infantry, glider troops, and parachute troops. Arnold assigned to the Ferrying Command the mission of operating, "either directly or by contract, all transport lines extending beyond the Western Hemisphere," but in time only military personnel served with the Ferrying Command outside the United States.

More important, perhaps without intention, General Arnold's decision essentially separated military air transport operations into either troop carrier operations or logistics operations. His decision mirrored the prewar organization which placed "tactical transport" airplanes under the GHQ Air Force and logistical transport airplanes under the Materiel Division. This was a watershed in doctrinal issues that has been present in questions of airlift since that time: the definitions and dichotomies of strategic and tactical airlift. Strategic airlift has long been defined as "the continuous or sustained air movement of units, personnel, and materiel between area commands; between the Continental United

Snatch pick-up of a glider in Wesel, Germany. 17 April 1945.

States (CONUS) and overseas areas: within an area command when directed." Strategic airlift, therefore, can be broadly termed as the logistical movement of troops and material and their resupply over intercontinental routes. These resources, General Arnold and the senior leadership of the Army Air Forces understood, must be organized into a single command with worldwide responsibilities. The mission of the Ferrying Command and its successors represented the logical development of these doctrinal ideals.

Tactical airlift was a different issue, however. It has been defined as those forces: organized, equipped, and trained to move combat forces and sustaining material under widely varying situations ranging from small movements in battle to large movements over long distances. Tactical or theater airlift is the employment of airlift within a theater to support combat operations; it includes airborne assaults, airdrop resupply and logistical support between theater bases.

Since tactical airlift was viewed as a combat resource in a manner that logistical airlift never had been, it was assigned to the air component of the theater commands, not to a single command with a global mission.

This factionalism proved burdensome for Ferrying Command personnel on numerous occasions during World War II. As the command expanded its operations, problems arose from the traditional concept that a theater commander controlled all resources within his area. Since a Ferrying Command route might pass through several theaters, the command's aircraft were often viewed as additional theater assets and subjected to redirection. Faced with critical airlift shortages, theater commanders understandably diverted scheduled Ferrying command flights for their own needs. This caused havoc with the efficient operation of the total air transport system. In June 1942, the War Department formally directed the theater commanders to refrain from diverting the transports, but the practice, although diminished, continued throughout the war.

Creation of the Air Transport Command

Despite the new organizational arrangement of March 1942, the Ferrying Command and the Air Service Command still duplicated too many functions. In addition the United States Navy had created its own Naval Air Transport Service (NATS) on December 12, 1941, to provide logistic airlift for the Navy's fleets and far-flung bases. L. Welch Pogue, Chairman of the Civil Aeronautics Board, recommended that General Arnold create a new command, separate from the Army and Navy, to control military air transportation. At the very least, Pogue believed, all transportation within the Army should be consolidated under a single command. Pogue suggested in a memorandum in the spring of 1942:

It would constitute a great step forward if the air transport services of the Army could be consolidated and placed under one command, provided all other commands and branches of the Army were required to present their demands for services of the airline organizations to such a unified Air Force Transport Command and to abide by its decisions.

Taking the problem under consideration in June 1942, General Arnold concluded that the responsibility for air transportation must be assigned to "permit the most efficient utilization of aircraft, facilities and personnel by the elimination of dual responsibility and duplication of services" He also believed that the Army Air Forces had to "provide transport operations by military personnel, rather than by civilians under contract, on routes that enter combat areas or are likely to become combat areas."

Motivated by these considerations, General Arnold issued AAF General Orders Number 8 on June 20, 1942, creating both the Air Transport Command (ATC) and the Troop Carrier Command. Effective July 1, the Air Transport Command received the following responsibilities:

A. The ferrying of all aircraft within the United States and to destinations outside the United States as directed by the Commanding General, AAF.

B. The transportation by air of personnel, materiel, and mail for all War Department agencies, except those served by Troop Carrier units.

C. The control, operation and maintenance of establishments and facilities on air routes outside the United States which were under the control of the Commanding General, AAF.

General Arnold's recommended course of action limited the Air Service Command to continental operations and allowed ATC to handle all other air transport save that reserved for support of theater combat forces. In addition, to the Troop Carrier Command went the responsibility of providing air transportation for parachute troops, airborne infantry and glider units; and for conducting local air transport missions within the theaters of operation. Colonel Harold L. George, who had assumed command of the Ferrying Command in April 1942, became ATC's first commander, a post he held until the end of the war.

The Navy, never interested in any plans for unification of airlift forces, continued to operate the Naval Air Transport Service as its own long-range airlift organization throughout the war. The Air Transport Command and Army Air Forces staffs studied the question of duplicated effort by ATC and

Many different aircraft have performed airlift functions.

NATS and presented their findings to the Joint Chiefs of Staff repeatedly during the early 1940s, without results. In early 1944, however, the Joint Chiefs agreed that NATS would restrict its operations to serving the naval establishment. To monitor duplication, the Joint Chiefs of Staff created the Joint Army-Navy Air Transport Committee (JANATC). Because of the decisions made by General Arnold, the Joint Chiefs of Staff, and the Navy, consolidating all military air transportation under a single command would not be achieved until many years later.

The Ferrying Command and its predecessor ACFC had existed as a separate entity for only 55 weeks, but during that time compiled an impressive record. Besides pioneering the global military airways, pilots ferried 13,513 aircraft to final domestic destinations and 638 airplanes to final foreign destinations. Additionally, the Ferrying Command transported 1,920 tons of cargo, mail and passengers before the reorganization.

Despite several false starts, the United States possessed a rudimentary military airlift service when it entered the Second World War. Although the Air Corps Ferrying Command existed for only 13 months, it created a strong base for the expansion of the wartime Air Transport Command. Growing from an organization comprised of two officers and a civilian secretary in May 1941 to a force of 11,000 by the time of ATC's activation, the command also demonstrated an exemplary level of service. Its pilots had ferried over 14,000 aircraft. All of this was but a prelude to the remarkable development of airlift during the Second World War, when the value of airlift became apparent.

Military Air Transportation in World War II

American usage of military air transportation before the Second World War consisted primarily of demonstrations of airlift's potential. During World War II, military airlift grew into a vital aspect of American air power. The necessity of fighting a mechanized, global war forced the nation to create an extensive military air transport system that stretched around the world. Because of the vast distances involved, the speed at which war resources could be transported became essential to the war effort. In this environment, there arose within the American military structure two distinct divisions: logistical airlift, or air transport, to move critically needed personnel and supplies over intercontinental distances and tactical airlift to support combat operations in the theater areas. The nation's budding airline industry joined with the Air Transport Command to perform the former task with the latter airlift remaining under the control of the theater air force commanders.

Douglas C-47 "Sky Trains," 12th AF Troop Carrier/Wing loaded with paratroopers on their way for the invasion of Southern France. August 15, 1944.

Especially within its first year, the Air Transport Command, underwent several organizational changes, adjusting as best it could to its massive growth in resources and span of control. In four years of war, the command's ferrying operations delivered 282,537 aircraft; its air transport operations, both military and contract carriers, flew over 8.5 billion passenger and 2.7 billion ton miles. Among the billions of miles flown, one air transport operation, the "Hump" Airlift, ranked as the command's most significant accomplishment.

Organizational Structure and Mission of ATC

At the time of its redesignation as the Air Transport Command on June 20, 1942, the command's mission had evolved from a limited one of ferrying aircraft for the British - from American factories to transfer points near the eastern seaboard - to one of ferrying aircraft to American and Allied forces wherever located.

17th Airborne Division troops assemble by their Curtiss C-46 Commandos. Rhine Crossing. March 24, 1945.

Specifically, the Army Air Forces' June 20 order made ATC responsible for:

1. the ferrying of all aircraft within the United States and to overseas locations as tasked

2. the transportation by air of personnel, material and mail for all War Department agencies, excluding troop carrier units

3. and the control, operation and maintenance of AAF facilities along overseas air routes.

The order further directed ATC to "utilize to the fullest extent possible, the services, facilities and personnel of the civil air carriers."

Initially, ATC established two major subordinate units that roughly corresponded to the command's primary mission responsibilities: the Ferrying Division for the delivery of aircraft and the Air Transportation Division for the shipment of resources to the theaters. The divisions' functions were to coordinate the command's activities. At the time of the redesignation, the command was already in the process of reorganizing its units and had received approval to form five independent foreign wings to manage aircraft over the various routes.

Headquarters Army Air Forces had designated the wings, the 23d through the 27th AAF Ferrying Wings. Air Transport Command officials believed the numbered designations might inadvertently, through typographical error, result in the shipment of cargo or personnel to the wrong theater. Accordingly on July 5, the wings were renamed after the geographical areas they served.

Upon the establishment of the five wings, the functions and resources of the Domestic and Foreign Wings of the old Ferrying Command were merged with the Ferrying Division. Colonel William H. Tunner, commander of the Domestic Wing since its activation, assumed command of the division. At the time of its inception, the Air Transportation Division consisted of the Contract Air Cargo Division, formerly under the Air Service Command. Colonel Robert J. Smith, previously vice president of Braniff Airlines, headed the Air Transportation Division. Later, in March 1943, the Domestic Transportation Division would replace the Air Transportation Division. Many of its functions had already been assigned to Headquarters ATC. The Domestic Transportation Division oversaw flight and ground crew training and the domestic air transportation service which was performed by civil carriers under contract until 1944. In light of the administrative difficulties involved in overseeing thirteen different carriers, ATC officials decided to militarize the domestic transport service gradually throughout the summer and fall of 1944. Officially redesignated "military air transport" with the abbreviation of MAT, the new MAT service was at first a joint responsibility of the Ferrying and Domestic Transportation Divisions. But dual control proved difficult, and the Domestic Transportation Division became a wing of the Ferrying Division on November 1, 1944.

Growing demands for its ferrying and transport services forced ATC to expand its routes and operations throughout the war.

The Air Transport Command formed new wings wherever needed. Between October 1942 and December 1943, the command added four more wings. The Alaskan Wing, established in October 1942 at Edmonton, Alberta, took over the Northwest route to Alaska, previously a Ferrying Division responsibility.

On December 1, 1942, the India-China Wing was activated and assumed from the 10th Air Force the responsibility for resupplying the Chinese over the Himalayan Mountains. The wing would become one of the command's largest. As war activities increased in the Pacific, there followed more requests for air transport and ferrying services. In January 1943, ATC renamed the South Pacific Wing the West Coast Wing and established a new Pacific Wing at Hickam Field, Hawaii, to manage the route between the base and Australia.

Anticipating the invasion of Europe, ATC formed the European Wing at Prestwick, Scotland, on January 14, 1943, but soon moved the wing's headquarters to London.

Allied activities in Africa eventually caused the Africa-Middle East Wing to split along the lines of its two sectors, forming the North African and Central African Wings in December 1943.

Essentially wherever the allied forces took back territory from the Axis forces, ATC followed, extending its air logistics and ferrying services.

By the summer of 1944, the overseas wings had become so large that two of the wings - India-China and North Atlantic - exceeded the personnel strength of an infantry division. Moreover, the European Wing with over 7,500 personnel was expected to grow considerably as a result of the Normandy invasion. Accordingly, on June 27, 1944, ATC inaugurated a major reorganization of its units, redesignating the nine existing wings as divisions and forming intermediate wing headquarters from the sectors of old. By the end of the war, in August 1945, there were eight foreign divisions and one domestic division.

Reliance Upon Civil Carriers

Especially early in the war, the Air Transport Command did not have sufficient resources to accomplish its diverse mission. Although Army Air Forces and command officials had envisioned a fully militarized organization as the ideal from the beginning, and while military personnel increasingly rendered overseas airlift services from 1943 on, the use of contract services continued until the end of the war. Virtually all of the commercial carriers operating in 1941 had some piece of the airlift pie during the war. The original agreements allowed the government to purchase the airlines aircraft and equipment and then operate them using civilian pilots and support personnel.

Beginning early in 1943, however, the War Department began to change its strategy and adopted "on-call" contracts which bound the airlines to render any service to the government within the general limits of the carrier's capabilities. While ATC did not write these agreements, it established the airlift requirements and had a supervisory responsibility.

Curtiss C-46 "Commando" in the snow-capped Himalayas between India and China. Every two and a half minutes, the India-China Division of ATC flys a plane across this route.

"Rooster Movement." Chihkiang. April 25, 1945.

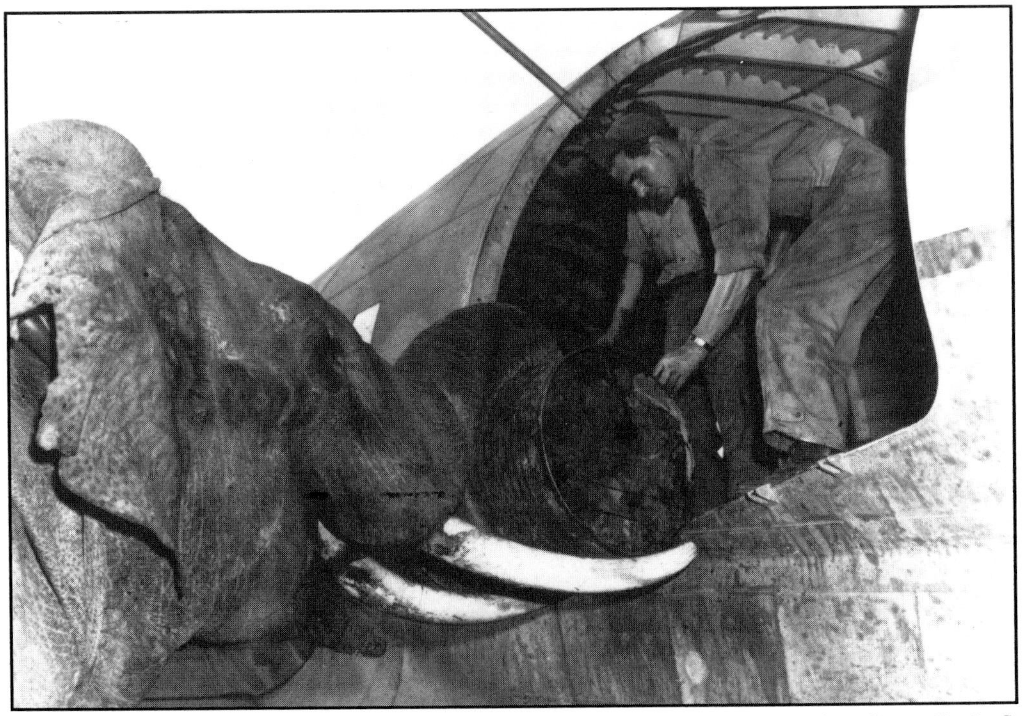

"Move over, boys, you're doing it the hard way." says Elmer the elephant, as he loads a drum of gasoline on Curtiss C-46 "Commando" of the ATCs India-China Division, somewhere in upper Assam. Feb. 1945.

Making 'Anything, Anywhere, Anytime' A Reality

Editor's Note: The information contained in this article was extracted from Anything, Anywhere, Anytime: An Illustrated History of the Military Airlift Command, 1941-1991 compiled by the Military Airlift Command History Office and Over the Hump by Lt. Gen. William H. Tunner. Additional information and picture support was provided by the AMC Office of History.

Air transportation of the people and materials of war became a vital aspect of American air power during World War II. The need to move personnel, equipment, and supplies quickly around the globe, highlighted the need for both military and civilian airlift capabilities. Air transportation methods and equipment have continuously evolved.

Nearly 50 years ago, Lieutenant General William H. Tunner took command of the India China Division. His two-fold mission was to increase the tonnage of supplies airlifted to the Chinese and decrease the number of aircraft mishaps. Within a year of taking command, Lt. Gen. Tunner was well on his way to achieving these goals. The airlift policies and methods he helped pioneer are still in evidence today. This is the story of how he and his men successfully flew over the Hump, the Himalayas between India and China.

This description of flying, from *Anything, Anywhere, Anytime,* paints a vivid picture of the varied weather conditions and geographical regions the area of the Hump covered:

To the men who served in the India-China Wing, the resupply missions over the Hump were just as difficult and dangerous as the strategic bombing missions in Europe. Out of the steaming, tropical valleys of India's Assam Province, the airlift crews flew fully-loaded transports northeastward over the Himalayas into southwestern China. They usually landed at the principal American airdrome at Kunming. If aircrews took a direct flight in about four hours at a maximum altitude of about 16,000 feet during optimum weather conditions. This route, however, necessitated flying over northwestern Burma, territory patrolled by Japanese fighters. To avoid them, many aircraft commanders detoured to the north flying a distance of 720 miles and crossing the Hump at the northwestern end, where its lowest peaks reached 16,000 feet. But over this route, aircrews had to operate at altitudes approaching 20,000 feet, a formidable task given the limitations of the aircraft.

At the end of the war, the Air Transport Command had 13 bases in India and 6 in China, a marked expansion from 1942 operations when Hump pilots shuttled between only a couple of airfields. Lt. Gen. Tunner described the phenomenal effort required to build runways in these remote regions in his memoir *Over the Hump*.

"We landed on a runway of gravel, which had been broke out of its quarry by sledges, chipped down to size by hand hammers, transported to the site by hand in small baskets, and then laboriously spread by hand. Thousands on thousands of Chinese working with only the rudest of equipment had leveled this runway, then covered it with hand-crushed stone and rolled it with hand-pulled rollers—two hundred men to a roller. It was rough, bumpy and hard on tires, but it was a runway, and I brought the C-46 down on it..."

If the runways seemed primitive, the living and working conditions were worse. Tunner made a preliminary trip around the wing several months before he assumed command. What he saw motivated him to address the problems of low tonnage levels and high mishap rates in what might seem a unique way. Tunner pinpointed morale among the people serving all along the Hump as a major contributor to the other difficulties. Living conditions were dismal. Tunner, though prepared, was shocked by what he saw:

Men off duty lounging aimlessly around reflected their living conditions .... "Their uniforms showed the lack of laundry and were badly worn .... Few showed common courtesy, much less military courtesy, as I walked by."

Tunner resolved to change the situation. When he took command, he brought military bearing and pride to the Hump. He ordered daily inspections and Saturday parades. He instituted a consistent rotation policy for all who served in the China Burma India Division (CBI). To his credit, Tunner managed to bring needed daily supplies to the Hump operations. No longer were the Americans without basic necessities such as soap, toothpaste, and toilet paper. Cleanliness, courtesy, and patriotic pride began to return to the men serving throughout the area. Many of his changes and programs caused griping, but Tunner learned, in a unique way, that his people understood he was looking out for their needs and that they were each important to the mission's outcome.

The day after Christmas, 1994, Lt. Gen. Tunner received a report from the Misamari airdrome of a record breaking number of trips over the Hump in 24 hours. Misamari's normal number of Hump trips was 30 per day. In part, the report read:

There was no pressure; no forcing men to do the job. Every effort was voluntary, like spontaneous combustion followed by wildly sweeping prairie fires. Enthusiasm concerning the project burst forth in every section of the field. A goal of 55 trips was set and the numerals painted on a banner. That the numerals would have to be changed to higher ones four times during the 24-hour period none of us ever dreamed.... The 81st trip left the runway one minute before the deadline.

With the living and working conditions improving, so too the tonnage and mishap rates improved. Tunner knew the way to control such a widespread operation was to use big business techniques. His hand-picked senior staff met each morning to discuss and decide how to handle the problems arising in the command. While commanding the Ferrying Division of ATC, Lt. Gen. Tunner instituted the use of statistical management of aircraft, loads, and personnel. He brought those statistical methods to the Hump. He wanted to know where every pilot, plane, and parcel was—at all times. His stated intention was to have each and every aircraft flying, undergoing maintenance, or being unloaded or loaded every minute of every day.

Sometimes, the lack of equipment gave rise to inventive ways to load and unload the aircraft. For instance, elephants were used in India to load 55-gallon drums of petrol onto the aircraft. Another method of unloading those drums of petrol was to use civilian labor. The civilian workers would roll the drums out of the cargo door and let them drop onto truck tires where other civilian workers loaded them by hand onto waiting trucks. With the aircraft fully utilized, the availability of aircrews became an ever-increasing critical issue. Again, Lt. Gen. Tunner turned to statistical management and good business practices to manage his personnel. He saw to it that the aircrews received training and proper amounts of crew rest. He empowered his operations officers with the scheduling of the aircrews.

With the mass movement of Chinese troops into the area around Kunming, Lt. Gen. Tunner was able to establish C-46 and C-47 units with flight crews and maintenance personnel within China. These units were instrumental in the movement of supplies in that area. As a result, tonnage throughout the Hump increased. In addition to improving the movement of supplies across the Hump and around China, Lt. Gen. Tunner deserves credit for the development and the use of production-line maintenance (PLM) and a fully organized search and rescue operation.

Production-line maintenance was just being established in the civilian airline industry when Lt. Gen. Tunner became a convert to the system. The basic idea of PLM was to move the aircraft through different maintenance stations. Each station would perform a specific maintenance task as the aircraft moved along the line. By breaking the complex maintenance process into more simplified steps, Lt. Gen. Tunner was able to augment his limited maintenance personnel with civilian workers.

Originally pilots flew search and rescue missions without official, command organization and support. One of the most colorful search and rescue pioneers in the CBI was Captain John E. Porter. Know as "Blackie" by pilots throughout the Hump, Porter was a former stunt pilot. He and his crew, other barnstorming types, would borrow or actually steal planes to go out on search missions. It was during such a rescue mission that Blackie and most of his crew were lost to enemy fire. A flight of Zeros attacked the aircraft. With communications gone, Blackie's bailout command failed to reach his crew. Blackie, piloting the bullet-ridden, burning B-25, ordered his copilot, Lt. Spain to bailout. Lt. Spain got stuck in the escape hatch. Blackie, holding the plane steady, stood and pushed him through the escape hatch. Lt. Spain was the sole survivor.

After taking command of the CBI, Lt. Gen. Tunner formally organized the search and rescue efforts along the Hump. One of the first jobs the newly-organized Search and Rescue organization accomplished was to pinpoint every existing sign of wreckage along the Hump. This task was difficult. Over the course of the airlift, 701 major mishaps resulted in the loss of 460 aircraft and 792 men— over 60 percent of the command's total mishaps and fatalities. They then educated the aircrews to report all signs of wreckage, which helped reduce the duplication of rescue efforts. The Hump, littered its length and breadth with aircraft wreckage, was nicknamed "the aluminum plated trail."

Perhaps, Lt. Gen. Tunner's observation summed up the importance of airlift, not only during World War II, but throughout all the conflicts which followed:

"...begun when air transportation of heavy cargo was in its very infancy, carried on with steady increase in spite of the enemy and formidable weather conditions, and over the most menacing terrain, all this a half a world away from home, the Hump Airlift proved, forever, the efficacy of air transportation. After the Hump, those of us who had developed an expertise in air transportation knew that we could fly anything, anywhere, anytime."

Still later, by the end of 1944, the military had taken over all domestic transport services performed by the commercials. Certainly the difficulty of coordinating and working with several different airline companies influenced the decision to gradually phase out contract operations, first overseas and then in the domestic air transport service. A completely militarized system greatly improved airlift's flexibility and responsiveness.

The military's use of civil carriers had been one of necessity. The relationship, although beneficial to both, had moments of friction. As a measure of the command's reliance on civil air carriers, nearly 88 percent of the transport work managed by ATC was performed under contract by commercial airlines during 1942. In 1943 this rate dropped to 68 percent, and by the end of the next year it had fallen to 33 percent. At the end of the war, ATC's military forces were operating all but 19 percent of the air transport missions.

The Hump Airlift to China, 1942-1945

During World War II, the Air Transport Command provided critical support to Allied forces around the world, but nowhere did this support reach the size of the "Hump" Airlift over the Himalayan Mountains in the China-Burma-India (CBI) Theater. In February 1942, President Roosevelt's commitment to aid the Chinese in their fight against more than a million Imperial Japanese troops precipitated the most extensive airlift ever undertaken by the United States.

Resupplying forces in China had tremendous significance both for the war effort and future airlift doctrine. For much of the war, the air route over the Himalayas formed the only link between the outside world and Major General Clair L. Chennault's Fourteenth Air Force; General Joseph W. Stilwell's mission to China; and Generalissimo Chiang Kaishek's Chinese National Army. This air route provided the logistical support necessary for the defense of China. Air Transport Command aircrews involved in the Hump Airlift, the nickname given to the several hundred mile aerial pipeline, made 167,285 trips to China, delivering almost 740,000 tons of war material. Of that amount, over 75 percent was delivered in the last year of operation and included moving entire armies from India and Burma to the battlefronts of China. Over the course of the airlift, ATC would record 701 major accidents, losing 460 aircraft and 792 men. Of these, enemy action destroyed seven aircraft and killed 13 airmen. Against this background, the airlifters accomplished their mission, and by the end of the war, the Hump Airlift was operating with business-like precision, largely due to the direction of Brigadier General William H. Tunner.

As Commander of the India-China Division, Tunner had a two-fold mission: increase tonnage while decreasing accidents.

Tunner, a superb administrator, initiated several improvements to the airlift system. First, he acquired additional personnel and aircraft.

Second, Tunner and his staff instituted a comprehensive safety program. They prepared a statistical tracking procedure to determine the causes of aircraft failure, the airfields where the most accidents took place, the type of weather involved, the model of aircraft most prone to accidents, and maintenance deficiencies. The information gathered, coupled with more rigorous flight checks, aircrew physicals and an efficient safety awareness program, proved most useful in combating accidents. The major accident rate decreased from .34 per 1,000 flying hours in August 1944 to .15 by August 1945, a 55 percent decrease.

Third, General Tunner improved aircraft reliability while decreasing maintenance time by introducing Production Line Maintenance (PLM). This procedure required an aircraft to be towed through seven different stations. At each station, crews performed specific maintenance operations. To make the PLM work, each base specialized in one type of aircraft. Consequently, repair operations were more efficient.

Each of Tunner's actions increased tonnage and decreased accidents. The Hump Airlift delivered 44,098 tons in January 1945; by July this had been increased to 71, 042 tons. Meanwhile, the accident rate dropped from 23 accidents and 36 fatalities in January to only eight accidents and 11 fatalities in the last full month of the war. After the end of hostilities in August 1945, Hump operations declined swiftly; dropping to 53,315 tons in August and 1,429 tons in November, the month it ended.

Unquestionably this airlift made it possible for China to continue its resistance against the Japanese. It demonstrated the viability of using large-scale airlift operations to deliver war material and combat troops. Between 1942 and 1945, 81 percent of all supplies entering China came via the Hump. Without these supplies, the Chinese defenders would have been unable to fight. As a result, the Japanese Imperial Army was forced to maintain 1.2 million troops on the Chinese mainland. Had it achieved a quick victory there, Japan could have left a small occupation force in China and moved the remainder of its forces to the Pacific, making the island-hopping campaigns more costly than they were.

The Hump Airlift also influenced the development of American foreign policy. Researchers preparing the Strategic Bombing Survey following the war duly recognized its importance:

The major significance, for the future, of all air operations in CBI was the development of air transport operation. During the first year of the war, the magnitude to which air transport operations could be developed was not appreciated. However, the terrain of Burma and China and the absence of land lines of communication forced all agencies in the theater to turn to the airplane - initially as an afterthought and an emergency last-chance measure. The inherent flexibility of air power permitted it, without adequate preplanning, to meet the exigencies of the various situations. Air transport operations expanded beyond the wildest prediction of 1942 - expanded because it was the one agency which could succeed.

Airlift Support of Special Operations

A final aspect of airlift in World War II as it relates to the future missions of the Military Airlift Command was the support given to special operations. The Air Transport Command, troop carrier, and other airlift organizations provided airlift for a wide variety of special operations. The most common was the clandestine single-ship airdrop or airland operation conducted in every theater of the war. These were often used to insert or extract special operations teams or agents and to resupply small guerrilla bands or insurgents behind enemy lines.

Airlift supporting special operations also included activities designed to demoralize the enemy. Psychological operations involved leaflet drops, electronic jamming and broadcasting.

There were other notable types of special airlift operations as well, among them emergency evacuations and glider insertions. Airlift units accomplished numerous emergency evacuations of both large and small numbers of Allied personnel.

The most celebrated special air operation units of World War II were the VIII Air Force Composite Command's units involved in Operation Carpetbagger designed to resupply insurgent bands in France and the low countries during the months leading up to the invasion at Normandy; and the 1st Air Commando Group, activated on March 29, 1944, in India, to support Wingate's Raiders, guerrilla forces operating in Burma behind Japanese lines.

Wartime operations firmly established the validity of dedicating aircraft for logistical supply and troop deployments. As Major General Robert M. Webster, ATC Commander from 1946-47, related to a National War College class in 1947, the United States had entered World War II "with only the basic types of military aircraft, the bomber and the fighter." Remarking further, Webster stated:

"I feel that we have come out of that war with an additional type, the transport plane, and that we should think in terms of bomber-fighter-transport -

Mobile Air Traffic "Control Tower" (Hot Dog Tower) Berlin 1948. Airlift. Tegel Field, French Sector, 1946th AACS Squadron. (Courtesy of Jack D. Fellman.)

since they are all equally important - and they must be properly balanced to each other it we are to be prepared to conduct successful war operations.

The Air Transport Command and other airlift organizations had proven this throughout World War II, responding to the needs of the moment in a myriad of ways and in a variety of situations to deliver cargo and passengers where and when they were required. The war years had also shown the necessity of consolidating duplicative airlift functions as well as the value of drawing upon the commercial airline industry. After the war, Military airlift underwent organizational refinement to accommodate the airlift lessons of World War II.

Demobilization and Cold War, 1945-1953

The Air Transport Command had built an impressive system for moving people and cargo during World War II; the command's transports had crossed the Atlantic every 13 minutes, the Pacific every 37 minutes in July 1945. However, when peace finally came, ATC joined other military elements in a rapid demobilization. While Americans had willingly heeded the call to arms, they had no intention of remaining overseas once the war ended. Isolationism had the potential, so it appeared in 1945, to become once again an important objective of the United States' foreign policy, just as it had been in the 1920s and early 1930s. Even those national officials who were committed to ensuring that the United States played a central leadership role in the world did not agree that a large standing military was necessary. After all, "the Bomb" had finally ended World War II in the Pacific, and its holocaustic potential would deter future aggressors at a moderate cost to the United States government.

It seemed like a time for the nation's military forces to go home, back to civilian life. Americans, however, would discover that the world remained a dangerous place. Some former enemies were becoming friends, while one former ally began looking very much like an adversary.

Postwar Organization and Mission

Victory in Europe meant a shift in emphasis for the Air Transport Command, but not less work. When Germany surrendered on May 7, 1945, the Army Air Forces transferred some forces to the Pacific but sent most of its people and resources home. Air Transport Command personnel, however, had to delay their own homecoming to assist with the troop and equipment withdrawals. From May through mid-September 1945, ATC participated in the "Green" (troops) and "White" (aircraft)Projects - as the demobilization and transfer of men and equipment from Europe and the Mediterranean theaters were called - returning more than 250,000 personnel and 5,900 aircraft.

The Japanese surrender in August 1945 following the atomic bomb attacks meant an even larger mission for ATC - moving occupational forces into Japan. To execute this assignment, known as "Special Mission Number 75," ATC pulled 154 C-54s from scheduled Pacific Division operations and placed them on standby for the deployment. The command diverted another 95 C-54s from points as far away as North Africa and the North Atlantic ("Purple Project") and rushed them to Okinawa, where they joined aircraft from the India-China Pipeline and the United States. These few hundred transports as well as another 360 B-24 bombers, used as cargo aircraft, made Special Mission Number 75 one of the greatest concentrations of troop carrier airlift.

Following the Japanese surrender, the United States began the usual process of dismantling its military forces.

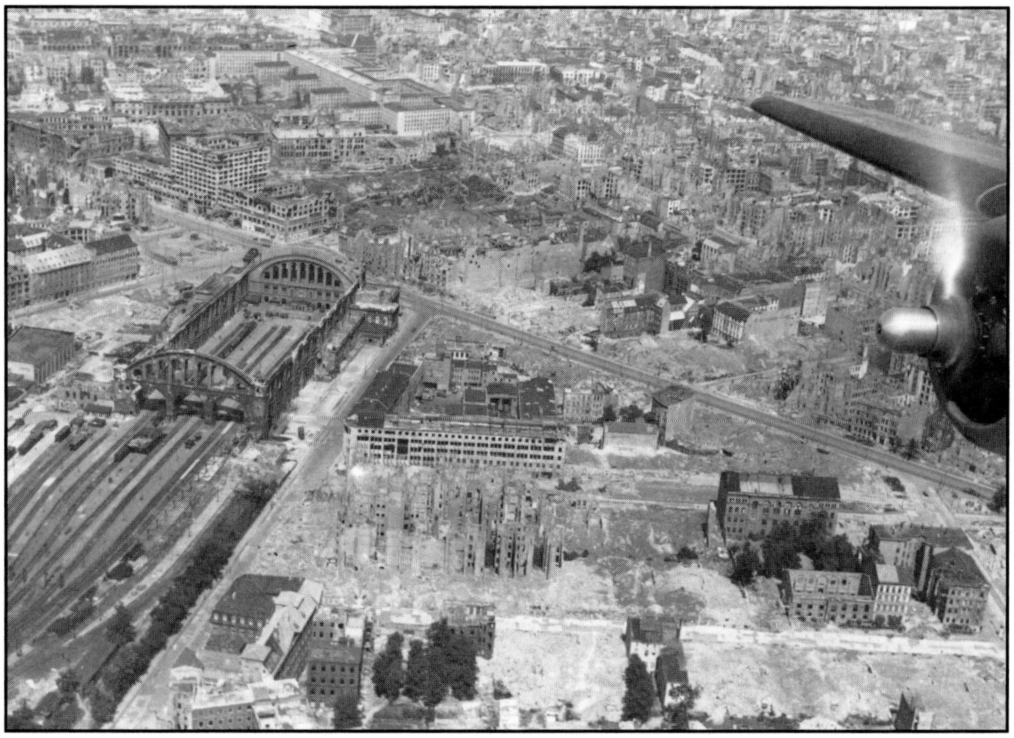

From this C-54 over Anholter Bohunhof Berlin, RAF and USAAF bombing damage was still evident. July, 1948. (Courtesy of Gail and Alta Halvorsen.)

The Air Transport Command had finished the war with one domestic and eight overseas divisions. A year later, there were three left in operation - Atlantic, Pacific and European.

Fortunately, ATC had some highly-placed advocates who made sure the command was not completely disbanded. When General Hap Arnold was preparing to pass command of the Army Air Forces to General Carl Spaatz, he asked American Airlines executive C.R. Smith for advice on the best structure for ATC. Cyrus Rowlett Smith had finished the war as ATC's Deputy Commander.

Smith was obviously proud of ATC's wartime accomplishments and asserted the command had the potential to be an even better airlift organization. Recounting his session with General Arnold to Lieutenant General Harold George, Smith beat ATC's drum:

"Army, through Air Transport Command, should be the preeminent air line operators in the world, better than any other military establishment and better than any airline organization. This is possible, on account of lack of civil restrictions (upon military), availability of equipment and the military necessity of always being ready on a highly skilled basis."

On December 6, 1945, eleven days after talking to Smith, Arnold wrote a "Dear Tooey" letter to his successor, offering General Spaatz advice on a variety of topics. In words that would seem prophetic, General Arnold stressed the difficulty of deploying and establishing air forces in combat areas. "During times of peace," he wrote, "we are apt to retain our combat units and sacrifice the essentials to their successful deployment and immediate operation. We must retain our bases and our means of deployment." Arnold then urged Spaatz to preserve essential airlift capability for future crises since it appeared that the nation's forces would be pulling back from forward locations overseas. Without airlift, Arnold reasoned, the nation could find itself in serious trouble in future confrontations.

Arnold also told Spaatz that ATC during peacetime would need to operate a regular air transport service between bases as well as serve other government agencies.

"The operation of these services will give us well-trained officers and men experienced in world-wide, all-weather flying. It will give us an opportunity to develop and to test new equipment, communications facilities and other airway aids. More important, the scope of the operations I visualize would be such as to present unusual opportunities for collecting general international intelligence. This potential has never been fully realized."

Although General Arnold clearly expected the civil airlines to carry on the American preeminence in aviation during peacetime, he felt the military should take the leadership role in developing routes, equipment, techniques and procedures. This would result in technical developments in long-range aircraft.

Above all, General Arnold realized that without a rapid deployment capability inherent in ATC, the deterring forces of the Army Air Forces were hollow. He stressed planning for emergency overseas deployment.

Consolidation of Airlift: The Birth of NATS

After Pearl Harbor, the Navy had organized its own Naval Air Transport (NATS), and throughout the war years both the Army's ATC and the Navy's NATS operated extensive airlift systems independently. With peace, however, Congress and other government officials began questioning the need to maintain two airlift services that were overlapping. In May 1946, the Joint Chiefs of Staff asked the Joint Army-Navy Air Transport Committee (JANATC) to work with the Joint Logistics Committee to produce a draft agreement delineating the respective responsibilities of the two transport services. The committee ultimately recommended a temporary joint task force, with the goal of withdrawing the Navy from all land-based air transport activities and having ATC pick up all overseas routes of "common interest." Lieutenant General George soon found that the Navy and Army Air Forces leaders could not agree on which routes were common interest, something that was essential to defining the interim joint task force's areas for cooperation. Headquarters NATS representatives insisted that the Navy need not turn over responsibility for operating a route of common interest to ATC if the Army Air Forces used contract

27 January 1949. Kearney AFB, NE. Operation Haylift. Loading hay on a C-82 to drop to cattle in Nebraska. This aircraft was from the 4408th Air Base Sqdn, Lawson AFB, Ft. Benning, GA. ***Inset:*** *Crew of the C-82. Lt. Charles R. Smith, T/SGt. Victor Rogers, Lt. Max J. Gardner, Capt. Phillip R. Heinmiller, T/Sgt. James E. Spell, Lt. James L. Hill, Lt. Don Spearel and Lt. Stan Rundell. (Courtesy of Donald W. Spearel.)*

carriers on the route. In 1946, Lieutenant General George wrote to the Assistant Secretary of War for Air, Kenneth C. Royall, stating that he expected the Navy to make high demands for air transport service just so it could claim nonsupport from ATC and maintain its NATS airlift structure permanently. George suggested that the only way to deal with this was to have the War Department, rather than any of its service components, determine the strategic air transport requirements for both the Army and the Navy and then assign the total mission to the Army Air Forces.

The Army Air Forces got nowhere with consolidation until civilian leaders directed the military services to organize the air transport mission under a single command. The issue became part of the give and take surrounding the legislation that created a separate, independent United States Air Force.

On July 18, 1947, President Harry Truman established a temporary Air Policy Commission "to make an objective inquiry into national aviation policies and problems," and to assist him in formulating an integrated national aviation policy. The tasking went beyond the military, encompassing all aspects of aviation in the country. The President's Air Policy Commission - also known as the Finletter Commission - held formal hearings between September and early December 1947. It would be difficult to find air power experts the Finletter commission neglected to interview. Besides the heads of aircraft corporations and government aviation agencies, the commission heard from General F. Trubee Davison, the former Assistant Secretary of War for Air; Colonel Thomas O. Hardin, Air Transport Command, who was formerly the Chairman of the Air Safety Board; Vice Admiral Emory S. Land (Retired), President of the Air Transport Association of America; Air Force Chief of Staff General Carl Spaatz; and Air Force Secretary W. Stuart Symington. Major General Laurence S. Kuter, who had co-authored the basic plan for the organization and employment of air power in World War II as a member of the War Department General Staff, also spoke to the commission.

C-54 assigned 317th Troop Carrier GP being loaded at Celle, Germany, from British lorries, April 1949, during Berlin Airlift. (Courtesy of Morley Yagen.)

The Finletter Commission addressed military and commercial transport services in a single section of its report. The commission revealed that estimates by the military establishment indicated both ATC and NATS, to include the commercial aircraft, would be unable to meet the nation's wartime needs. "For this," the commission's report stated, "we must increase our commercial fleet."

At this point, the military was not the only group with scattered transportation resources. The report also disclosed that the United States did not yet have a single agency to consider transportation policy for the nation. The Finletter Commission recommended creating a Department of Civil Aviation within the Department of Commerce, which would later be upgraded to a Department of Transportation. The commission also noted that the military planned to "take over, as they did in World

War II, as much of the civilian lines, domestic and international, as circumstances permit" and suggested the preparation of prior agreements to specify what equipment and services the airlines would furnish to ATC. When the Finletter Commission submitted its final report at the end of December 1947, among its most significant recommendations was the creation of a single military air transport service.

Based upon the President's Finletter Commission, the new Secretary of Defense, James V. Forrestal, told the services to decide how - not whether - to consolidate air transport. From January through March 1948, Air Force and Navy working groups discussed consolidating air transport responsibilities. With agreements in general principle, the debate moved to higher levels between late March and early May 1948, as the specifics were decided. The Navy was particularly reluctant to give up its transport assets, but Air Force Secretary Symington consistently carried topics of disagreement to Defense Secretary Forrestal for resolution. Accordingly, on May 3, 1948, Forrestal signed a memorandum that created the Military Air Transport Service (MATS) as the single organization to manage strategic airlift operations. This new command, which began operations June 1, 1948, took over all ATC assets. The Naval Air Transport Service gave up fewer resources; 446 officers and 2,372 enlisted personnel from three Navy squadrons which flew C-47s and C-54s, went to MATS effective July 1, 1948. Other transport squadrons dedicated solely for fleet support remained a part of the Navy. No naval air facilities transferred to MATS, and the Navy only considered the aircraft "on loan," liable for recall at any time. A rear admiral became the deputy commander of MATS, while another rear admiral commanded MATS' Pacific Division. To the extent that Air Force and Navy forces were incorporated into a single command, MATS represented the first attempt at "jointness" within the Department of Defense. The Military Air Transport Service also included three technical organizations that had been part of ATC since March 1946: the Air Weather Service, the Air Rescue Service and the Airways and Air Communications Service.

The consolidation of strategic airlift came none too soon. Less than a month after the Military Air Transport Service was created, the Soviet Union blockaded land and water routes to West Berlin, and the United States and her allies inaugurated what would become a historic airlift to resupply the city. The Military Air Transport Service and its airlift structure would face a difficult test.

Helping Hands to Berlin

In the closing days of the war, American forces waited on the Elbe River while the Russians captured Berlin. The Americans joined them in occupying Berlin on July 2, 1945, when the first American troops landed at Tempelhof Central Airport in battle-scarred C-47s flown in from Halle, Germany, by the 301st Troop Carrier Squadron. Eventually, the victorious Allies divided Germany and Berlin into four zones, one each for France, Great Britain, the United States and the Soviet Union.

Relations with the Russian allies were so good at this early stage that the 301st Troop Carrier put on an airshow at Tempelhof the end of August 1945.

Soon, however, relations between the United States and the Soviet Union began to decline, and jointly-occupied Berlin offered the Soviets an ideal opportunity. In early 1948, the Soviets began exploiting the vulnerability of the western Allies' position in Berlin by stopping coal deliveries to their sectors of the city. A larger crisis arose in April 1948 when the Soviets restricted rail traffic into Berlin in a "mini-blockade" to a proposed West German

Top to Bottom: *Lt. Don Spearel in front of the control tower at K-46 airfield (Hoeng Song) Korea. June 1951. (Courtesy of Donald W. Spearel.)*

Air EVAC patients being loaded at Tokyo. Operation Big Switch.

Ambulance bus being loaded aboard a C-124.

Air Rescue Service unloading a patient off a H-19.

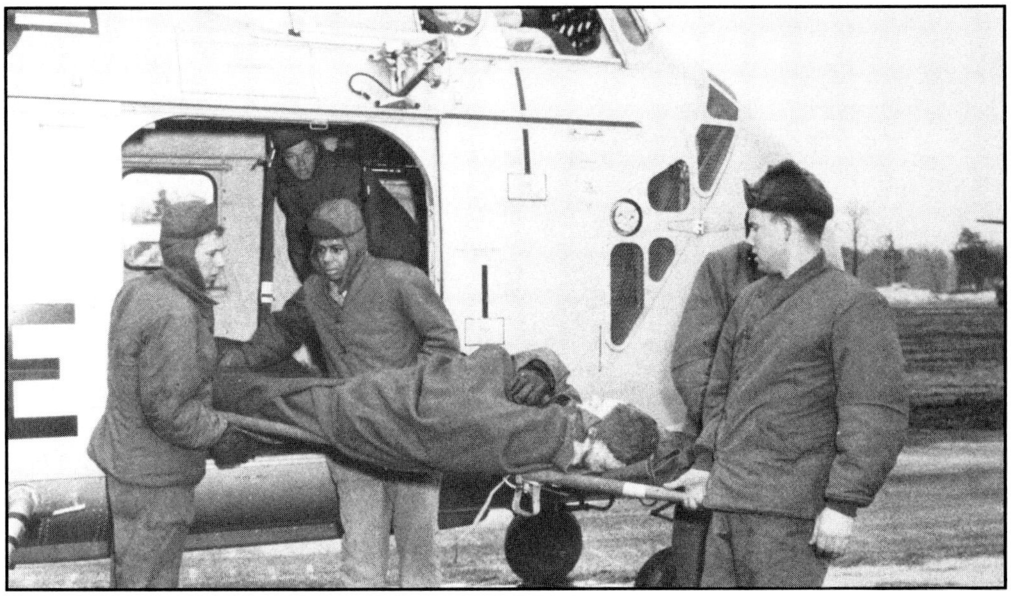

A Brief Chronology of Significant Airlift Events of the Korean Conflict 1950-1953

Office of AMC History, Scott AFB, IL

25 Jun. 1950 The North Koreans crossed the 38th Parallel and invaded South Korea. The first aircraft destroyed in the assault was not a fighter or a bomber. It was a C-54 transport operated by the Military Air Transport Service (MATS), Military Airlift Command's (MAC) immediate predecessor, on the ramp at Kimpo airfield, near Seoul.

25 Jun. At the outbreak of hostilities, the intratheater airlift mission was not a part of MATS; it fell under the direct command of the theater air component commander. Accordingly, the Far East Air Forces (FEAF) depended upon the 374th Troop Carrier Wing to manage most of its intratheater airlift mission. In addition, FEAF had scattered 13 C-46s and 22 C-47s throughout the Pacific for other air transport activities.

27 Jun. This evening General Douglas MacArthur, directed the FEAF commander to airlift 150 tons of ammunition from Japan to Korea on 28 June and to move an additional 200 tons per day until 1 July.

28 Jun. The first intratheater airlift operations in Korea took place on this date when 7 C-54s, 4 C-46s, and 10 C-47s, protected by an armada of fighters, evacuated 748 people from Korea to Japan. Simultaneously, these airlifters began an emergency movement of ammunition and other critical supplies to forces in Korea. These efforts did not stave off the North Korean advance, however; in the process FEAF lost two C-54s, the first transports destroyed by enemy action in the air since World War II.

28 Jun.-1 Jul. The 374th Troop Carrier Wing flew elements of the 24th Infantry Division from Japan to Korea to deploy these forces along the thinly defended Pusan Perimeter, the last remaining stronghold of the allies on the southern tip of Korea. The wing flew 757 hours in C-47s and C-54s to support this effort.

30 Jun. The North Koreans captured the airfield at Suwon, the last base in Korea where C-54s could take off and land. This forced intratheater airlifters to employ smaller aircraft, such as C-47s, which could be operated into airstrips still in allied hands.

1 Jul. The 21st Troop Carrier Squadron, part of the 374th Troop Carrier Wing, moved from its permanent headquarters at Clark Air Base, the Philippines, to Japan so that its C-54s could augment the airlift forces already operating in Korea.

17 Jul. MATS' Pacific Division had less than 60 modern C-54 aircraft when the war began, so Headquarters MATS directed its Continental Division to add another 40 C-54s to the Korean Airlift. One of those was MATS' VR-3, a Navy transport squadron flying R5Ds, the equivalent of the C-54, which had been providing Alaskan airlift support. In addition, Headquarters USAF directed that two troop carrier groups from the Tactical Air Command also augment the airlift force. One of these units was the 62d Troop Carrier Group which flew C-54s over the Northern Pacific Route from McChord AFB, Washington.

20 Jul. The Royal Canadian Air Force Squadron No. 426, operating six Canadian-built Douglas North Star transports, a version of the C-54, was placed under operational control of MATS at McChord AFB, Washington, and began to fly Korean Airlift missions.

23 Jul. By this date FEAF's airlift forces were capable of sustaining the movement of more than 150 tons between Japan and Korea per day.

26 Jul. The 61st Troop Carrier Group, operating C-54s and belonging to the Tactical Air Command, was officially placed under the operational control of MATS and began flying strategic airlift missions between McChord AFB, Washington, and Japan.

28 Jul. The first Korean Airlift mission of the Royal Canadian Air Force departed for Japan from McChord AFB, Washington.

28 Jul. The first four Air Rescue Service (ARS) SA-16 amphibious airplanes arrived in Japan to provide air search and rescue coverage for United Nations air operations over Korea.

31 Jul. In response to General Douglas MacArthur's call for a C-119-equipped paratroop group to support a proposed offensive in Korea, Tactical Air Command's 314th Troop Carrier Group (Medium) received an order on this date to deploy to the theater. The unit could only manage to send four C-119s to FEAF by 3 August, however, and the main part of the group did not arrive until 16 September.

31 Jul. By this time MATS was averaging 186 round trips to Japan per month over the Great Northern Pacific Route, almost all of them passenger missions. In addition, its monthly airlift total was 1,300 tons moved between the United States and Japan.

5 Aug. On this date the first open-sea rescue of the Korean Conflict took place, when an ARS SA-16 picked up a Navy Corsair pilot off the western coast of Korea.

10 Aug. A Belgian Air Force C-54 assigned to support the strategic airlift to Korea arrived at Fairfield-Suisun AFB, California, in preparation for central Pacific missions.

11 Aug. The First Belgian Air Force C-54 mission departed for Japan from Fairfield-Suisun AFB.

14 Aug. Canadian participation in the strategic airlift to support the Korean conflict was augmented with the addition of three four-engine transports from Canadian Pacific Airlines.

15 Aug. In one of the speediest saves during the war, a downed F-51 pilot, who bailed out over the Sea of Japan, was rescued by an ARS SA-16 amphibious plane less than five minutes after he had crashed.

22 Aug. FEAF organized the 1st Troop Carrier Task Force (Provisional) at Ashiya Air Base, Japan, to provide airlift for the projected invasion of Korea in mid-September 1950.

26 Aug. Lieutenant General George Stratemeyer, FEAF commander, organized Combat Cargo Command (Provisional) under the command of Major General William H. Tunner, an airlift expert from the Military Air Transport Service. This organization reported directly to Headquarters FEAF.

10 Sep. MATS' Pacific Division had in operation for the Korean War, by this date, a strategic airlift total of 41 C-54s; 14 R5Ds, 11 C-97s; and 9 C-47s. In addition to these 75 aircraft assigned to the Pacific Division, it was augmented by 75 Troop Carrier C-54s from the Tactical Air Command; 8 C-54s from the Royal Canadian Air Force; and 66 commercial aircraft.

10 Sep. The largest increment of C-119s, from the 314th Troop Carrier Group (Medium), arrived in Japan on this date.

15 Sep. The Inchon amphibious invasion took place as assault elements of the X Corps went ashore supported by air and naval forces.

16 Sep. The last elements of 314th Troop Carrier Group (Medium) were deployed to the Far East and assigned for FEAF's Combat Cargo Command. This brought to the command a total of 96 C-119s for intratheater airlift operations.

19 Sep. At 1426 local, the first C-54 landed at Kimpo field since the opening days of the Korean War. During the afternoon, 8 other C-54s and 23 C-119s landed with 208 tons of supplies for ground troops, lighting equipment, and 280 men of the 1st Combat Support Unit (Provisional) to provide aerial port functions.

20 Sep. FEAF Combat Cargo Command began an around-the-clock airlift between Japan and Kimpo which delivered in excess of 226 tons per day to ground forces. On their return trips, the transports handled aeromedical evacuation for casualties. In accomplishing this, Major General Tunner emulated the Berlin Airlift by setting up a tight airlift schedule with landings and takeoffs every two minutes and by directing that pilots use instrument flight rules exclusively to ensure that the flow and route were meticulously accurate for every mission. By the end of the month, the Combat Cargo Command was successfully airlifting between 800 and 900 tons per day into Kimpo.

21 Sep. Nine C-119s made emergency airdrops of ammunition and rations to front-line troops.

20 Oct. In one of the major operations of the war, FEAF's Combat Cargo Command conducted an airdrop in the Sukchon-Sunchon area about 30 miles north of the North Korean capitol of Pyongyang. At 1400 local time, 71 C-119s and 40 C-47s dropped 2,860 troops from the 187th Airborne Regimental Combat Team and 301.2 tons of supplies between the retreating North Korean soldiers and their sanctuary in the north. By day's end, the paratroopers had secured the drop zones, established positions at strategic sites, and blockaded roads and railways.

21 Oct. As a continuation of the previous day's airdrop, at 1000 local time 40 C-119s delivered 1,093 additional paratroopers and 106.8 tons of supplies. That afternoon the 187th linked up with United Nations forces moving north. In all, allied forces killed or captured about 6,000 North Koreans and rescued several allied POWs.

8 Nov. Major General Tunner increased Korean intratheater airlift capability by acquiring additional forces. Less than 36-hours after its C-46s bedded down at Brady Field, Kyushu, Japan, the 437th Troop Carrier Wing began shuttling cargo into Korea.

26 Nov. The Royal Hellenic Air Force Flight No. 13 arrived in theater with transport aircraft and was attached to the 21st Troop Carrier Squadron of the 374th Troop Carrier Wing for support.

27 Nov. The Red Chinese Army intervened in the Korean War, beginning a harsh assault on United Nations troops which forced them back near the 38th Parallel. In the process some of the UN troops were cut off from support in the South.

28 Nov. Elements of the X Corps were trapped at the Chosin Reservoir deep in North Korea. One of the most savage campaigns in American history, it pitted about 15,000 allied troops against an estimated 120,000 Chinese and North Korean troops. During this operation airlift offered the only reliable means of supply. The first aid to reach the troops came during an airdrop by 16 C-47s from the 21st Troop Carrier Squadron. These aircraft dropped 25 tons of ammunition.

29 Nov. The same 16 C-47s dropped 35 tons and 15 C-119s dropped another 80 tons of ammunition to the troops at Chosin.

Dec. The 61st Troop Carrier Group was hurriedly dispatched to Japan on a temporary duty basis to provide addition intratheater airlift support during the Chinese offensive.

4 Dec. Lieutenant General Stratemeyer instructed Combat Cargo Command to use all of its C-46s, C-47s, and C-119s in support of X Corps at the Chosin Reservoir.

4-10 Dec. Elements of the 1st Marine Regiment, which had been cut off at Hagaru-ri and Koto-ri during operations to assist troops withdrawing from the Chosin Reservoir, hacked out two rough airstrips, and C-47s made 240 landings, most of them by the 21st Troop Carrier Squadron, to bring in 273.9 tons of supplies and take out 4,689 sick and wounded. In addition, planes from the 1st Marine Air Wing made 56 sorties, and C-47s from the Royal Hellenic Air Force flew 30 sorties carrying additional cargo to the Marines.

5 Dec. The Eighth Army decided to evacuate as many patients in hospitals in Korea as possible, lest they be captured by advancing Chinese forces. C-54s from the 801st Medical Air Evacuation Squadron moved more than 4,000 patients to Japan in this one 24-hour period.

7 Dec. The cut-off forces at Hagaru-ri and Koto-ri joined together and tried to fight their way out of enemy encirclement and move to Hungnam. They soon found that a critical bridge had been blown over a 1,500 foot gorge and requested an airdrop to bring in eight spans of an M-2 treadway bridge, complete with plywood planking. These were packaged, one span per bundle, and weighed two tons each. The FEAF C-119 detachment at Yonpo made an unsuccessful test with six small chutes, but then decided to use two large parachutes per bundle. A single C-119 flew from Yonpo to Koto-ri and dropped the bridge from 800 feet in a valley on an unmarked drop zone. Although one span fell into enemy hands, the remainder were serviceable and used to get the troops over the gorge.

8 Dec. After 13 days of isolation, the cut-off Allied Forces met up with troops from Hungnam. Only with the aid of airlift did they withdraw with most of their equipment. Since 27 November, FEAF Combat Cargo Command had supported the cut-off force with 313 C-119 and 37 C-47 airdrop missions, delivering 1,580.3 tons of supplies and equipment. In all, airdrop and airland, the FEAF Combat Cargo Command delivered about 2,000 tons of supplies to troops during this operation.

14-17 Dec. Combat Cargo Command undertook a maximum effort air evacuation of allied forces from Yonpo. During this operation, it flew 393 sorties, moving 228 patients, 3,891 passengers, and 2,088.6 tons of cargo. Yonpo was overrun by enemy troops at 0900 local time on 17 December.

20 Dec. In the midst of the Chinese advance, in Operation CHRISTMAS KIDLIFT, 12 C-54s from the 61st Troop Carrier Group flew 989 Korean orphans from Kimpo Airfield to safety at Cheju-du Island, off the southern coast.

1-24 Jan. With roads and railroads clogged by bad weather and retreating UN forces, airlift became an increasingly important means of supplying troops in Korea. FEAF's Combat Cargo Command lifted 5,041 tons of men and material for the Fifth Air Force and 7,445 for the Eighth Army throughout the theater. On return trips, the transports evacuated 10,489 combat casualties to rear hospitals. In addition, on 406 C-119 sorties, FEAF airdropped 2,004 tons of supplies to forces, most of it to X Corps.

2 Jan. A C-47 loaded with flares spent five hours over a target area north of Seoul dropping a 129 Mark VIII flares to provide illumination for B-26 bombardment operations. This was the first such use of the aircraft, but not the last. Soon, FEAF modified three C-47s, which it called the "Firefly," for flare dropping and assigned them to the Tactical Flight Section, 3d Air Base Group. These modified C-47s stood alert at Taegu Air Base. Troops all over Korea soon began referring to the C-47s performing this mission as "The Old Lamplighter of the Korean Hills."

2-6 Jan. The 21st Troop Carrier Squadron airlanded 115 tons of cargo in C-47s, and the 314th Troop Carrier Group's C-119s dropped an additional 460 tons of supplies to embattled ground forces at Wonju during the Chinese assault.

30 Jan. The 61st Troop Carrier Group delivered 270 tons of cargo to Army forces at Suwon.

4 Feb. In an effort to interdict enemy convoys, during Operation TACK the flare-dropping C-47s were called up to drop tacks on sections of enemy-held roads. When the trucks stopped to change punctured tires, allied fighters would destroy all stopped vehicles. On this night several C-47s flew at a height of 10 to 20 feet, scattered 8 tons of roofing nails along 4 highways south of Pyongyang. One of the C-47s nearly collided with three Chinese tanks during this operation and called in B-26s to attack the tanks. In addition, FEAF netted 28 vehicles destroyed in this operation. It was moderately successful, but not sufficiently so to continue as a normal part of operations.

15-16 Feb. When American soldiers were completely surrounded by the enemy at Chipyong-ni, 20 miles east of Seoul, the 314th Troop Carrier Group's C-119s airdropped to them 87 loads of ammunition, gasoline, and rations. In addition, H-5 crews from Detachment 1, 3d Air Rescue Squadron, evacuated 52 critically wounded soldiers. This effort, as well as other air support, allowed the troops to hold out until UN forces could link up with them.

23-28 Feb. When OPERATIONS KILLER, a UN counterattack in central Korea, bogged down in the thawing muck and impaired the lines of communication, the 314th Troop Carrier Group began supporting advancing forces with airdrop operations. During this period it delivered by airdrop 1,358 tons of cargo on 256 C-119 missions to ground troops at Wonju.

1 Mar. The C-47 "Fireflies" began to team with Marine F4Us and F4Fs on interdiction missions less far north than the deeper penetrating B-26s. This enhanced the survivability of the C-47s and proved to be a most productive marriage. The two services cooperated in this way throughout the rest of the war.

23 Mar. In Operation TOMAHAWK, at 0900 local time, a force of 72 C-119s dropped 2,001 paratroopers and 204 tons of supplies, while 48 C-46s dropped 1,436 paratroopers and 15.5 tons of supplies. The troopers were from the 187th Airborne Regimental Combat Team and were delivered on two drop zones near Munsan-ni where they were used to help trap an enemy force which was retreating in the face of UN ground advances. During this operation, five C-119s incurred damage from small-arms fire. A sixth sustained greater damage, exploding in the air, and although five crew members bailed out, the pilot and copilot were killed.

24 Mar. Following this airdrop, the 187th Airborne Regimental Combat Team attacked eastward. Far outstripping the Army's surface logistics capability in the region, it called on the 315th Air Division for airlift support. On this date 36 C-119s airdropped 40 men and 187.7 tons of supplies to Munsan-ni to sustain the paratroopers.

24 Mar. General MacArthur announced that the United Nations would probably be willing to carry the war from the Korean peninsula into mainland China. This was a striking statement, made by a soldier rather than by a person charged with formulating foreign affairs policies, running counter to President Harry S. Truman's official position.

26 Mar. The 315th Air Division used four C-46s to drop 10 tons of supplies to the 187th Airborne Regimental Combat Team during its advance eastward.

27 Mar. The 315th continued to support the advance of the 187th, airdropping 65.8 tons of cargo from 12 C-119 missions on this date. This airdrop was especially important because the Army forces were receiving almost no supplies by surface transport and this effort resupplied them with food and ammunition.

11 Apr. President Harry S. Truman relieved General Douglas MacArthur from all his commands in the Far East, because MacArthur did not agree with the military strategy of the United States in Korea: "to repel attack...to restore peace...to avoid the spread of the conflict." He replaced MacArthur with Lieutenant General Matthew B. Ridgway.

22-27 Apr. During a major enemy counterattack in western Korea, the 315th Air Division supported UN forces trying to hold the line. It airlifted more than 4,500 tons of cargo from Japan to Korea, airlanding about 1,700 of these tons at Kimpo. The remainder was airdropped or airlanded at other locations.

30 Apr. During April 1951, the 315th Air Division delivered in Korea, by both airdrop and airland, a total of 15,900 tons of cargo.

16-25 May On each of these days, the 315th Air Division airlifted at least 1,000 tons per day throughout the Korean Combat theater, to support ground forces under a general enemy assault.

11 Jun. First Lieutenant John J. Najarian successfully rescued a downed fighter pilot from behind enemy lines on the Taedong River in North Korea. He executed a night landing and takeoff without lights on the river in his SA-16 amphibian. He received the distinguished service cross for his action.

30 Jun. During this month, the 315th Air Division delivered a total of 22,472 tons of cargo to Korea, by both airland and airdrop methods. In addition, its 6127th Air Terminal Group loaded 10,938 aircraft sorties with 60,475 passengers, 17,146 tons of cargo, 899 tons of mail, and 10,520 aeromedical evacuation patients.

16-17 May The 315th Air Division was notified at 0900 local time of the requirement to move elements of the 187th Airborne Regimental Combat Team from Ashiya and Brady Air Bases, Japan, to Pusan East airfield. The last of the 160 C-119s moving 2,361 troops and 889.1 tons of equipment landed at Pusan by noon the next day. Included among the tonnage moved were mortars, vehicles, weapons, and ammunition. The 187th then moved by ship to Koje-do where it put down a riot by prisoners of war.

Jun. Only 28 of 71 C-119s in the Korean theater were in commission during this month, and none of those 28 were counted on as actually safe for flying. They had been worn out in the difficult operations of the first part of the war; this caused the 315th Air Division's leadership on 19 June 1952 to request their replacement.

8-29 Jun. During this three-week period, the 315th Air Division handled the single largest airlift of an Air Force unit in the war. It moved the entire 474th Fighter-Bomber Wing from Misawa Air Base, Japan, to Kunsan Air Base, Korea.

5-15 Oct. As part of a ruse designed to test the strength and determination of the enemy, General Mark Clark, commander of forces in the theater, directed his key subordinates to plan and give every sign that they would participate in an airborne and amphibious assault at the coastal city of Kojo. Lieutenant General O.P. Weyland, commander of Far East Air Forces, directed his key air leaders to do the same. His fighters and bombers stepped up strikes in the area. Only the top leaders understood that the plan was a hoax.

15 Oct. At 0300 local time this day, crews from the 403d Troop Carrier Wing were awakened and told to fly from Ashiya Air Base, Japan, on a feint in Korea. Before dawn 32 C-119s were in formation heading for Chorwan. They leveled to a paradrop altitude of 800 feet and headed for enemy territory. Just before crossing the lines in airdrop formation, however, they turned and recovered to Taegu where they loaded more paratroopers for return to Japan. From the reactions of the enemy, military analysts assessed their readiness and capabilities to react to this type of operation. In addition, the 315th Air Division had received some valuable refresher training in airdrop operations.

1 Feb. The 315th Air Division moved the 461st Troop Carrier Squadron and the Royal Hellenic Air Force detachment to Seoul, Korea, to serve as a feeder line for cargo and passengers airlifted from Japan to Korea on C-124s.

Mar Following the loss of a C-119 in an accident, the 315th Air Division prohibited the old and rapidly degenerating aircraft from carrying passengers in Korea, although they could still carry up to six tons of cargo and engage in airborne training with the 187th Airborne Regimental Combat Team.

6-11 Apr. Negotiators at Panmunjom worked out the details of a prisoner of war exchange.

5 May The last group of allied exchanged prisoners of war were flown from Haneda Air Base, Japan, to the United States by a MATS C-54.

29 May A C-124's number-two engine caught fire in flight over Korea, raising doubts about the fuel system in the aircraft.

11 Jun. Another C-124 had an engine fire during a mission in Korea. This prompted an investigation, which found that the engine fires were the result of faulty generators.

18 Jun. The generator problem led to the worst air transport disaster in the Korean War, when a C-124 on this evening lost power from an engine on takeoff and spun into the ground near Tachikawa Air Base, Japan, killing all 129 passengers and crew. This led to a grounding of the fleet and overhaul and repair of all generators.

21-23 Jun. General Mark Clark directed that the 315th Air Division move the 187th Airborne Regimental Combat Team from Japan to central Korea as a defensive measure before the final truce. This was accomplished on 53 C-54 and 249 C-119 missions. Daytime flights landed at Chunchon and night landings used the ground control approach capability at Seoul. Moving 1,770.6 tons of cargo and 3,252 troops, the operation was completed on 23 June 1953.

28 Jun.-2 Jul. Using C-46, C-54, and C-119 aircraft the 315th Air Division moved elements of the 24th Infantry Division from Misawa and Tachikawa Air Bases, Japan, to Pusan and Taegu Air Bases, Korea. The airlift moved 898 soldiers and 284.2 tons of equipment from Misawa and 3,039 troops and 943.27 tons of cargo from Tachikawa. These troops were moved to fill in defensive positions prior to the final truce.

8 Jul. Seventeen C-124s resumed operations after undergoing extensive examination and repair to generators. Eight other C-124s remained grounded because of a lack of spare parts.

27 Jul. An armistice was executed between the United Nations allies and the North Koreans and Chinese ending the three-year long Korean War. During the conflicts MATS, using its fleet of C-47, C-54, C-97, C-119, and C-124 aircraft, had airlifted 214,000 passengers and 80,000 tons of cargo from the United States to staging areas in Japan. Within the combat area, the Combat Cargo Command (later redesignated 315th Air Division) under the control of the Far East Air Forces, with an average of 210 aircraft, flew 210,343 sorties and carried 391,763 tons of cargo, over 2.6 million passengers, and more than 307,000 patients. It airdropped 15,000 tons of supplies and equipment during such critical operations as the allied withdrawal from the Chosin Reservoir. About 386,000 patients were airlifted to medical treatment centers, 320,000 in the theater and 66,536 to the United States. The swift evacuation resulted in the lowest death rate from wounds experienced by United States forces in any war to date. Finally, the Air Rescue Service recovered 9,898 United States troops, of which 996 were from the combat zone.

C-124 52-1073 20th ATS, Dover AFB, 2 Dec 1963. Lightning strike over Med/Enroute Incirlik AB, Turkey. Chateroux AB, France. (Courtesy of Robert W. Schorr.)

state. The western Allies responded with the "little airlift" as local transport aircraft, mostly C-47s carried enough supplies into Berlin to over-come the short-lived blockade. All of this portended the blockade and airlift that followed between June 1948 and September 1949.

When the western Allies did not end their efforts to establish an independent West German state, the Soviets on June 24, cut off all food supplies and electricity to Berlin, making the excuse that it was the result of "technical difficulties."

General Lucius D. Clay, the American commander in Germany, and several other leaders immediately anticipated using aircraft to overcome the blockade. Clay ordered Lieutenant General Curtis E. LeMay, Commander of the United States Air Forces in Europe, to organize his fleet of C-47 transports into an airlift to resupply Berlin. LeMay placed Brigadier General Joseph Smith in charge, and Smith began the first airlift flights using the 102 C-47s and two of the newer C-54 aircraft of the 60th and 61st Troop Carrier Groups on June 26, 1948. The British also furnished *Dakota* (C-47) transports. That first day, American crews flew in 80 tons of milk, flour, medicine and other high-priority cargo on 32 C-47 flights. Thereafter, daily runs from Wiesbaden and Rhein-Main Air Bases to Tempelhof Field occurred. Operations VITTLES, as the airlift was called, at first only resupplied American forces stationed in the city. Later, President Truman expanded the airlift to include necessary food, medicine, fuel and clothing for all Berliners.

Within a month, the airlift was proving successful but not spectacular. Originally, General Clay had estimated that the city needed 4,500 tons of supplies daily. A month into the operation, however, Americans were moving only around 1,000 tons per day, the British about 750 tons more. To achieve the minimum tonnage required, General Clay expanded the operation, pushing personnel and aircraft resources to their limits. Although the airlift was a USAFE operation, the Air Force also brought in the resources and expertise of the Military Air Transport Service. Major General William H. Tunner, an airlift expert of proven ability and senior official in MATS, was called in to take over the operation. He headed a multinational Combined Airlift Task Force (CATF), formed in October 1948, largely from C-47 and later C-54 transport units as well as Navy R5D aircraft and personnel assigned to MATS. Through the CATF, the British and Americans merged their efforts; RAF Air Commodore J.W.F. Merer served as Tunner's deputy.

Major General Tunner adapted the lessons learned while operating the Hump Airlift in World War II to beleaguered Berlin, always reminding the world that the seemingly impossible had been done before. Mass strategic airlift was already a proven commodity by the time of the Berlin Airlift. He expanded the number of bases dedicated to the airlift, exchanged the small C-47s for the larger C-54 transports, increased the number of aircraft and personnel assigned, streamlined the size and complexity of the airlift support system, and, most important, greatly enhanced efficiency through a number of management innovations. Tunner emphasized using all 1,440 minutes of the day for the airlift. He envisioned an airplane landing every minute. Although this was impossible, it indicated the degree of proficiency he sought. He settled for the more practical rate of one every three minutes that his predecessor had established. This rate, he noted, "provided the ideal cadence of operation with the control equipment available at the time." He explained, "At three-minute intervals, this meant 480 landings at, say Tempelhof, in a twenty-four-hour period. The planes that came in had to go out again, of course, and with the take-off interval also set at three minutes, this meant that a plane either landed or took off every 90 seconds." in an understatement he noted, "There was little time wasted sitting at the ends of runways."

The Berlin Airlift was significant for many reasons. In foreign relations, it demonstrated the United States' resolve to meet a Cold War challenge. America's Allies around the world regarded the airlift as a triumph of will. It also impressed the Soviet Union. At no time in their history prior to 1948 could the Soviets have mounted such an extensive operation. The airlift affected Air Force doctrine as well. It showed, for example, that virtually any amount of cargo could be moved anywhere in the world, if proper support were available. The airlift provided valuable experience in operational techniques, air traffic control, and aircraft maintenance and reconditioning. Furthermore, as the editor of *Air Force* magazine reflected in September 1948, "For the first time in history, the United States is employing its Air Force as a diplomatic weapon." The Berlin Airlift proved what has been confirmed many times since: airlift is a more flexible tool for executing national policy than either fighter or bomber aircraft.

Airlift and the Korean Conflict

Most histories of the Air Force stress that when the orders came to airlift troops and supplies to Korea, a force seasoned by the lessons of World War II and the Berlin Airlift responded. However, when the Combined Airlift Task Force finished the Berlin Airlift in the fall of 1949, the transport fleet was in rough shape. The demanding flying conditions and seepage of coal dust and flour into aircraft seams required the fleet to undergo extensive rehabilitations. Flying hours were reduced to minimum levels. The Air Force and Military Air Transport Service were again shrinking to a core from which they planned to expand in wartime. Between the conclusion of World War II and June 1950, more than three quarters of the airlift capability of the command had been demobilized, mothballed, or placed in the reserves.

On June 25, 1950, at 0400 hours, the North Koreans invaded South Korea.

The United States was completely unprepared for the North Korean attack. Its forward military forces had been drawn down, and the nation had neither the strategic nor the tactical airlift capability to recover quickly. During the first eight months, only outstanding efforts in all of the combat arms, coupled with considerable luck, prevented Allied troops from being pushed off the Korean peninsula.

The Military Air Transport Service was responsible for providing the strategic airlift necessary to sustain the war effort in Korea. Unfortunately, MATS was ill-equipped for the task at hand. It was symbolic, perhaps, that the first American aircraft lost in the Korean War was not a fighter of a bomber but a MATS C-54 supporting the United States Korean Military Advisory Group. The C-54, grounded at Kimpo Airfield near Seoul because of a damaged wing, was destroyed on the first day of the war by two Yak fighters strafing the field. It represented the state of an airlift system which a few years before had sustained the two million inhabitants of Berlin.

If MATS' C-54s were Berlin-worn, the real limiting factor for Korea was the number of aircrews. There were only enough men to achieve an aircraft utilization of 2 and 1/2 hours per day. The war was 30 hours flying time away from the United States for a C-54 via the shortest route - the Great Circle - and 34 hours by the more practical mid-Pacific route. Fortunately, the airlines were manned for high utilization rates. During the 75 days it took to recall and train personnel, the civilian airlines operated 66 aircraft under contract, flying an average of 10 hours per day. Orvis M. Nelson's Transocean Air Lines, a supplemental carrier, was the first commercial airline to fly Korean airlift missions for MATS. Transocean used seven DC-4s, the commercial version of the C-54, to handle nearly 14 percent of the entire Korean Strategic airlift.

During Korean airlift operations, civil air carriers flew more than 40 percent of the missions on the United States-Japan shuttle.

Airlift operations during Korea were divided. Strategic airlift, which went from the United States to Japan, was the responsibility of the Military Air Transport Service's Pacific Division, based at Hickam Air Force Base, Hawaii, and commanded by Navy Rear Admiral William G. Tomlinson. The trans-Pacific airlift had to be augmented by a second, intratheater airlift between Japan and Korea. The Military Air Transport Service's Deputy Commander, Major General William H. Tunner, was once again called upon. Headquarters USAF assigned him to organize for FEAF a Combat Cargo Command (Provisional) on September 10, 1950. This unit handled all kinds of airlift, including airlanding supplies and troops and airdropping combat troops, equipment and supplies.

Within nine days of its creation, the Combat Cargo Command was flying men and supplies to Inchon. Air Force cargo-handling teams, another Tunner innovation, speeded unloading. Return flights provided aeromedical evacuation for casualties, carrying them from the beachhead to hospitals in Japan.

Tunner's Combat Cargo Command performed a variety of missions including airdrop and airland in combat environments, logistical airlift in areas where the threat of enemy action was not great, and, increasingly, aeromedical airlift of sick and wounded.

The Military Air Transport Service's aerial cargo delivery from the United States to Japan began on a modest scale - only 2.5 tons per day - but rose to an average of 106 tons daily. By the time of the truce of July 27, 1953, MATS had airlifted 214,000 passengers and 80,000 tons of cargo to Japan. Together, the theater and strategic airlift systems flew 210,343 sorties and carried 391,763 tons of cargo, over 2.6 million passengers, and more than 310,000 patients. They airdropped 15,000 tons of supplies and equipment, often in situations where it was impossible to bring supplies in any other way. During the course of the war, MATS transported 43,196 Korean War casualties to the United States for further hospitalization or specialized medical treatment.

The Berlin Airlift was not a combat operation; its lessons were more managerial in nature. On the other hand, the Korean Conflict taught some difficult lessons about conducting modern combat airlift. Tactics, organizational structure, operations planning, and a myriad of other issues of a more subtle nature were elucidated as a result of the Korean experience. As with the Berlin crisis, American military airlift, especially tactical airlift, emerged from the conflict with its reputation enhanced by its many successes. Following Korea, however, the military was once again neglected. The concept of nuclear deterrence, combined with emphasis on air defense and tactical fighters, left little room for the aerial resupply of troops.

The Civil Reserve Air Fleet

One of the lessons learned during World War II and confirmed in Korea was that the nation could not maintain enough airlift capability in its military to respond to wartime requirements. This provided the genesis for the inauguration of the Civil Reserve Air Fleet (CRAF), a partnership between the commercial airlines and military airlift to ensure that sufficient airlift was available for deployments in the event of contingencies or wars. This concept was developed during the latter 1940s after both government and civilian airline officials realized that they had been overly optimistic about the volume of passenger traffic to expect after the war.

The Finletter Commission had maintained that the only valid reason for subsidizing commercial cargo airlift was to develop a fleet of aircraft that would act as a "pool" for military emergencies.

Nearly three years after the Finletter Commission delivered its findings, a more specific study was conducted concerning wartime airlift requirements. The study was performed by the Committee on Wartime Airlift Requirements and Capabilities, chaired by James H. Douglas, and therefore called the Douglas Commission. The study recommended establishing a three-tiered reserve of four-engined transports in civil airlines. First, the committee members stated that by the end of June 1951, airlines would have the equivalent of 350 C-54 airplanes that could be converted to extended military operations within 48 hours. Second, by December 1951, another 400 C-54 equivalents would become available to augment the military within 48 hours. Finally, the commission recommended designating another 100 C-54 equivalents, to include training crews and modifying the aircraft's range for Europe. The Douglas Committee recognized that the required military modifications, making the aircraft heavier, would increase the operating expense of commercial airlines. The committee calculated what the added costs would be and suggested that military pay the difference.

In 1951, the Douglas Commission's report became the basis for organizing the commercial carriers to augment for the military airlift system. The Department of Defense and the Department of Commerce jointly approved the basic concept late in the year, calling it the Civil Reserve Air Fleet. Headquarters MATS worked with the Department of Commerce and the civil airlines planning how to incorporate 300 four-engine airplanes into MATS' operations following the declaration of a national emergency.

Fully mobilized, CRAF participants were expected to airlift 95 percent of the passengers and 35 percent of the cargo required by overseas theaters. The CRAF program has been an important element of defense airlift planning since its inauguration. Except during the recent Middle East crisis, Operation Desert Shield, there was never a need to activate the CRAF, for the airlines have always made aircraft available when crises and contingency operations required more airlift than the military was able to provide.

Following World War II, the military in general and the Military Air Transport Service in particular learned some valuable lessons. Strategic airlift had been consolidated in the military, at least on paper, and had proved its worth as a diplomatic tool in Berlin and as a combat force in Korea. At the same time, the drawdown in resources and the search for an optimum organizational structure for airlift forces continued to challenge both the command and the Air Force.

The Interlude, 1953-1964

During the period before Vietnam, the Military Air Transport Service with its mission of military airlift became an increasingly important element of the national defense posture. This was largely due to a series of Cold War crises that began in the mid-1950s and required immediate military action short of nuclear confrontation. As the Military Air Transport Service rose to meet its tasking, it gradually acquired a combat airlift mission in addition to its air logistics responsibilities. None of this, however, was discernible in 1953.

In the post-Korean environment, the Military Air Transport Service found itself under serious attack by the commercial air carriers who had secured favorable support from congressional leaders. Two government reports, in particular, fueled the ensuing debate over the proper role and function of military airlift. In the end, the controversy brought fundamental changes to MATS. Most significantly, it substantiated the need for military airlift, strengthening the command's peacetime and wartime missions. At the same time, the commercial carriers, through the Civil Reserve Air Fleet Program, assumed a greater portion of the government's airlift business. The debate also resulted in an extensive aircraft modernization program for MATS.

Loading a jeep aboard a C-54.

Airlift Modernization

Besides working out the relationship between military and civilian air transportation agencies, the extensive congressional airlift hearings also determined what kind of transport aircraft the military would procure. Prior to this congressional direction, the Military Air Transport Service in the early 1950s had hoped to replace its aging World War II fleet with two types of turboprop aircraft: a pure cargo aircraft capable of carrying 50 tons a distance of 3,500 miles and a passenger-cargo aircraft capable of transporting 15 tons or 100 passengers as needed over the same distance. The requirement of 3,500 miles was based on the realization that many en route air bases would in all probability not be available in wartime. But as the mission of the Military Air Transport Service evolved to include combat airlift, command officials advocated procuring jet aircraft.

Although Congressman Flood had raised the aircraft issue during the 1956 House Defense Subcommittee hearings, no congressional directive was forthcoming until 1958 when the Holifield subcommittee recommended that the Military Air Transport Service concentrate on airlifting outsize and special cargo, leaving the passenger and conventional cargo business to the commercial carriers. Consistent with this division of airlift, the subcommittee also stipulated that the Air Force take action to modernize the MATS fleet by procuring a large, long-range cargo aircraft.

Congressman L. Mendel Rivers in his subcommittee investigations of 1958 also expressed great displeasure at the Air Force's air transport modernization program. Rivers proposed jet aircraft as the ideal for MATS so that it could keep up with the strike forces was supposed to support. At a minimum, the Rivers subcommittee advocated procuring DC-8, B-707, C-133, and /or C-130B aircraft. Accordingly, Congress in the military appropriations for 1959 directed the Air Force to modernize its military air transports.

Responding to the congressional recommendations, Defense Department officials stated their concurrence on modernizing airlift, provided it was not placed ahead of other military programs. Demonstrating their commitment, they disclosed on-going plans to retire MATS' piston-powered C-54 *Skymaster* and C-97 *Cargomaster.* Military Air Transport Service and Air Force planners also studied future airlift requirements and recommended acquiring Lockheed's C-130B *Hercules* and a "swing-tail" jet cargo aircraft to complement the C-133, envisioning then the C-135 *Stratolifter* as a swing-tail aircraft. They also proposed developing a cargo jet for the 1966-1970 time period, this became the C-141 *Starlifter.*

By 1960, MATS' airlift mission had evolved from solely supporting the nuclear forces of the Strategic Air Command and the fighters and bombers of the Tactical Air Command to include deploying Army and Navy combat forces worldwide and maintaining an aerial resupply system. John F. Kennedy had made the airlift issue part of his presidential campaign and spoke of the need to obtain "additional air transport mobility - and obtaining it now" in his State of the Union address in January 1961. Kennedy also disclosed that the United States defense policy would adopt a flexible response strategy versus swift and massive nuclear retaliation. Rapid mobility became a key element of the Kennedy Administration's posture of deterring the full spectrum of warfare. Support for MATS' airlift modernization program had never been more certain.

Incorporating President Kennedy's new defense strategies, Secretary of Defense Robert S. McNamara secured several changes in the airlift modernization program. He asked Congress to increase the procurement of the longer range C-130Es from 50 to 99; sanctioned the modification of 17 KC-135 tankers under production into transport configurations; requested the purchase of 13 other C-135s for a total of 30; and proposed the development of a modern, long-range cargo aircraft, the C-141. The new national security posture had made an interim modernization program for the Military Air Transport Service of the utmost importance. The command would use the new C-130 and C-135 aircraft to fill the void until the arrival of the C-141. President Kennedy inaugurated the C-141 program on March 13, 1961, when he announced the Lockheed Company as the winner of the design competition. Less than three years later, on December 17, 1963, the C-141 *Starlifter* soared.

MATS Acquires Combat Airlift Mission

As national leaders came to endorse the flexible response concept, they acknowledged a shortage in airlift assets. Increasingly, the Military Air Transport Service found itself tasked to support contingency operations, often at short notice. Airlift, in this period, evolved from a purely air transport function to one of providing combat airlift. World crises and the White-Lemnitzer Agreement of March 1960 had essentially brought about these changes for MATS. Ever since the Military Air Transport Service had gained troop carrier assets in 1957, the command had the means to support the Army's tactical requirements for airdrops although MATS had eliminated formation training. Despite the precise statements in MATS Regulation 23-1 (April 15, 1959) that troop carrier operation included "airdrop supply" and "airborne operations," the command initially used its troop carrier aircraft for logistical resupply missions.

Aware of the changing environment and the increased emphasis on airdrop and airborne operations, the Military Air Transport Service sought to improve its ability to deliver troops and supplies to forward areas when it issued a directive in January 1961, requiring all of its C-124 units to qualify in the computed air release point (CARP) aerial delivery procedures. Motivating its aircrews to this end, the command initiated annual competitions in April 1962. In this first event, seven troop carrier and air transport wings flew their C-124s in day and night formations, testing their accuracy in airdropping paratroopers and cargo. The competition underscored the need for MATS aircrews to participate in more realistic aerial delivery training.

The command, however, soon received more explicit tasking. In May 1962, following a planning discussion held at Headquarters Commander in Chief Atlantic (CINCLANT) in December 1961 over the types of aircraft to be used for airdrop requirements, Headquarters United States Air Force directed MATS to develop a formation flying capability for C-124 aircrews. Military Air Transport Service troop carrier units were to become fully qualified in formation flying by March 1963, with the air transport wings to follow thereafter. A few months later, CINCLANT's massive airdrop plans for the Cuban missile crises led to an accelerated training program.

Accordingly, based upon this new guidance from higher headquarters, the Military Air Transport Service issued its own regulation changes on July 20, 1962, which incorporated airdrop of personnel and cargo in mission statements of both troop carrier and air transport units, specifically stating the directive applied to C-124A, C-124C, and the new interim C-130E aircraft. Highlighting the command's relative inexperience in aerial delivery, the new change further stipulated that "Operations of this type will employ delivery by single ship or by three-ship element formation techniques." When Headquarters United States Air Force issued a revised mission statement for MATS in July 1963, Air Force Regulation 23-17 defined airlift to include the "aerial delivery of troops, equipment and supplies." It also directed that MATS would train and equip its airlift forces in "all airlift tasks, consistent with the capabilities of the aircraft assigned." In addition, it stipulated that the command's airlift forces be mobile and flexible. In May 1964, the regulation was amended, replacing the words "aerial delivery" with the more specific "airborne assault and airborne operations."

In actuality, the regulation recognized a de facto situation. In 1957, MATS had gained troop carrier and aircrews capable of performing airdrop missions but had generally used them for the command's logistical resupply mission. Given the present airlift shortage and the fact that the command now possessed the C-130E and would soon have the C-141 also capable of airdrop/airborne operations, it was only natural that the Military Air Transport Service assume the full spectrum of airlift operations. Use of these aircraft, however, for both theater and intertheater operations clearly eroded the time-honored distinctions between the strategic and tactical airlift missions.

Related to these significant mission developments were Headquarters MATS' efforts to redesignate its units and congressional attempts to rename the Military Air Transport Service. In October 1961, Lieutenant General Joe W. Kelly, MATS Commander from 1960-1964, clearly understood the tremendous challenges facing the Military Air Transport Service when he remarked:

"The increased emphasis on limited war capability, the Presidential Approved Courses of Action, and the language of MATS' modernization legislation all point to a reorientation of MATS' activity from a predominantly scheduled operation to a posture responsive to the requirement for rapid global deployment of limited war forces as well as the requirements of general war."

In June 1962, Congressmen L. Mendel Rivers (D-SC), Melvin Price (D-IL), Durward G. Hall (R-MO), and William G. Bray (R-IN) had introduced bills which attempted to strengthen MATS' single manager charter by proposing greater centralization of the airlift forces. The bills also sought to rename the command the Military Airlift Command in recognition of the Defense Department's growing reliance upon airlift to deliver troops and equipment to areas of conflict. In effect, the name change would acknowledge that MATS was a major Air Force command on the same level as the Strategic Air Command and the Tactical Air Command. By January 1963, the Air Force had formulated its position on the proposed legislation. Although Headquarters Air Force supported consolidating strategic airlift assets further, it clearly stipulated the exemption of assault airlift, the airlift mission assigned to the Tactical Air Command. More importantly, the Air Force opposed the redesignation of MATS to include the assumption of specified command status as unnecessary. Undaunted, Representatives Rivers, Price, and Hall reintroduced the measure during the 1963 session and eventually secured its passage in 1965.

Effective January 1, 1966, MATS changed its name to the Military Airlift Command. After MATS became MAC, Headquarters USAF asked MAC to draft a new version of the almost ten-year-old command charter, which would outline MAC's relationship to each agency of the Department of Defense. Command leaders used this opportunity to update all aspects of the subject. When Headquarters USAF issued the new mission statement, Air Force Regulation 23-17, the charter stated in part that both the Secretary of the Air Force and the Executive Director (the MAC commander) could deal directly with all Defense Department and other government agencies on airlift matters. The MAC supplement to the new regulation noted, "MAC is the single Air Force Agency to provide movement control for airlift forces engaged in deployment and redeployment operations." Later that year, MAC also assumed two new functions: the operation of the Single Passenger Reservation System, a worldwide passenger reservation system for all international travel; and aerial port management which controlled the volume and rate flow into the military airlift system. The 1960s set the stage for the consolidation of all airlift resources within a specified command that followed in the 1970s.

C-54G prepares for take off.

Going to work in "Veet Nam." A U.S. Air Force C-130A aircraft in flight near Cam Ranh Bay Air Base, Vietnam. 1966.

Tactical Airlift in Vietnam

Forces And Organization

Article taken from The United States Air Force in Southeast Asia 1961-1973: An Illustrated Account Office of the Air Force History USAF Washington, D.C. 1984

The first USAF transports sent to Vietnam were four C-47s, which arrived at Bien Hoa Air Base on November 16, 1961 as part of the Farm Gate detachment. The C-47 airlifters performed diverse missions—support flights for Farm Gate, airdrops of Vietnamese paratroops, and night flareship operations. Their most demanding task, however, was to resupply United States Army Special Forces detachments at remote locations throughout South Vietnam. Such deliveries were often by airdrop, either by parachute or free-fall, the C-47s escorted by fire-suppressing Farm Gate or VNAF strike aircraft. In 1963 the number of C-47s increased to six, but airlift tasks gradually yielded to gunship roles.

The Vietnamese Air Force also used C-47s, organized in a two squadron group commanded in 1962 by Lieutenant Colonel Nguyen Cao Ky, future VNAF commander who later served as Prime Minister and Vice President of his country. A shortage of VNAF pilots in early 1962 led the Air Force to assign USAF pilots to the two squadrons. Thus, in April, 30 American officers arrived in Saigon to serve as copilots with otherwise all-Vietnamese C-47 crews. The "Dirty Thirty" Americans soon appreciated the flying skills of their VNAF counterparts and acquiesced to their own copilot roles. Some at first criticized the informality of Vietnamese flying methods, but all came to agree that these were often the surest, given the difficult flying environment. Ky's tactful leadership preserved good relations between the two groups. A second contingent replaced the original Dirty Thirty officers in the spring of 1963. The project ended in December 1963, although the idea was temporarily revived in modified form in 1965. The VNAF airlift squadrons remained healthy, frequently transporting the Vietnamese airborne battalions in their nationwide reserve role.

The larger, twin-engine C-123 Providers gradually took over the role of USAF C-47s in South Vietnam. The Provider had been tagged as "obsolescent" as early as 1956 and in 1961 was scheduled for retirement from the active inventory. Instead, in December 1961, a squadron of 16 C-123s began deploying to Vietnam (Project Mule Train) to provide "tactical airlift support of South Vietnamese armed forces." The first four ships reached Tan Son Nhut on January 2, 1962. A second C-123 squadron entered Vietnam in the spring of 1962, a third in 1963, and a fourth in 1964. All eventually were assigned to the 315th Air Commando Wing. The C-123s thus became until 1965, the principal airlift element in South Vietnam. Their ability to land on short and rough fields proved most valuable, and the four squadrons remained in South Vietnam until 1970. During 1967-1968, the aircraft underwent major modifications—a supplementary jet engine was installed under each nacelle, and improved brakes, flaps, and landing gear were provided—all of which enhanced C-123 payload, climb, and short-field performance.

Dominating airlift operations in Vietnam after early 1965 was the C-130 Hercules. Equipped with four turboprop engines and first flown in 1954, the C-130 was far superior in performance to earlier tactical transports. Its 15-ton payload, for example, was three times that of the C-123. In 1961 three squadrons of C-130As were stationed in the Western Pacific under the 315th Air Division in Japan; a fourth squadron deployed in the spring of 1962. These units assisted in deploying a U.S. ground and air task force to Thailand in May 1962 and regularly airlifted personnel and cargo into South Vietnam from offshore locations before 1965. They occasionally made deliveries between points in South Vietnam, making extra stops while transiting from offshore or deploying to an in-country base for short periods. When President Johnson ordered United States ground units into South Vietnam, the C-130s during March 8-12, 1965 deployed a Marine battalion landing team from Okinawa to Da Nang. On May 4-7, 1965, the C-130 fleet carried the Army's 173rd Airborne Brigade from Okinawa to South Vietnam in 140 flights.

A U.S. Air Force C-123 sprays herbicides on jungle growths and thickets in South Vietnam. Defoliants of this kind are those used for agricultural purposes in the United States. Missions of this kind were always aimed at specific and limited areas of jungle vegetation.

C-130 in-country missions from South Vietnamese bases became routine in April 1965. Thereafter, aircraft and crews rotated into South Vietnam on one or two week cycles from home bases in the Philippines, Taiwan, Okinawa and Japan. By the end of 1965, the in-country force had grown to 32 ships, positioned at four bases. Airlift requirements expanded with the scope of ground combat, the C-130 force in South Vietnam reaching 96 ships in February 1968. The rotational scheme was retained—offshore basing exploited existing maintenance facilities, reduced in-country construction and manpower requirements, and reduced exposure to enemy shelling. In an emergency, the in-country force could be quickly augmented.

In the spring of 1965, the offshore C-130 force increased to eight squadrons, including four on temporary duty from TAC. The force grew to 12 permanently assigned squadrons by mid-1966 and reached a peak of 15 (including three temporarily deployed TAC) in early 1968. The 315th Air Division also controlled a C-124 squadron based at Tachikawa; employing these four engine craft for overwater transport and occasional hauls of outsized items within Southeast Asia.

Within Vietnam, the C-123s and C-130s were centrally controlled through the Common Service Airlift System (DSAS) and its Airlift Control Center (ALCC) at Tan Son Nhut. In the spring of 1966 MACV established the Joint Movements Transportation Board and the Traffic Management Agency for theater-level management. The Board reviewed monthly forecasts of airlift requirements and capabilities and made allocations as necessary, while the Agency gave daily attention to priorities and controlled the flow of cargo to aerial ports. For emergency moves, the MACV Command Operations Center could levy special flights, bypassing the transportation agencies. The ALCC scheduled the aircraft flights and controlled the daily missions.

The ALCC functioned countrywide through local airlift control elements, liaison officers, field mission commanders and mobile combat control teams—all linked by often troublesome radio and land line communications. The whole apparatus was placed under the 834th Air Division upon its activation on October 15, 1966. Also assigned to the 834th were the 315th Wing and its C-123 squadrons, along with three in-country aerial port squadrons and numerous subordinate aerial port detachments. Expansion of the aerial port system tended to lag behind workload requirements, a condition intensified by shortages and breakdowns in cargo handling equipment.

Also assigned to the 834th Air Division (but employed outside the centralized scheduling system) were six squadrons of C-7A Caribou transports. These twin-engine reciprocating aircraft had been purchased by the United States Army in the early 1960s to support its airmobile forces. An Army Caribou company deployed to South Vietnam in the spring of 1962. By 1966 the force had expanded to six companies and operated under the scheduling and mission control of specified Army corps and divisions. In April 1966 the Army and Air Force chiefs of staff agreed to transfer the Caribous to the Air Force. Later that year, USAF air and ground crewmen entered the Army companies as trainees and replacements. On January 1, 1967, the six companies officially became Air Force squadrons, based at three locations and assigned to the 483rd Tactical Airlift Wing at Cam Ranh Bay. For the most part, the squadrons continued to operate under Army scheduling. The Air Force acquiesced in this "dedication user" procedure, although it was a departure from its doctrine of centralized control.

Airlift In The Ground War

Basic to Allied strategy were search-and-destroy operations, penetrating remote regions used by the enemy. Pivotal in these ventures were the fixed-wing transports, which lifted multi-battalion task forces to forward airheads and resupplied them with POL and ammunition. The Army helicopters, which performed local assaults and short-haul distribution of supplies, often refueled and rearmed at the C-130 airhead. Highway lines of communication supplemented air delivery when feasible, but airlift avoided the problem of enemy road ambushes, enabling the Allies to operate (like communist forces) in a "war without fronts."

Before 1965, during the period of Mule Train C-123 operations, such combat applications of tactical airlift were only glimpsed. Each day several C-123s and VNAF C-47s formed an alert element ready to move Vietnamese paratroops on instant notice. Although the airlift force was rarely exercised in this fashion, it did conduct a series of paratroop assaults, some preplanned and others in response to enemy ground attacks. These operations were uniformly disappointing, due to poor coordination, inaccurate drops, or the troopers' inability to close with the enemy.

By contrast, the C-123s won praise in airlanded troop deployments, hauling multi-battalion Vietnamese forces for reinforcement or offensive purposes. Efforts to mesh the best of C-123 and helicopter assault qualities had gratifying results, as exemplified during Allied operations in the Mekong Delta in March 1964, when the Providers delivered fuel and ordnance for Army helicopters at forward dirt strips. More significant than the assault roles, however, was the C-123's vital service in assisting Farm Gate C-47s in resupplying Special Forces camps.

Demands for airlift increased substantially in 1965, after the arrival in South Vietnam of large U.S. Army contingents. Among the first were two paratroop brigades, seen by General Westmoreland as a central reserve force quickly available for offensive or reaction operations. In August 1965 the 173rd Airborne Brigade was airlifted from Bien Hoa to Pleiku in 150 C-130 flights for just this purpose. Then, after returning to its home base, the 173rd began a series of sweeps over the Saigon plain, relying on a mix of truck, helicopter, and air transport support. During Operation New Life 65, which began on November 21, 1965, the 173rd made a helicopter assault into a dirt airstrip 40 miles east of Bien Hoa. The first C-130 landed within one hour and another 70 C-130 sorties followed during the next 36 hours to deliver both troops and cargo. Overland communications established by the third day enabled the C-130s to reduce their effort to 10 sorties daily.

Capt. John C. Flournoy logs 1000 hours in the HH-43B at Moody AFB, GA. Capt. Pfadenhauer (left) Det CC looks on. (Courtesy of John Flournoy.)

The key role played by air transport also was seen in the operations of the 1st Brigade, 101st Airborne Division, during the spring and summer of 1966. The brigade made five successive moves, each requiring some 200 C-130 lifts and each operation largely sustained by air resupply. The C-130s first airlifted the brigade from Tuy Hoa to Phan Thiet early in April, next to the highlands airstrip at Nhon Co later that month, then north to Cheo Reo in May, then to Dak To soon after, and finally back to Tuy Hoa in July. In the fall of 1965, soon after its arrival at An Khe, the 1st Cavalry Division undertook operations requiring a major airlift effort by the Air Force. When strong NVA forces attacked Plei Me camp south of Pleiku, General Westmoreland committed the division. Its assault helicopters consumed large amounts of fuel in the battle area and almost depleted available stocks, despite the efforts of the Army's Caribous and Chinook helicopters to sustain an air line of communications. Faced with this emergency, MACV called for USAF airlift forces. On the morning of October 29, the C-130s began an extended stream from Saigon, hauling POL and munitions to the main Pleiku field for redistribution by helicopter to the battle area. After several days, the C-130s landed closer to the battleground on a dirt strip at Catecka Tea Plantation, where the division's helicopters refueled. Over a period of 29 days, the Air Force delivered to the 1st Cavalry Division an average of 186 tons per day, most of it POL. These operations indicated that in the future airmobile forces would require substantial logistical airlift support from the Air Force, probably greater than that required by conventional ground forces.

Airlift support for the conventional units in search-and-destroy operations was exemplified during Operation Birmingham, a four week air deployment into Tay Ninh province beginning April 24, 1966. The C-130s flew 56 D-day sorties into the 4,600 foot dirt strip at Tay Ninh. C-123s lifted other forces to smaller, nearby strips. An around-the-clock resupply airlift into Tay Ninh followed to sustain the two-brigade force. The Air Force averaged 424 tons daily the first week, close to the forecast requirements. A land line of communications, opened on May 1 reduced the burden on the C-130s which were also supporting the airborne brigade at Nhon Co. Army helicopters carried supplies from Tay Ninh to field units, while Caribous provided courier service between Saigon base camps and the battle scene. During the second week of Birmingham, the runway at Tay Ninh deteriorated from heavy rains, which also halted ground movements. By the time the operation ended on May 17, the C-130s and C-123s had flown nearly 1,000 sorties and delivered nearly 10,000 tons of cargo for the division.

C-130 parachute capabilities were exercised during Operation Junction City early in 1967. The operation got under way on February 22, with airmobile assaults using all available helicopters. Executed simultaneously was the war's only battalion-sized parachute assault by American troops. Thirteen C-130s departed from Bien Hoa on schedule; the troopers of the 173rd Brigade jumped as planned at 0900. All landed in the drop zone near the Cambodian border. There was negligible opposition, although one C-130 received a single hit. At 0927, ten C-130s dropped the brigade's equipment, returning at 1300 to make additional cargo drops. Some of these loads landed in swamp areas, complicating retrieval. Five cargo-carrying Hercules sustained hits, none seriously.

On the second day, the Air Force launched 38 C-130 resupply sorties, but bad weather reduced drop accuracy. During the next five days, coordination and accuracy improved with daily drops averaging 100 tons. During the final stages of the operation in late March, the C-130s made airdrops to a "floating brigade," using drop zone locations which the ground unit provided by radio. The drops in support of Junction City totaled 1,700 tons. The sustained effort strained the Army's parachute rigging capability, but also eased the workload of the Army's resupply helicopters and provided USAF crews with invaluable experience.

Air landed operations in support of subsequent Army offensives followed the pattern of the earlier ones. In November 1967, C-130s in some 250 sorties flew the 173rd to Dak To. In the three weeks of heavy fighting that followed, daily streams of C-130s delivered more than 5,000 tons, landing on a much-patched and busy 4,200-foot asphalt strip. Enemy shelling on the 15th destroyed a 1,300 ton ammunition dump and two parked C-130s. Crewmen courageously taxied a third Hercules from the inferno and received Silver Stars for their deed. Although land deliveries supplemented the air resupply, General Westmoreland later remarked that "along with the gallantry and tenacity of our forces, our tremendously successful air logistic operations was the key to the victory."

Combat operations in the northern portion of South Vietnam created special transportation problems for Allied forces. In 1966 highway communications north of Da Nang were in disrepair and highly insecure, while port facilities and airfields near the DMZ were primitive. In July the Marines deployed their forces immediately below the DMZ, using more than 250 trips by Marine KC-130s (with some aid from USAF aircraft) into the red dirt airstrip at Dong Ha. The Allies later surfaced the strip and also opened a second all-weather field at Quang Tri. These projects required large quantities of scarce airstrip matting, as did the Khe Sanh strip improvement to the west. A prolonged C-130 airdrop supported operations and airstrip construction at Khe Sanh during the fall of 1967. The effort required cooperation among Navy suppliers at Da Nang, Army parachute riggers, Air Force crews, and the Marines at Khe Sanh.

The air supply of Khe Sanh during the first four months of 1968 represented a remarkable achievement, enabling 6,000 Allied defenders to survive under heavy NVA pressure until reopening of land communications in April. Favored by absolute air supremacy and proximity of Khe Sanh to Da Nang (30 minutes), the Khe Sanh airlift nevertheless faced major problems stemming from difficult weather and the enemy's imaginative use of firepower. For much of the period, communist fire forced a halt to C-130 landings and limited severely those by C-123s, thus necessitating reliance on airdrops. The Air Force overcame its hitherto chronic weakness in dropping supplies during poor visibility, by using ground radar to guide the C-130s to precise points over the Khe Sanh runway. The foul weather helped screen the transports from enemy gunners, although it also hindered fire suppression efforts by allied fighters. On most days the ability of Marine retrieval parties to clear loads from the rough drop zone limited the volume of supplies. Late in the siege C-130s began bulk deliveries within the Khe Sanh main perimeter using low-level extraction techniques.

Between January 21 and April 8, the Air Force delivered 12,400 tons to Khe Sanh. The C-130s accounted for more than 90 percent of the total in 496 drops, 67 extractions, and 273 landings. C-123s made 105 drops and 179 landings; C-7 Caribous made eight landings. During the early weeks, Marine KC-130s also delivered cargo. Throughout the siege, Marine helicopters resupplied hill positions outside the perimeter. During February alone, they hauled well over 1,000 tons. Supply levels never became dangerously low. Three C-123s and one Marine C-130 were destroyed during the battle.

The 1968 Communist Tet Offensive severely challenged the airlift crews. The early attacks at Tan Son Nhut and many of the up-country airstrips temporarily dislocated the airlift system. However, transport crews managed to fly numerous emergency troop and supply missions on behalf of hard-pressed garrisons. By February 3, the fourth day of the offensive, countrywide airlift sorties regained former levels. Sortie totals increased thereafter as crews flew to the limit of their capacity and as additional planes and crews arrived from offshore. With surface lines of communication disrupted throughout the month, demands for airlift remained high. The most critical area was the northern provinces, where winter monsoon weather and airspace congestion hampered resupply of expanded allied forces. But in all regions, the recovery from the enemy offensive depended upon the air transport system. Missions into Khe Sanh were unaffected by the Tet attacks.

Efforts of airlift crews were extraordinary in two other 1968 operations. The first was Operation Delaware, the air invasion of the A Shau Valley in April. Once more the C-130s faced a gauntlet of enemy fire and bad weather, this time without the assistance of ground radar. In daily supply drops to the 1st Cavalry Division at A Luoi, the C-130 crews used their radar and doppler equipment to navigate up the cloud-filled valley, breaking out just before visual release of cargo. During the nine day period starting April 26, the C-130s in 165 flights dropped 2,300 tons, nearly all of it ammunition. Drops continued even during one period when the weather was so bad that helicopters could not operate. One C-130 and its crew was lost to enemy fire; four others received major battle damage. The cavalry division commander termed the C-130 effort "one of the most magnificent displays of courage and airmanship that I have ever seen." A rebuilt A Luoi airstrip received its first Caribou on May 2 and its first C-123 and C-130 two days later. Before rain closed the airstrip on the 11th, USAF transports made 113 landings, more than half by C-130s.

The air evacuation of the allied garrison at the Kham Duc mountain post on May 12, 1968 was equally spectacular. Intermittently through much of the day, Army and Marine helicopters lifted out survivors, while allied air strikes held off the enemy on all sides. During the morning a C-130 landed but received heavy damage, then departed with only three passengers, fuel steaming from shrapnel holes and tires stripped away. A C-123 made a successful morning pickup. In early afternoon, three C-130s attempted pickups. Enemy fire destroyed one after it took off with more than 100 civilian passengers; another, crippled in landing, was abandoned. Only the third made a successful landing and evacuation. Then, late in the afternoon, three C-130s succeeded in bringing out the last of the garrison. Of 1,500 survivors of Kham Duc, the Air Force flew out more than 500, nearly all in the final crucial minutes before the outpost fell.

One final act of valor ensued. Under orders to bring in a three-man USAF control team and unaware that the evacuation was already completed, a final C-130 landed at the now hostile Kham Duc strip. Radio conversations soon after the Hercules took off made known the blunder. A C-123 made a touch-and-go landing and spotted the three men. A second C-123, piloted by Lieutenant Colonel Joe M. Jackson and Major Jesse W. Campbell, landed under heavy fire and successfully removed the team. For his role in the rescue effort Jackson received the Medal of Honor, the only airlifter of the war so honored. Jackson wore the award with dignity in the years that followed, representing the thousands of USAF airlifters who shared pride in his recognition.

ARVN paratroopers board a C-130 prior to dropping into a Viet Cong area.

The small payload capacity of the C-7 Caribou normally discouraged its use in major tactical operations. On the other hand, excellent maneuverability at low altitude and slow airspeed enabled the Caribou to make accurate drops into small places. Furthermore, its small payloads were appropriate for garrisons lacking heavy recovery equipment; finally, simplicity of construction minimized the C-7s vulnerability to ground fire. Thus, the Caribou was used frequently for emergency drops, of which three were particularly noteworthy: the resupply of Duc Lap in August 1968, Ben Het in the spring of 1969 and Dak Seang a year later. The garrisons at all three survived heavy enemy pressure with the help of air-delivered supplies; hostile fire was severe in each instance, necessitating special tactics. At Duc Lap, the Caribous flew in at tree-top level, popping up to 300 feet at the last moment for the release. At Ben Het, the C-7s made run-ins at intervals of about 20 seconds, in coordination with preplanned fire suppression from covering fighter aircraft. Similar tactics were less successful at Dak Seang, where three aircraft were lost in the first week of operations. The force then went over to night drops, the drop zone marked by signal fires and airborne flareships. During the battle, the C-7s made 125 drops over Dak Seang, releasing 250 tons; the garrison recovered 94 percent of these supplies.

Airlift efforts during 1969-1971 included massive support of the allied incursions into Cambodia and southern Laos. During nine weeks commencing on April 28, 1970, aircraft of the 834th Air Division delivered 60,000 tons to more than 20 airfields in South Vietnam immediately adjacent to Cambodia. Ammunition lifts from Bien Hoa for a short period approached 1,000 tons daily. Missions into Cambodia were less extensive, but included more than 150 C-7 flights into the border strip at O Rang and several C-130 ammunition drops to U.S. forces near O Rang. Lam Son 719 was preceded by 250 C-130 flights, hauling a Vietnamese airborne division and other ARVN forces from Saigon to Dong Ha and Quang Tri. Over a seven week period during the Laotian operation, the C-130s delivered more than 14,000 tons to a reconstructed logistics base at Khe Sanh. During the same period, C-141s of the Military Airlift Command flew some in-country movements to reduce aerial port backlogs.

The major North Vietnamese spring offensive in 1972 led to frequent Air Force supply flights in support of the defenders of Kontum and An Loc, two primary objectives of the enemy. Sustained air resupply into isolated Kontum began in April, with day landings halting on May 17 after enemy shells damaged several C-130s, burned two VNAF C-123s and destroyed a C-130E. C-130 landings resumed exclusively at night on May 18-19 with 17 aircraft overcoming extreme hazards to make successful deliveries. Resupply continued nightly under cover of allied gunships, although in one instance a C-130 delayed its departure past dawn and was destroyed by ground fire. On May 25 the enemy seized a part of the runway, closing the field to landings. The supply burden momentarily fell upon VNAF and United States Army helicopters, but on the 28th C-130s began drops into one corner of the airfield. Although bundle recovery was difficult and slow, more than 2,000 tons were dropped in 130 sorties before C-130 night landings could be resumed on June 8.

Aerial resupply to the 20,000 defenders and refugees at An Loc began in April, with deliveries by VNAF and United States helicopters and VNAF C-123s approximating 20 tons daily. Loss of a VNAF helicopter and a C-123 to enemy fire caused successive suspensions of these efforts. Air Force C-130s began daylight drops on April 18. The first four aircraft successfully parachuted supplies but each received battle damage; a fifth ship went down west of the city. To reduce exposure, altitudes were raised above 6,000 feet, dropping with ground radar guidance. "High-altitude, low-opening" parachute techniques proved unsuccessful, however, as numerous chutes failed, in part because of inexperienced Vietnamese packers. The C-130s resumed low-level conventional drops even though all ships continued to receive hits.

On April 26, after another C-130 was lost, the Air Force switched to night drops. Conducted visually with light signals, these too proved unsatisfactory, as many bundles missed the mark and some fell into enemy hands. The low point occurred the night of May 2-3, when the C-130s failed to make a single successful deliver. Loss of a third C-130 on the following evening caused the cancellation of further night missions. High altitude, daylight drops were resumed on May 5 but now with the assistance of American parachute riggers brought in from Okinawa. Half of the parachutes opened properly and only one of the 24 bundles dropped that date fell into enemy hands. On May 8 the An Loc defenders recovered 68 of the 88 tons dropped. The drops became increasingly successful thereafter, the airlifters exclusively resupplying the defenders until the end of the siege on June 18. In all, C-130s dropped 7,600 tons into the beleaguered city in more than 600 sorties, a dramatic and fitting climax to the 10 year history of the USAF tactical airlift arm in Vietnam.

Other Applications of Tactical Airlift

Air Force Transports performed numerous noncombat transportation roles during the war. Passengers, mail, perishable foods, and high value equipment were moved between aerial ports throughout South Vietnam. Such loads frequently originated at airfields served by Military Airlift Command and the principal seaports: Saigon, Da Nang, and the Qui Nhon, Nha Trang, Cam Ranh Bay complex. For routine bulk hauling within Vietnam, air transport remained secondary to the surface modes. An exception was the Army Special Forces logistic system, which depended primarily on airlift for the routine supply of its camps. The C-130 force also made overwater hauls between off-shore points and fields in Vietnam, averaging 7,000 tons monthly in the peak year, 1966.

The war in Laos required substantial air transport, of which USAF helicopters based in north Thailand provided a part. Civilian contract firms were primarily responsible for fixed-wing transport in Laos. The 315th Air Division operated a passenger and air logistic service within Thailand, linking the principal Air Force bases there. In September 1965, a rotational C-130 detachment at Bangkok took on this role, replacing C-123s previously rotated from South Vietnam.

Casualty evacuations in South Vietnam were almost entirely by air. Army helicopters performed most battlefield evacuations, while USAF transports hauled between the 18 in-country hospitals, mostly located at C-130 airfields. Transports carried unrigged litters at all times for emergency evacuations. The 903rd Aeromedical Evacuation Squadron, organized in 1966, provided flight medical personnel and operated a control center at Tan Son Nhut to monitor patient movements. Mobile casualty staging teams entered South Vietnam in early 1967, thereafter performing patient care at certain forward airheads. The 315th Air Division C-130s and C-118 performed patient evacuations from Vietnam to offshore points. Specifically configured for patient evacuations, the C-118s were assigned to the 6485th Operations Squadron, based originally in Japan and later at Clark in the Philippines. In 1968 a C-118 detachment began operating at Cam Ranh Bay to support in-country patient movements. All-jet C-9s finally replaced the C-118s in 1972 as the PACAF specialized aeromedical evacuation force.

Late in 1965, the Air Force deployed a squadron of CH-3 helicopters to South Vietnam, each able to carry a three-ton payload about 200 miles. The squadron performed a variety of tasks, including delivery of supplies to USAF radar and communication sites in South Vietnam, and occasionally joined Army and Marine units in air supply operations. Both the CH-3s and the Air Force UH-1s later deployed played major roles in special warfare operations in Laos, Cambodia and remote areas of South Vietnam. Management remained outside the Common Service Airlift System (CSAS). Because of interservice agreements, Air Force operation of transport helicopters was quite limited. Techniques of air delivery were advanced. Introduced in 1966 was the container delivery system, which permitted supply drops from 600 rather than 1,000 feet. Thereafter, C-130s employed the system in all drops except for the heaviest items. Low-level extraction systems developed in the early 1960s proved useful for pinpoint aerial deliveries. One of these, the ground proximity extraction system (GPES), used a hook-and-cable arrangement which required prepositioning of an arrestor gear. The other, the low altitude parachute extraction system (LAPES), required a large extraction parachute and a long cleared space on the ground (often an old airstrip). The Air Force used both systems at Khe Sanh with mixed success.

For most airlifters, flight operations in Southeast

C-130 using ground proximity extraction system.

This C-123 lands on a PSP landing strip in Vietnam.

Heavy equipment supply drop in RVN.

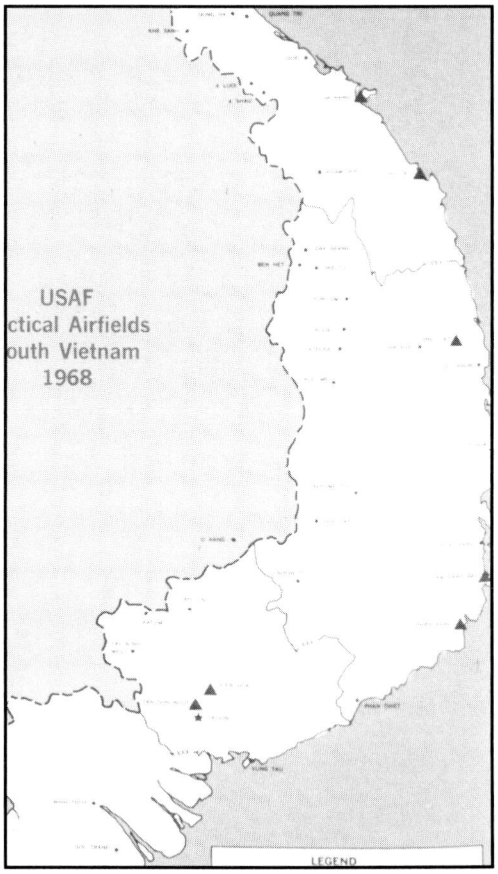

Asia were an abrupt change from the methods taught and practiced in the United States. In peacetime flying, crews adhered to written regulations, regularly attended flying safety meetings, and practiced endlessly the mechanical techniques of instrument flight. In Southeast Asia, however, crewmen quickly learned to rely on their own wits and judgment. Prescribed criteria of ceiling and visibility were generally overlooked. Crews flew visually whenever possible, looked for breaks in overcasts, and stayed underneath low ceilings except when over hostile areas. Squadron commanders frequently had to curb the enthusiasm of their crews and caution them against unnecessary risks—a difficult message in view of the heady sense of mission accomplishment generally felt.

Special hazards existed at the forward airheads, where the fixed-wing transports shared crowded ground and air space with seemingly uncontrolled helicopters, trucks, pedestrians, and assorted hazards. Landings required the concentration of the entire crew to avoid other traffic, friendly artillery, and high terrain—all while maneuvering clear of clouds and setting up precision final approaches. Joint working groups set up in late 1968 belatedly addressed the hazardous conditions.

Beginning in 1970, the Air Force substantially reduced its airlift forces in the Pacific. Remaining by the end of 1971 were only five C-130, three C-7, and one C-123 squadron; the 834th was inactivated in November. In 1972 the Caribou and C-123 wings went out of existence, leaving only four squadrons of C-130Es based in Taiwan and Thailand. On the other hand, the VNAF airlift force expanded from two squadrons in 1970 to eight in late 1972, including three C-7 and three C-123 squadrons. Two squadrons were equipped with C-130As just before the 1973 ceasefire, making available the C-123s for the air forces of Laos, Thailand and Cambodia.

Over the years between 1962 and 1973, the Air Force delivered more than seven million tons of passengers and cargo within South Vietnam. By comparison, United States and British transports carried slightly more than two million tons during the Berlin Airlift and about 750,000 tons during the Korean War. The Air Force lost 53 C-130s in the Southeast Asia war, more than half of them in 1967 and 1968. C-123 losses (including one spray aircraft) also totaled 53 and C-7 losses numbered 20. Of these 126 aircraft, enemy action accounted for 61, including 17 destroyed by sapper or shelling attacks. The other 65 were lost from "operational" causes, mainly associated with the difficult conditions at forward airstrips. All but 10 of the losses occurred in South Vietnam. Those airlift crewmen killed or missing numbered 269. In Vietnam, the USAF tactical airlift arm lived up to its proud history, adding a new, illustrious chapter.

If you doubt any of this, talk to any U.S. Army or U.S. Marine "Grunt" with whom we were so proud to serve our country.

Strategic Airlift in Vietnam

Even though airlift played an important role in Vietnam before 1964, this was the year the Military Air Transport Service became heavily involved. The command's primary task throughout the war was to provide intertheater or strategic airlift, the movement of military assets over intercontinental distances. Transporting personnel and critical supplies halfway around the world by ship had become too slow in the rapid response environment of the nuclear age. United States military officials, therefore, increasingly relied on airlift to support military projections worldwide. Before MAC's participation in the Vietnam War had ended, the size and scope of the strategic airlift to Southeast Asia had grown to historic proportions. Command leaders used all their resources and creativity to meet the ever-expanding airlift requirements.

One of the first problems confronting the MATS staff, as they strove to support the airlift needs of the theater commanders, concerned the age and capability of the airlift fleet. Although command officials had already planned for a new transport to replace the aging C-124, they had to wait until the 1964 build-up of conventional forces in Vietnam necessitated the acquisition of additional capability. At that time, MATS possessed the following squadron strengths: 21 C-124 *Globemaster*, 3 C-133 *Cargomaster*, 7 C-130 *Hercules*, 3 C-135 *Stratolifter*.

However, the Military Air Transport Service's new jet-powered C-141 *Starlifter*, with a 4,000 nautical-mile range and the means to airdrop both troops and cargo, had already entered the production phase in 1964 and would serve as the backbone of the airlift force throughout the conflict. The C-141 literally revolutionized intertheater airlift by tripling MATS' capability, the most significant increase in airlift capacity in the history of the Air Force. As the first jet aircraft designed specifically for military airlift, the C-141 boasted a cruising speed of 422-440 knots, heavy cargo load capacity, ocean-spanning range, and a complete aerial delivery system. Interior rails and rollers, that were part of the 463L materials handling system, folded up to leave a smooth floor for loading vehicles and passengers.

In 1964, the Military Air Transport Service also submitted a request to Headquarters USAF for a huge, high-speed, cargo carrier to replace the aging C-133 and to reduce the time needed to transport outsized cargo around the world. Revised Army strategy and tactics had changed airlift requirements. When engineers designed the C-141 in 1961, Army leaders envisioned airlifting small, rapid-reaction paratrooper forces. Three years later, military strategists were planning larger-scale movements of heavy infantry units supplemented by mechanized and armored battalions. A large portion of the heavy equipment these Army units used was classed as "outsized" and would not fit into the *Starlifters*.

Defense Secretary Robert S. McNamara announced on December 22, 1964, the administration's plan to develop the C-5A *Galaxy*. In 1965, President Lyndon B. Johnson told Congress, "We must further improve our ability to concentrate our power rapidly in a threatened area so as to halt aggression early and swiftly." To enable the United States to deploy a fighting force overseas in a matter of days, instead of weeks, the President supported the C-5A, which would "represent a dramatic step forward in the worldwide mobility of our forces and in American leadership in the field of aviation."

The Vietnam era was a revolutionary one for the Military Air Transport Service/Military Airlift Command. The principal role of airlift changed dramatically, shifting to a greater emphasis on direct combat support. To meet these new challenges, the command used its available resources plus those of the Air Force Reserve, Air National Guard, and commercial airlines to build the national airlift capacity to a wartime level. In the process, the command's mission expanded from strictly intertheater airlift to include intratheater airlift. Additionally, the command streamlined its organization to make the entire operation more efficient. An increase in the number of aerial ports, prepositioned crews at enroute stations, and innovative maintenance procedures were just a few of the creative ideas that improved efficiency. Planners dusted off several of General Tunner's innovations from World War II and the Berlin Airlift to improve airlift operations and reduce cargo delivery time. During the Vietnam period, the command also received its first aircraft specifically designed as a military airlifter the jet-powered C-141; the giant C-5A joined the C-141 a short time later to help convert the Military Airlift Command's intertheater airlift fleet from propellers to jets. Military airlift came of age in Vietnam. Airlift's expanded role proved that military airlift should consolidate under one commander and presaged the growing sentiment for the Military Airlift Command to become a specified command directly under the control of the Joint Chiefs of Staff.

Capt. Myles A. Rohrlick (center, foreground), and his wife, Joy (right), are congratulated by Brig. Gen. Richard H. Ellis (left), 315th Air Division commander, after the general decorated Rohrlick with the Legion of Merit in a recent flight-line ceremony at Tachikawa Air Base, Japan. A C-130 Hercules instructor pilot of the Division's 815th Troop Carrier Sqdn., Rohrlick distinguished himself as the originator and test project officer for the new parachute Low Altitude Delivery (PLAD) system. In the background are the two 815th loadmasters who were Air Force Commendation Medal recipients for their work in the project. Left to right are SSgt. Wayne A. McFarland, SSgt. James O. Oxford, and Oxford's wife, Edna. (Courtesy of W.A. Rohrlick.)

Evolution and Expansion, 1973-1981

For nearly a decade, the Military Airlift Command had transported thousands of soldiers and countless tons of war material to Southeast Asia to sustain American and South Vietnamese military operations. This had been the command's primary focus. As the Vietnam cease-fire agreement went into effect on January 27, 1973, MAC began a period of transition. Initially, the command supported the withdrawal of military forces and evacuated civilians from Vietnam. These operations required considerable airlift resources and proved the value of airlift's responsiveness. Thereafter, the nine-year period extending from 1973-1981, was characterized by several developments that shaped the course of military airlift well into the command's fiftieth anniversary year. World events such as the Arab-Israeli War, the oil crises, Iranian Revolution, and Soviet intervention in Afghanistan dramatically underscored the need for military airlift to support American diplomatic initiatives or assist, if required, contingency operations. Additionally, the command continued its long-standing tradition of transporting food, supplies, and medical equipment during times of natural disasters. Besides demonstrating the United States' good will, these humanitarian airlifts helped many people around the world to overcome devastating circumstances.

A Decade of Change, 1981-1991

Near the end of the 1970s, dramatic world events affected American foreign policy, and in the process, the Military Airlift Command increasingly served as an instrument for the execution of the United States' national security objectives. First, the 1979 Iranian revolution and the escalating Soviet threat roused Congress to commission a mobility study to investigate the military's ability to respond to various crises throughout the world. The *Congressionally Mandated Mobility Study* recommended rebuilding American airlift capability. Second, the failed attempt in March 1980 to rescue the hostages held captive by Iranian students dramatized the limitations of the United States' special operations forces. Military operations in Grenada further highlighted organizational shortcomings among the Services. As a result, the military and Congress expended much time and effort in the 1980s determining the optimum organization and force structure of Special Operations Forces and the need to better integrate global air, land, and sea transportation resources. These reviews brought about a reorganization of the Department of Defense and the creation of unified commands for special operations and transportation.

MAC Airlifts 101st Airborne Division to Vietnam

HQ Military Airlift Command—The Pentagon announced today (Dec. 13, 1967) that two brigades of the 101st Airborne Division, the famed Screaming Eagles, are being airlifted by Air Force transports from Fort Campbell, Ky., to Vietnam in the largest aerial deployment of combat troops in the war.

All-jet C-141 Starlifters and turboprop C-133 Cargomasters of the Military Airlift Command have already carried 6,475 troops and 4,123 tons of cargo from the 101st's home base to Bien Hoa, South Vietnam, in the giant airlift — termed Operation Eagle Thrust — which began Nov. 17.

In all 10,386 troops and 5,118 tons of cargo will be transported over the 10,000-mile aerial pathway.

"The airlift demonstrates the tremendous strides MAC has made in strategic airlift capability in very recent years," said General Howell M. Estes Jr., MAC commander.

"Our aircrews are trained to carry combat-ready troops and their equipment anywhere in the world in a matter of hours and accomplish it on a tight, fixed schedule with the able backing of our support and maintenance personnel.

"There is no question that this ability to move division-size combat units to war zone quickly is a vital deterrent to general war."

The airlift of the Second and Third Brigades of the 101st is MAC's largest and longest deployment of combat-ready troops into a war zone. The main body of troops began moving Dec. 1. The airlift is scheduled to end late this month. The First Brigade of the 101st was previously shipped to Vietnam in 1965.

A total of 373 C-141 missions and 20 C-133 missions were scheduled for the move of the Second and Third Brigades.

MAC military and commercial contract airlift of cargo, troops and passengers to other global destinations, including Southeast Asia points, has continued without interruption during Operation Eagle Thrust.

The Military Airlift Command's distinguished history dates back to the Berlin Airlift of 1948-49. Organized just three weeks before the Russians blockaded the city, MAC — then the Military Air Transport Service — helped deliver more than 2 million tons of supplies to Berlin over a 15-month period to keep the city's population alive.

The long-range C-141 fanjet, which cruises at 500 miles an hour, is carrying a normal troop load of 94 or a combination of troops and cargo. Paratroopers wear combat clothing and carry their weapons aboard the aircraft.

Aircrews from Air Force bases at Travis and Norton, Calif.; McChord, Wash.; Dover, Del.; Charleston, S.C.; Robins, Ga.; and McGuire, N.J., are taking part in the airlift.

Normal flying times for the C-141 through the mid-Pacific is 24 hours 20 minutes. Total elapsed time is 30 hours 20 minutes for the 10,472 mile flight.

Starlifters make the flight through Alaska and Japan over a modified polar route in 22 hours 55 minutes flying time with a total elapsed time of 26 hours 56 minutes for the 9,783-mile flight.

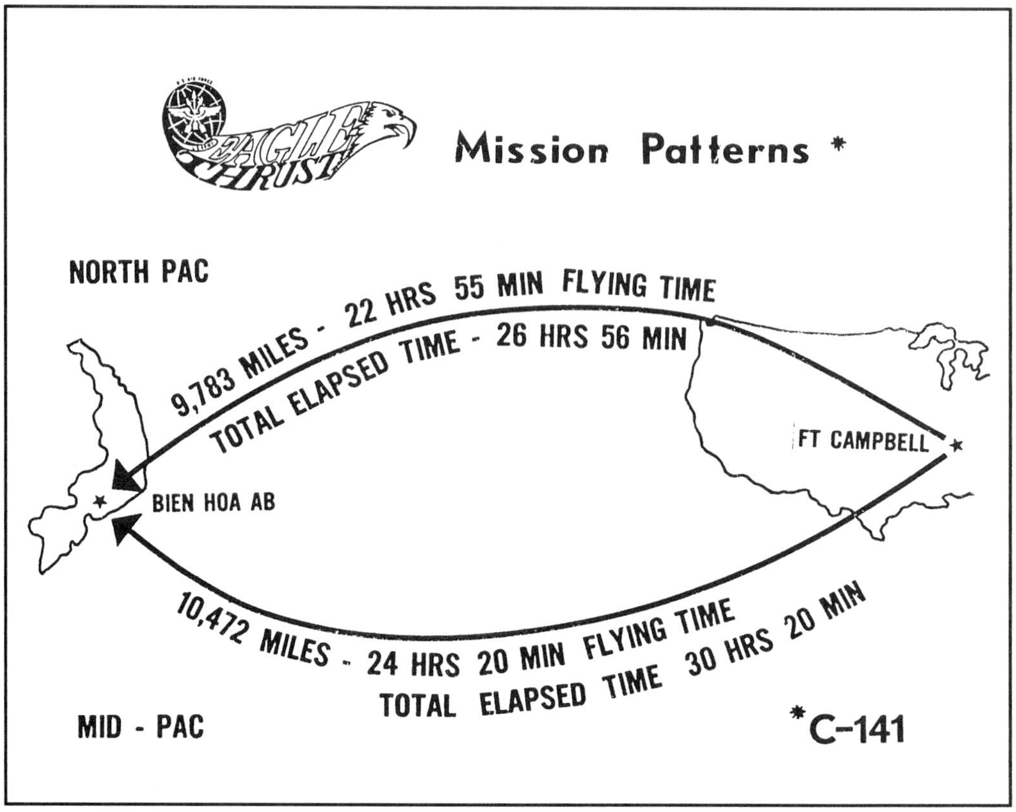

Middle photo: General Westmoreland (left) with MAC Commander, General Estes. Both seem pleased with how well "Eagle Thrust" was accomplished. Dec. 13, 1967. Bien Hoa AB, RVn.

Bottom photo: An Army vehicle is loaded aboard an Air Force C-141 before departing Fort Campbell, KY, for Vietnam.

Finally, although the Military Airlift Command received praise for its performance during a variety of difficult missions, even relatively small contingencies, such as those in Grenada and Panama, underscored the need for additional airlift capability. At the end of the command's fifth decade, as the Cold War ended, events in the Middle East overshadowed the new order dawning in Europe.

MAC and the United States Transportation Command

In 1977, the Joint Chiefs of Staff had authorized the *Transportation Operating Agency Organization Alternatives Study* to consider the propriety of a special command to manage all Department of Defense transportation functions. The study had concluded, however, that MAC should remain a specified command because of its global and interagency mission.

Although Defense Secretary Casper Weinberger and some members of the military community continued to support a unified transportation command initiative, they made little headway until the Packard commission issued its interim report on February 28, 1986. These findings recommended creating a "single unified command to integrate global air, land and sea transportation." The final report included the same suggestion and on April 18, 1987, President Ronald Reagan approved the activation of the United States Transportation Command (USTRANSCOM) at Scott Air Force Base, Illinois.

General Cassidy acted as the executive agent for USTRANSCOM from its establishment until its formal activation on October 1, 1987, when he became the "dual-hatted" United States Transportation Command Commander in Chief and MAC Commander in Chief. The Navy's Military Sealift Command joined MAC and the Army's Military Traffic Management Command as service components of the new unified command.

The Military Airlift Command retained its specified command status until October 1, 1988, when United States Transportation Command became fully operational. With the loss of MAC's specified command status, three areas of responsibility transferred to the new unified command. First, USTRANSCOM assumed operational command of MAC's C-5, C-141, United States-based C-130, WC-130, and WC-135 aircraft. Second, the new command also exercised operational command of mobilized Air Reserve Component transport assets. Third, mobilized HC-130 and combat search and rescue helicopter aircraft transferred either to USTRANSCOM or USSOCOM.

The loss of specified command status, however, had little impact on many MAC activities. Several functions, such as the 89th Military Airlift Wing, Air Weather Service, Aerospace Audiovisual Service, operational support airlift assets, and aeromedical evacuation forces, which were "service-unique" and not part of MAC's specified command mission, continued without change. The command's theater-based forces, including intratheater airlift, special operations, operational support airlift, weather support, visual information, combat rescue, and aeromedical airlift forces assigned to the operational control of theater commanders or the senior Air Force commander, also remained unchanged at this time.

While these organizational changes brought new challenges to the Military Airlift Command, the command continued to fulfill its mission responsibilities. There were several widely-publicized humanitarian relief efforts as well as the Urgentfury contingency. Changing relations with the Soviet Union again demonstrated airlift's uniqueness as an instrument of diplomacy. The command's resources would also support two major contingencies: Operation Just Cause, an incursion into Panama to depose strong-man Manuel Noriega; the Operation Desert Shield, into Saudi Arabia to oppose Iraq's aggression in the Middle East.

Events of the 1980s helped to restructure the organization and mission of the Military Airlift Command. Following the congressionally directed cut-backs after the Vietnam War, crises in the Middle East drove Capitol Hill to reevaluate the need for airlift as well as the organization of the Defense Department. The *Congressionally Mandated Mobility Study* of 1981 concluded that the United States lacked the mobility capacity to meet any one of the number of likely contingencies and recommended adding 20 MTM/D to MAC's projected airlift capability for a total of 66 MTM/D. The command spent the remainder of the decade striving to attain that goal. The unstable conditions in the Middle East and two important studies prompted Headquarters United State Air Force to approve the consolidation of special operations forces into Twenty-Third Air Force under MAC's control in 1983. Leading Congressmen, however, were determined to give special operations more autonomy and enacted the Defense Reorganization Act of 1986, creating the unified United States Special Operations Command, with Twenty-Third Air Force as its air component, in 1987. Three years later, the Air Force converted Twenty-Third Air Force into the Air Force Special Operations Command, an independent major command. The Reorganization Act also established the United States Transportation Command composed of MAC, the Navy's Military Sealift Command, and the Army's Military Traffic Management Command. The Military Airlift Command lost its specified command status when USTRANSCOM became fully operational on October 1, 1988. The end of the Cold War and presidential and congressional efforts to bring the federal deficit under control meant that in the 1990s the Military Airlift Command would have to streamline some of its organizational elements still further.

Several significant military operations tested the command's ability to respond during its fifth decade: Urgentfury, Rubber Wall, Just Cause, and finally, Operation Desert Shield. Although MAC performed well, these contingencies taxed the command's capacity and documented the need for additional airlift resources. Military Airlift Command crewmembers also flew several historic missions into the Soviet Union in support of nuclear underground testing verification, the INF Treaty, and Armenian earthquake relief efforts. During each of these operations, command crewmembers flew to areas previously shielded from the West. The period was also characterized by a transformation in super-power relations, and the Military Airlift Command was at the forefront of that change.

General Robert E. Huyser, CINCMAC, makes acceptance remarks following the presentation of a C-141B model by Lockheed-Georgia's president, Robert B. Ormsby (left). The presentation was made Dec. 4, 1979 at Charleston AFB, SC where General Huyser flew the first C-141B delivered to the Air Force.

Congratulations on a great record spector, C-130 gunship. L to R: Col Harry Williamson, 16 SOS Commander; Col F.P. Humphries, 8th TAE Fighter Wing Commander and Col Nelson L. Neil, Deputy Commander Special Operations. (Courtesy of Nelson L. Neil.)

Urgent Fury

by Sgt. James L. Davis, AMC News Service

Operation Urgent Fury

Ten years have come and gone since the operation to restore political stability and rescue 709 American students from the battle-torn Caribbean island of Grenada.

The operation was a shining moment in the history of the Military Airlift Command, a testament to the Global Reach capability MAC and now the Air Mobility Command continue to provide the United States.

The memories of that early morning on Oct. 25, 1983, when Urgent Fury began with low-level airdrops and gunfire, are still vivid for Maj. Gen. James L. Hobson Jr., AMC's director of operations and transportation.

A decade ago the general was a lieutenant colonel and commander of the 8th Special Operations Squadron, Hurlbure Field, Fla., who found himself piloting the first MC-130 to drop U.S. Army Rangers over Point Salines Airfield, Grenada.

Urgent Fury was an operation born of the Cold War. Grenada, the tiny, spice-producing island of 108,000 people is located 90 miles north of Venezuela. Because of its strategic location, American leaders became concerned over the Soviet and Cuban presence there, including the construction of a large airport complex. If Soviet aircraft operated from the island, they could pose a serious threat not only to shipping in the Caribbean, but to Central America, northern South America, and the southern United States.

When the Organization of Eastern Caribbean States asked the United States to join in a military action to restore democracy to the island, then President Ronald Reagan approved a plan with three primary goals: To evacuate American and third-country nationals; to neutralize the Grenadan military and any military force on the island that might interfere with the evacuation; and to restore political stability and democracy to Grenada.

The special operations capability that would put the first troops on Grenada would be the responsibility of today's Air Force Special Operations Command. But in 1983, MAC owned the capability and the command's aircraft and people would be the first to put the plan into action.

"The plan was for two MC-130s to airdrop a runway seizure team 30 seconds apart over Point Salines at 0530. Thirty minutes later the third MC-130, which I commanded, would make the first landing and drop off the Ranger command and control element," Hobson said.

That was the plan. But on the way to Grenada things suddenly began to change.

No Longer A Plan

MAC MC-130s were aerial refueled en route to Grenada, along with the command's AC-130 gunships, which allowed enough fuel to loiter in the area, if necessary. It was necessary.

Just prior to arriving in the area of Grenada, an AC-130 Gunship was tasked to fly over the Point Salines airfield and report on the condition of the runway. Using low-light level television, the crew reported the runway was obstructed with vehicles and 50 gallon drums.

The mission commander on board the C-130 command and control aircraft made the decision that all of the first wave of C-130s would airdrop due to the condition of the runway. The original plan to secure the field with Rangers and then airland the remaining force was scrapped.

"We were all in a holding pattern 75 miles west of Grenada. The first two airplanes departed holding to make their 0500 drop time," Hobson said. "The weather was bad, with thunderstorms in the area."

En route to the drop zone (Point Salines runway), the number one MC-130 lost its inertial navigation system. Considering the severe weather in the area and afraid that with a degradation in navigation aids there was a possibility of either compromising the operation or dropping Rangers in the ocean, the aircraft commander made the decision to call "no drop."

"Since the first two Talons were a package, the air mission commander directed them to hold south of the initial run-in point (the IP) and called us to see when we could make it to the runway. We were now sequenced to make the first drop," Hobson said.

This wasn't the best news for the battalion commander on Hobson's aircraft since he was not heavily armed and had been briefed to expect stiff resistance.

With the runway as the drop zone, Hobson took his plane in.

"A search light from the tower shined on the aircraft at about a mile from the end of the runway and we readily knew that this was not a surprise operation. As soon as the Rangers started out of the aircraft we started to take small arms and 20mm fire. I told the loadmaster to tell me when the last guy was out of the airplane so we could get the hell out of there," Hobson said.

They dropped from 500 feet, giving the Rangers about 10 seconds of "hang time," - all that they wanted under the circumstances. When the last Ranger was out of the aircraft Hobson turned his plane and descended sharply to the right rather than continue down the runway.

"As we went by the IP the number one and two aircraft joined behind us. After I pulled off the Cubans turned everything on the number two aircraft. He received damage and subsequently pulled off without dropping," the general said.

Circling AC-130 gunships were waiting to lend a hand and Hobson called the mission commander for help.

"The rules of engagement were that we wouldn't fire unless we were fired on, so I called the airborne mission commander and said 'we're getting a lot of fire, we need to bring in the gunship,'" Hobson said.

The gunship poured 20mm and 40mm cannon fire into the enemy positions and the remaining MC-130s in the first wave completed their airdrop missions, putting the first squads of Rangers on the ground by 6 a.m.

Although they didn't execute the original plan, both the aircrews and the Army troops found ways to adapt to complete the mission.

C-141s at Clark AFB, Philippines during Operation Homecoming. (Courtesy of Ralph Leske.)

Making Things Better

Even though Operation Urgent Fury was a success, Hobson said that many valuable lessons were learned as a result of Grenada with communications as one of the most important.

"We learned that communications on the cheap is no way to go to war. We only had one channel on the SATCOM (satellite communication) radio. When the original plan changed to where everyone was going to air drop the Army needed to get on the SATCOM radio for coordination and it rapidly became saturated," the general said.

They had only one radio operator on each plane and he was working both Air Force and Army operations, which caused great confusion, according to Hobson.

"We're all using common radios now, so hopefully that's been resolved. We also have more access to satellite capabilities than we had in 1983. Today in a crisis we would have access to several SATCOM channels," Hobson said.

The C-17 Globemaster III will be even better, according to the general. The C-17 was built so that the Army would have their own access to SATCOM radios in the cargo compartment of the airplane. The C-17 would also have helped in the Grenada operation by putting more forces on the runway in a shorter period of time.

For the operation, Hobson was awarded the MacKay Trophy for the most meritorious flight of the year. He said that advances in communication, improvements in intelligence and the advent of the C-17 will surely make potential Grenada-type operations more successful.

Operation Just Cause: Panama

From 17 December 1989 through 14 February 1990, all branches of the United States Armed Forces took part in a large-scale, complex military offensive in Panama. Operation JUST CAUSE was the largest American combat operation since Vietnam. The United States considered the Isthmus

C-141s in formation.

of Panama and the canal that bisects that nation of vital interest since Panama gained its independence from Colombia in 1904. In recent years, animosity between the United States and Panama escalated, culminating in an American indictment against the Panamanian strong-man Manuel Noriega on drug trafficking charges in February 1988. The United States government's political and economic sanctions failed to force Noriega from power but instead drew threats and violence against Americans living and working in his country and in the Canal Zone.

Friction between the two countries became especially acute around the time of the Panamanian presidential election on 7 May 1989. Following the overwhelming defeat of his hand-picked candidate, Noriega nullified the elections causing thousands of Panamanians to march and riot in protest. Noriega's person army, the Dignity Battalion, reacted with force, including assaulting the victorious presidential candidate, Guillermo Endara and his running mates.

President George Bush recognized that the violence and coercion surrounding the election posed a serious threat to the safety of the Panama Canal and the 51,000 Americans who lived in Panama. Accordingly, he decided on 11 May 1989 to deploy a brigade-size security augmentation force to ensure the protection of American lives. Between 11 and 18 May 1989, MAC used 34 C-5, 39 C-141, and 2 commercial L-1011 missions, airlifting 2,679 soldiers and Marines and 2,950 tons of cargo to Panama, to complete the highly-publicized Operation NIMROD DANCER.

That same day, 11 May, President Bush ordered some 5,000 dependents living among the civilian community to move onto a military facility by 12 May or to return to the United States. During the seven-week period from 16 May through 29 June 1989, Operation BLADE JEWEL airlifted a total of 5,915 persons on 49 commercial, C-5, and C-141 missions, from Howard Air Force Base, Canal Zone, to Charleston International Airport, South Carolina.

Following a failed coup attempt against him on 3 October 1989, Noriega acted even more aggressively to eliminate opposition and consolidate his authority. As conditions in Panama deteriorated in the ensuing weeks, several events occurred in rapid succession, causing President Bush to order military action. On 15 December 1989, at Noriega's instigation, the Panamanian National Assembly declared that a state of war existed between Panama and the United States and enacted measures to repel foreign aggression. The following night, 16 December, a member of Noriega's Panamanian Defense Force, the country's combination military police/internal security force, killed an American Marine officer. The defense force also arrested, beat, and threatened a United States Navy lieutenant and his wife, who witnessed the killing.

The attempted coup prompted American leaders to solidify plans for military intervention. Planning for the Military Airlift Command's participation intensified when representatives of MAC, TAC, and SAC attended an interservice meeting at Fort Bragg, North Carolina, between 5 and 9 October 1989. The Air Force's major operational commands perfected a series of invasion plans that included simultaneous airdrop/airland operations. Headquarters MAC determined that the command's airlift system required 60 hours to marshal the crew force needed for the invasion: approximately 36 hours to locate the crews and get them to onload locations; and another 24 hours for mission planning, preparation, and en route flying time.

Fortunately, Operation JUST CAUSE began at the start of the Christmas holiday season, so MAC had fewer scheduled commitments. Most of the 63 C-141s, 21 C-130S, and 2 C-5s the command needed to support the Army's invasion force were in place at their home stations. In preparation for a coup de main in Panama, MAC aircraft from throughout the airlift system prepositioned troops and equipment the day before at the staging locations: Pope Air Force Base, North Carolina; Lawson and Hunter Army Air Fields, Georgia; and Charleston Air Force Base, South Carolina.

Despite complications from fog and icing in the loading and staging areas, 12 C-141s and 4 Special Operations Low Level (SOLL) II C-130s began airdropping paratroopers and heavy equipment from the 75th Ranger Regiment on Torrijos/Tocumen Airport near Panama City early on 20 December 1989. At the same time, 15 C-130s executed an airdrop on Rio Hato, a major Panamanian Defense Force base and airfield. Within a few hours, ground forces had secured Rio Hato, and SOLL C-130s commenced landing to deliver additional equipment and evacuate wounded.

Special operations forces then assigned to MAC's Twenty-Third Air Force also played a critical role in the successful execution of JUST CAUSE. At H-hour, when the operation began, Twenty-Third Air Force had 26 aircraft committed to offensive air operations: 7 AC-130s, 7 HC-130s, 3 MC-130s, 4 MH-60s, and 5 MH-53s. The C-130s attacked ground targets and provided tactical support for the Army Rangers while the MH-53 and MH-60 helicopters deployed and repositioned forces in Panama City. All of the United States-based AC-130s and 80 percent of the United States-based MC-130s participated in the operation.

Contingency plans called for two C-5s to land at the Torrijos/Tocumen airport, after ground troops had secured it, and to provide a refueling platform in the event enemy action rendered the Howard runway unusable. The *Galaxys* also deployed an airlift control element (ALCE) to Torrijos/Tocumen to manage the airlift flow of follow-on support forces. Since Panamanian Defense Forces made no assault on Howard and the base's runway remained open, the C-130s of the assault force refueled there. Consequently, the C-5s landed at Howard and

Midnight off loading at Mogadishu. (Courtesy of Geoffrey R. Douglas.)

deployed the ALCE to Torrijos/Tocumen airport a few hours later. Working throughout the night of 20-21 December, the MAC ALCE and air traffic controllers at Torrijos/Tocumen had the airport in full operation by the time the security forces arrived on the twenty-first.

Even as United States forces were securing Torrijos/Tocumen, Rio Hato, and other military objectives, President Bush made a television/radio broadcast to the American people. He outlined events in Panama over the past two years and reminded the nation that he had not taken the decision to invade lightly. The President reported that the fighting was still underway and that some American lives had been lost. He went on to say that the United States had restored the legitimate government of President Guillermo Endara, duly elected on 7 May 1989. President Bush told the nation he had directed the Secretary of the Treasury to lift economic sanctions against Panama in respect for the Endara government. Finally, the President said he was returning the American Ambassador to Panama and taking appropriate action to achieve an orderly release of Panamanian assets frozen in the United States.

On 21 December, the day after MAC completed its participation in the pre-dawn assault, the command airlifted an additional 2,500 troops to Panama. Many of the soldiers were security policemen sent to help restore order amid widespread looting. Later that day, other MAC aircraft took augmentation forces assigned to the 7th Infantry Division to Torrijos International Airport. Military Airlift Command transports also evacuated military personnel wounded during JUST CAUSE from Howard to Kelly Air Force Base, Texas, after they had received their initial treatment in Panama. Aeromedical evacuation missions lasted from 20 December 1989 through 5 January 1990. During this period, 257 patients were flown from Panama to Kelly aboard 1 C-130 and 8 C-141 missions. Included among the wounded were 216 litter and 41 ambulatory patients.

Although having earlier boasted that he would never be taken alive, Noriega asked for and received asylum at the residence of the Vatican representative in Panama on Christmas Eve. American soldiers and tanks quickly surrounded the residence of the Papal Nuncio and blocked all approaches to the area. After failing to negotiate safe passage to a neutral country, Noriega left his sanctuary on 3 January 1990 and surrendered to American forces. An Army UH-60 Black Hawk helicopter flew him to Howard, where Drug Enforcement agents arrested him on the outstanding warrant. A MAC MC-130E then carried the deposed strong-man to Homestead Air Force Base, Florida, and delivered him to waiting federal officials, who whisked him away to the federal courthouse in Miami for booking and incarceration.

The Military Airlift Command's role in restoring democracy in Panama consisted of more than military action. Noriega was still in the Vatican embassy when the United States began its large-scale assistance program for the Panamanian people. The first of many humanitarian airlift missions commenced two days after Christmas. By the time this airlift ended on 3 January 1990, 8 C-5s and 14 C-141s had transported three tons of medical supplies, 10,000 blankets and sheets, several tons of baby food and food staples, and two million field rations to Howard Air Force Base and Torrijos International Airport from several continental United States locations.

Events in Panama had stabilized by New Year's Day to the point where MAC C-130s redeployed 56 troops to the United States. These were the first of the deployed forces to go home. By the end of January, MAC aircraft had airlifted approximately 10,000 troops back to the United States. The last returned to the United States on 14 February 1990, the day JUST CAUSE officially ended for MAC.

C-5 Galaxy over the Golden Gate Bridge.

From the time President Bush ordered the execution of JUST CAUSE on 17 December 1989 through 14 February 1990, C-5s, C-141s, C-130s, C-9s, C-12s, and civilian aircraft contracted by Headquarters MAC flew 775 missions to transport 39,994 passengers and 20,675 tons of cargo to and from Panama. Special operations units of MAC's Twenty-Third Air Force fulfilled a critical role in neutralizing pockets of resistance at H-hour and thereafter. Twenty-Third's AC-130, EC-130, HC-130, MC-130, MH-60, and MH-53 aircraft flew 796 missions during all of JUST CAUSE.

Media commentators correctly reported JUST CAUSE as being the largest United States military operation since Vietnam. Similarly, the early morning airdrop on Torrijos/Tocumen and Rio Hato Airfields constituted the largest personnel airdrop since Korea and the largest nighttime parachute assault in the history of air power. The restoration of democracy in Panama was a triumph for the concept of joint operations, as many intricately related assault operations had occurred simultaneously at H-hour. As measured by the number of separate organizations that participated directly in JUST CAUSE, MAC's role was greater than any other Air Force organization.

MAC Epilogue

Initially, the requirement to air transport military materials and passengers arose in the interwar years as the nation became air minded. During World War II, as the statistics attest, the air transport mission was critical to the war effort, and it proved its worth again during the Korean conflict. Following these wars, the need for a military air transport system during wartime remained undisputed; there was, however, great debate over whether the Defense Department should maintain such a system in peacetime, especially when the commercial airlines could handle the government's transpiration requirements. Thus, over the decades - driven by technological advances, political processes, and national security interest - the air transport mission developed into a highly responsive, world wide, military airlift system. These three major themes permeate the fifty-year history of the Military Airlift Command. Together, they capture the essence of how the command has evolved.

First, aviation technologies have continually influenced the evolution of military air transportation. Consequently, the history of airlift encompasses the endless search for larger, more advanced, and more capable airlift systems to include improving attendant systems that provide for ground handling, aerial port, maintenance, and command and control. In reality, this quest incorporated changes in political priorities, service requirements, and defense strategies as well as the social and economic climates of the times.

Cargo capacity, range and speed have always been central issues when deciding whether to purchase a particular transport aircraft. Starting in World War II with military versions of commercial aircraft such as the venerable C-47 (2.5 ton payload) and C-54 (6-7 ton), the command sought to develop its own transports to meet the armed services' specialized requirements. Although the prop-driven C-124 Globemaster was built specifically for military use, the shift from commercially-designed transports gained momentum when the Air Force moved from propeller to jet aircraft for its strategic airlift fleet.

The C-141 *Starlifter*, which entered service in the Mid-1960s, fundamentally changed the military airlift system. General Howell M. Estes Jr., the MAC Commander at the time, understood better than most that the acquisition of the C-141 ushered in a revolution in airlift. Estes identified nine overlapping technological limitations confronting an aircraft system: speed; range and payload tradeoffs; flexibility of employment; cubic capacity; ease of loading; self-sufficiency; fuel efficiency; direct operating costs; and terminal/airfield requirements. Although Estes asserted that no single aircraft had ever overcome all of these difficulties, he did conclude, however, that the C-141 had made a quantum leap forward by minimizing many of these historic limitations. The C-141 fostered a transportation revolution by its ability to airlift large loads over intercontinental distances to either a combat or non-threatening environment in a matter of hours. It opened up new avenues for the employment of airlift.

The C-5 *Galaxy* demonstrated a similar commitment to the never-ending search for technological developments that would increase airlift's capability. General Duane H. Cassidy, a former MAC Commander in Chief, suggested recently:

We started to build the C-5 and wanted to build the biggest thing we could build. We went to the engine manufacturers knowing that the power plants were going to determine how big the airplane could be. We said, "How big an engine can you build?" Herman the German, a guy working for GE at the time, Gerhard Neumann, ... said, "We can design you an 8:1 bypass ration engine." everybody rolled their eyes because that was twice as much as anybody had ever done before, but he designed it, and he built it. It was the TF-39 engine. That engine then has become a core for GE derivatives that are flying in every aircraft in the world today. Quite frankly, the C-5 program was a great contribution to commercial aviation. We'll never get credit for it, but we incentivized that industry by developing that engine. We figured how much four 8:1 bypass ration engine could lift and built an airplane around those four engines.

The C-5's wide-body design, massive cargo capacity, speed and range (especially with aerial refueling) represented a tremendous breakthrough in airlift capability. A future airlifter, the C-17, incorporates recent technology and the best features of both the C-141 and C-5; it promises to represent another technological leap forward.

Technological advances, however, have not been limited to aircraft, for virtually every aspect of the airlift mission has been affected. For instance, the packaging, handling and management of cargo has undergone substantial improvements since the pre-World War II era when everything was essentially hand loaded piece by piece. Even as late as the Berlin Airlift, it was not uncommon for items weighting 500 pounds to be loaded manually although forklifts did ease the burdensome loading procedures somewhat. When the C-124 was purchased in the early 1950s, it was designed not only for greater cargo capacity but also for easier loading. A former MATS Commander, William H. Tunner, wrote, "The front opens up like a clamshell, and ramps go down like the legs of a praying mantis. You can roll any cargo you want, even light tanks, right into this big beast."

In the mid-1950s, the Military Air Transport Service issued a contract for developing a rigid pallet, side guide rail system for the C-133. The contract resulted in the 463L materials handling system which has standardized cargo loading and management ever since. This gave rise to researching and developing special vehicles to move pallets and computerizing management tracking systems and the overall aerial port structure.

Another example of MAC employing technology expeditiously has been in the area of command, control, and communications. Virtually every aspect of the airlift mission has been affected by the advances in this field. From a simple system in World War II, which relied on rudimentary short-wave radios, radar, and telephones, the Military Airlift Command has developed an exceptionally capable system which allows near real-time flight-following of individual airlift missions world wide as well as the means to communicate with that aircraft from MAC headquarters. Implementation of this system in the 1980s has been one of the most significant events in the command's history and represents the immediate benefits of futuristic technology. The search for appropriate technology will continue to ensure a responsive airlift system.

310th MAS, Howard AFB, Panama, Dec. 13, 1987. (L to R); Back row: Ltc. Bruce Sutherland (P), Capt. Andy Alderson (CP), MSgt. Windsor Williams (LM), unknown, unknown. Front row: MSgt. Sager (FE), Capt. Neil Smith (N), unknown, unknown. (Courtesy of Andrew B. Alderson.)

The second major factor that has shaped the course of military airlift is the political process. Indeed, Congress' direction has been felt at every major organizational change. There also exists the political maneuvering within the Defense Department and Headquarters United States Air Force. Military airlift, although part of the aviation family, has always been regarded as a stepchild. The reasons for this are complex. Even though airlift has long been officially considered one of the primary missions of the service, airlift, in essence, has been largely viewed as an auxiliary capability, not contributing directly to the quest for air superiority or strategic bombardment. This impression that airlift is closely tied to an essentially unglamorous logistical effort has reinforced the step-child perception. In addition, airlift's ties to the Army have probably affected the standing the larger Air Force has given the airlift mission. Considering the divorce from the Army in 1947 had been a difficult one, the Air Force has sought to prove it has a mission and a significance beyond that of supporting ground operations.

This political environment ensures that the command's leadership has had difficulty obtaining the resources and priorities it required. This political process also drove the quest for ever greater control of airlift assets within the Defense Department. Lieutenant General Tunner, one of the foremost airlift strategists, argued eloquently throughout his career for placing all airlift resources under a single major command. "Airlift, rather than airlift airplanes, must be allocated to those commands which need it," he reflected. "To scatter thousands of expensive aircraft, complete with crews, maintenance, and spare parts throughout the entire military establishment has always been an inefficient and expensive waste of men, equipment, and money." Consolidating all airlift made sense not only on the basis of efficiency but also because of the inherent stature and authority a world wide organization commands. In many respects, the Defense Department's regard for the airlift mission improved dramatically in the mid-1970s with the consolidation of tactical airlift resources into the Military Airlift Command and the command's designation as a specified command under the Joint Chiefs of Staff. One of the important challenges for the future will be to maintain the single manger concept, for difficulties abound and the stepchild stigma remains.

Third, from an operational perspective the command's mission is one of the most significant because it daily supports the national security interests of the United States. Unlike virtually all other Air Force missions, airlift has an important role in both peacetime and combat environments. It has been used in every conceivable scenario across a spectrum of foreign policy options from sustaining combat forces overseas; to projecting military power to Korea, Vietnam, the Caribbean and Latin America, the Middle East, or Southwest Asia; to humanitarian airlift after disaster; to the movements of diplomats and foreign affairs advisors such as the shuttle-diplomacy missions of the early 1970s. All have been at the core of MAC's operations since the command's inception.

General Estes maintained that government officials began to recognize the importance of airlift early in the 1960s as a tool for executing the foreign policy objectives of the United States. He wrote that "global military airlift has been shown, throughout the era of the cold war, to be a principal medium of achieving maximum military flexibility." By the time of the Vietnam Conflict, Estes contended that MAC had become "the key element in a far-ranging change in national policy: to a strategy of multiple options for flexible, measured response to any situation in the spectrum of war." He called this linear progress; it was relatively straightforward, except to the most myopic and obtuse individuals, that the obvious advantages of airlift outweighed its limitations. The Military Airlift Command's unique, varied, and far-reaching responsibilities will continue to prove the importance of airlift into the twenty-first century.

Air Force Performance in Desert Storm, April 1991

Deployment, Sustainment and Resupply Efforts

Airlift, tanker support, prepositioning of supplies, and a large, modern, base infrastructure permitted movement of forces into the AOR, and provided the ability to operate quickly from Saudi Arabia. These forces assured we had the "global reach" and support to exercise "global power."

Strategic Airlift

Desert Storm was the largest airlift since World War II. Airlifters moved combat forces half way around the world. By the cease fire, airlift had moved over 482,000 passengers and 513,000 tons of cargo into the AOR — the equivalent of moving Oklahoma City, all of its people, all of its vehicles, all of its food, and all of its household goods halfway around the world.

Air Force C-5s and C-141s, moved 72% of the air cargo and about one-third of the people while commercial augmentation moved the rest. C-5s were 90% and C-141s 80% committed to Desert Storm. The remainder of Air Force airlift flew other high priority DOD missions to the rest of the world. This operation was the first time in its 38 year history that the Civil Reserve Air Fleet (CRAF) was activated. On 18 August, the first stage of CRAF, 18 passenger and 23 cargo aircraft, was activated. When fighting commenced, the second stage, 77 passenger and 40 additional cargo aircraft, were activated. These commercial carriers provided additional airlift capacity needed to meet CENTCOM's requirements. These commercial aircraft carried the majority of the troops to the AOR.

Strategic airlift forces have been going at full speed since Desert Storm started and will continue until our forces are redeployed home. Airlifters with Airlift Control Elements were the first to land in the Arabian Peninsula and will be the last to leave. Many of these people are Guardsmen and Reservists who have twice demonstrated their patriotism during the past 18 months — in Panama and now the Arabian Peninsula.

Air Refueling

The rapid deployment of fighter aircraft to Saudi Arabia resulted from Air Force tanker capability. Within 38 hours of the deployment notice, the first F-15 aircraft had landed in Saudi Arabia and were ready to defend the Persian Gulf area. The Strategic Air Command deployed 256 KC-135s and 46 KC-10s into the AOR during the war. In Desert Shield, tankers flew 4,967 sorties and 19,089 hours, refueled 14,588 receivers, including 5,495 Navy and Marine aircraft, and offloaded 68.2 million gallons of fuel. Tankers surpassed this effort during the six weeks of Desert Storm when they flew 15,434 sorties, logged 59,943 hours, refueled 45,955 aircraft and offloaded 110.2 million gallons of fuel. Approximately 20% was used for Navy and Marine receivers. Every aircraft — fighter, bomber, airlift, AWACS, or JSTARS — and every service and some allies used Air Force tankers to do their mission. One F-15 pilot commented about tanker accessibility: "There was more gas in the sky over Saudi than in the ground below" — a testimony to the tanker force. No other nation has such a capability.

Prepositioning and Base Infrastructure

Saudi Host Nation Support was superb. During the past decade, whole bases were built, stocked and prepared for war. These locations made it easy for forces to quickly move from a deployment phase to full combat operations. The Gulf Cooperation Council provided fuel and many other materials. In other locations, the Air Force operated from open runways. There, civil engineers literally built cities on the sand. During the past decade, the Air Force developed tents, hospitals, supply areas, ramps and other base essentials had been developed as modular buildings. This investment guaranteed operation from austere locations anywhere around the globe. Afloat and ashore prepositioning included bare base and fuels equipment, supplies, vehicles, and munitions which were required to sustain and project Desert Storm forces. These supplies saved an estimated 1,800 airlift missions to the AOR and provided prepositioning of supplies and base infrastructure for 21 principal airfields.

Theater Airlift

Over 145 C-130s deployed to support Desert Storm. These aircraft moved units forward as they deployed into the theater. From 10 August to the 2 April, C-130s flew 46,500 sorties, logged over 75,000 hours, and moved over 209,000 people and over 300,000 tons of cargo. Once the fighting started, they provided logistical support, medical evacuation of wounded, and battlefield mobility through airland and airdrop. During the ground campaign, C-130s flew over 500 sorties a day. During Desert Storm, C-130s had a mission capable rate of 84% — 6% higher than in peacetime.

Air Mobility in Operation Desert Shield/Storm

On August 2, 1990, 100,000 Iraqi troops overran and occupied Kuwait. President George Bush, on August 7, ordered United States forces to the Persian Gulf region to prevent an Iraqi invasion of Saudi Arabia, thus beginning Operation Desert Shield. When hostilities began on January 17, 1991, Desert Shield became Operation Desert Storm.

The first MAC aircraft to arrive on the Arabia peninsula, a C-141 from Charleston Air Force Base, South Carolina, landed at Riyadh Air Base, Saudi Arabia, on August 8 with an advance team from the United States Central Command Air Forces (CENTAF). Only airlift combined the speed and flexibility needed to deploy rapidly tens of thousands of American troops to a distant theater. An equally massive sealift, in concert with thousands of airlift missions, transported the vast quantities of material needed to resupply the forces deployed to the Persian Gulf area of responsibility (AOR). Airlift remained the principal means of transporting troops and for deploying time sensitive material. Along with the airlifters, tanker aircraft made up the other component of the air mobility equation. During the conflict, American refueling aircraft also supported the war effort.

Desert Shield soon became the most massive airlift in the history of air power. It was the first large strategic deployment of combat forces by air. By its sixth week, Desert Shield had surpassed the Berlin Airlift (1948-49) in total ton-miles flown. Air Force Chief of Staff, General Merril A. McPeak, later described MAC's contribution to Desert Shield "as the equivalent of a Berlin Airlift every six weeks." The statistics tell the story. During the Berlin Airlift United States aircraft moved 1.78 million tons an average distance of 300 miles, but the MAC airlift system had moved nearly 550,000 short tons of cargo between the United States and Europe to the Persian Gulf by the time most of the deployed forces had returned to their home bases by late April 1991. While the average distance from an European operating base to Berlin was 300 miles,

Resourcefulness is an airlift hallmark. (Courtesy of James J. Hoogerwerf.)

The first A-10 arrives at Edwards AFB for testing. 1975. (Courtesy of Ted Venturini.)

it was about 7,500 miles from the United States east coast to the Persian Gulf region. A comparison of Desert Shield/Storm with the Berlin Airlift follows.

Comparison of Desert Shield/Storm MAC Airlift with United States Participation in the Berlin Airlift

Desert Shield/Storm* STRATEGIC AIRLIFT	BERLIN AIRLIFT
Duration	
9 months	15 months
Distance (Average from onload)	
6,330 miles	300 miles
Average Ton Mile Daily	
13.6 million	1.2 million
Missions	
19,780	189,960
Short Tons	
548,000	1.78 million
Passengers	
590,956	62,750

•August 7, 1990 to April 19, 1991

Any comparison of the Berlin Airlift with the Persian Gulf War requires qualification, because the Berlin Airlift was also a remarkable accomplishment. During the 15-month Berlin Airlift, American C-47s and C-54s transported a daily average of 3,853 tons to Berlin. This was roughly 44 percent more cargo than the 2,149 tons MAC airlifted daily to the Arabian peninsula. Just three German airfields had supported the 1948-49 operation, but MAC organic transports and commercially contracted aircraft eventually were landing at a dozen airfields in the AOR. the relatively small C-47s and C-54s were manually loaded and unloaded. This made the Berlin Airlift a much more labor-intensive operation. (Rpt, MAC/XPY, Desert Shield/Storm Analysis, April 1991; R.D. Launius and C.F. Cross II, MAC Office of History, MAC and the Legacy of the Berlin Airlift, April 1989, pp. 58-60.)

In assessing the Desert Shield airlift, the magnitude of the operation and the long distance from the continental United States to the Persian Gulf region must always be kept in mind. Desert Shield reaffirmed the oft-made statement that airlift is a total system, dependent upon the smooth working of many integral components. These include but are not limited to the availability of aircrews, suitable operating bases, and the ability of aircraft to be maintained in a timely fashion. The various elements of the airlift system must operate smoothly to produce maximum efficiency. This assessment will focus on the planning and management of the Desert Shield/Desert Storm airlift. It will discuss the contributions of the Civil Reserve Air Fleet (CRAF), the total force policy, C-130 operations, and air refueling. Also examined is the extent to which the Gulf War might serve as a model for future operations.

General Hansford T. Johnson, who served as both Commander in Chief, United States Transportation Command (CINCUSTRANSCOM) and Commander in Chief, Military Airlift Command (CINCMAC), believed that one of Desert Shield's major lessons had already been learned from Operation Just Cause, the United States military operation in Panama in December 1989. Just Cause, he said, marked a turning point in United States military doctrine. It demonstrated that "once our nation decides to intervene, we intervene very quickly...with overwhelming force." So it was with Desert Shield.

Lessons applicable to air mobility arose daily, even hourly. Commanders, mid-level managers and technicians fine-tuned the airlift system to deal with new requirements, unanticipated difficulties and ever-changing circumstances. The need to move tens of thousands of troops quickly gave MAC no time to build an airlift system from scratch. Rather, the airlift's success resulted in large measure from MAC having perfected a seamless airlift system over five decades. In 1990 this system was mutually dependent upon strategic airlift, intratheater airlift, the commercial airlines and the integration of the active duty, Air Force Reserve and Air National Guard forces.

The pattern of operations begun in Desert Shield's first week continued until the redeployment airlift was largely completed in April 1991. Ninety percent of the strategic airlift missions staged through four European bases: Torrejon Air Base, Spain (31 percent); Rhein-Main Air Base, Germany (27 percent); Zaragoza Air Base, Spain (18 percent); and Ramstein Air Base, Germany (14 percent). Other European bases used by MAC were RAF Upper Heyford, United Kingdom; Rota Air Base, Spain; Lajes Air Base, Azores; and Sigonella Air Base, Italy. This relatively small number of en route bases made the aggregate airlift system more vulnerable to variables such as weather and ramp congestion.

Desert Shield validated the importance of having an airlift system in place that could expand as escalating requirements dictated. The intense August airlift would have been even more demanding without an established en route system at Torrejon, Rhein-Main, Ramstein and Zaragoza. The cadres of trained people, already present at these bases, were quickly augmented with temporary duty (TDY) personnel to support the surge of missions that stopped en route to the Arabian peninsula. Time was a valuable commodity in August 1990, when an Iraqi invasion of Saudi Arabia seemed likely. It was appropriate, then, that in the aftermath of the war, General Johnson emphasized to Secretary of Defense Richard B. Cheney the importance of the United States retaining both a Central European and an Iberian peninsula base. He asked that the United States Transportation Command be consulted on any base closure issues "affecting the global strategic mobility mission."

Unlike Operation Just Cause in Panama, no approved transportation plan existed for Desert Shield. Initially, Persian Gulf deployment requirements were based upon a United States Central Command (USCENTCOM) operation plan for Southwest Asia that was still being coordinated among the organizations that supported USCENTCOM. Members of the Crisis Action Team reviewed all of the operations plans existing for Southwest Asia. None fit the circumstances because a scenario involving an Iraqi invasion of Kuwait had never been used.

General Johnson described the mood of airlift customers in August 1990 as one in which "everybody wanted to rush off to the war. Delays and inherent difficulties arose because too many users of airlift wanted to move troops and cargo forward as quickly as possible. No Time-Phased

C-130/C-141 Snuggling Workhorses.

Force Deployment Data (TPFDD) existed for the Desert Shield scenario. Without an approved transportation plan, unit deployments were simply based upon gross estimates of requirements. The need for Desert Shield to begin at maximum speed and volume became more difficult without an approved transportation plan. If, for example, a Time-Phased Force Deployment Data List (TPFDDL) reported that a given unit was ready for airlift on day 5 with 500 tons of cargo and 500 troops, MAC scheduled whatever combination of C-5s, C-141s or contract commercial aircraft was needed to satisfy the requirement. The Crisis Action Team quickly discovered that the airlift requirements being submitted through the TPFDDLs by dozens of Army, Air Force, Navy and Marine Corps organizations were unreliable indicators of a unit's true airlift needs. Onload locations were often wrongly stated, and the terms "outsize" and "oversize" cargo were frequently applied incorrectly. These circumstances caused the following types of inefficiencies: too much airlift was scheduled, not enough airlift was tasked, or the wrong type of airlift was planned.

The problem of unreliable TPFDDLs was fixed early. A Requirements Validation Cell was created within MACs Crisis Action Team and staffed around the clock, usually with six to nine validators. Cell members telephoned every unit requesting airlift to confirm the departure dates and quantities of cargo and passengers reported in the TPFDDLs. This, too, did not always go as planned because it sometimes took several calls to find someone who could state with certainty a unit's true airlift requirements. Unreliable TPFDDLs remained a major problem throughout the operation. The telephone cell never disbanded. Even in the late stages of the redeployment, MAC still called the units to confirm the accuracy of the requirements stated in the TPFDDLs.

Changing requirements and priorities were no less troublesome than unreliable TPFDDLs. "More than anything else," said MAC's senior CAT director, "changing requirements were our greatest difficulty " in August 1990. The Joint Operation Planning and Execution System (JOPES), the services' automated deployment system, created the sequence for each unit to deploy. The JOPES, however, could not cope with the magnitude of rapidly changing requirements. One day the United States Central Command changed its airlift priorities seven times between the 1st Fighter Wing at Langley Air Force Base, Virginia, and the 82nd Airborne Division, Pope Air Force Base, North Carolina. Aircraft already on their way to one of these bases had to divert to other locations. During the first three days of September, the 101st Division Air Mobile, Fort Campbell, Kentucky canceled 28 C-5 and 25 C-141 missions. During the surge period of August and early September, the airlift system was often able to generate aircraft more readily than the supported units could mobilize personnel and cargo. MAC responded by sending aircraft to the major onload locations at one-hour intervals rather than at one-half hour intervals as was done during the deployment's first weeks.

The frequent changing of airlift requirements, especially in the first days of a major operation, is likely to occur again. Even with more accurate TPFDDLs, supported commanders will always be redefining their transportation requirements as circumstances dictate. Major General Vernon J. Kondra, who became MAC Director of Operations on August 23, 1990, made the point, "As requirements change, it becomes a dynamic, evolving, changing thing. You'd better understand it's going to happen, no matter how hard you try... It absolutely will not go away."

By the end of August, the MAC system had completed nearly 1,700 strategic airlift missions to the Persian Gulf region. General Kondra, at this time, found the greatest difficulties to be rooted in there not being a total airlift system firmly in place. The absence of a total system stemmed in part from the United States Central Command not assigning MAC a base on the Arabian peninsula from which aircrews could rest and fresh crews stage to fly the inbound C-5s and C-141s back to one of the European bases.

The absence of a stage base complicated aircrew management. For safety reasons, aircrews were normally not to exceed 125 flying hours in 30 days and 330 hours in 90 days. To support the lengthy Desert Shield missions, General Johnson increased these maximums to 150 hours and 400 hours respectively. He also increased the maximum crew day from 16 hours to 20 hours. Given the delays in loading aircraft, C-5 and C-141 missions from the United States to Europe were lasting up to 17 hours. Similarly, many of the round-trip missions between Europe and the Persian Gulf that should have lasted less than 20 hours sometimes stretched to 30 or more hours because of unloading and refueling delays. Some pilots accrued flying time so rapidly that they exceeded the monthly flying-hour ceilings and were grounded.

The absence of a stage base prevented aircrews from completing their cycle of mandatory crew rest in the AOR. This circumstance required augmenting the basic C-5 and C-141 crews departing Europe for Southwest Asia with an extra pilot (three pilots instead of two) so that the same aircrew could return the aircraft to a European base. The pilot augmentees helped the primary crewmembers with their flying duties or rested themselves when deploying between the United States and Europe and on the missions to and from the Persian Gulf region. Yet even when not at the controls of a strategic airlifter, the pilot augmentees still accumulated flying time that counted toward their 30 and 90 day flying hour maximums.

In September 1990, to help slow the rapid accumulation of flying time, Headquarters MAC implemented a system of pilot pools at the four major European bases. Approximately 200 C-5 and C-141 pilots were apportioned among the four bases on three week temporary duty assignments. The pilot pools enabled a significant number of pilots to remain in Europe for an extended period and conserved flying time by reducing the need for pilot augmentees to shuttle between the United States and Europe so frequently.

Jeddah New Air Base, Saudi Arabia (with its plentiful supplies of fuel and considerable ramp space) was MAC's first choice for an in-theater stage base. Had Jeddah New become MAC's stage base, refueling time in the AOR would have been reduced greatly. On September 20, 1990, United States Central Command offered MAC Cairo West Air Base, Egypt, as an in-theater stage facility. Cairo West, however, could not support extensive refueling operations and would still have required the cargo transports to refuel at their offload points in Saudi Arabia. The offer also carried the proviso that the airlifters had to depart Cairo West within 24 hours after the outbreak of hostilities to make room for other United States Air Force assets. By this time, the pilot pools were working so well that MAC declined the offer. Having a stage base in a distant theater became one of Desert Shield's most important lessons for air mobility. A Rand Corporation study later reported that the absence of an in-theater stage base had reduced the strategic airlift capability by 20-25 percent during Desert Shield/Storm.

During the deployment of the defensive force that was completed about September 20, MAC averaged 100 to 105 onloads daily from approximately 35 United States locations. While there were delays loading in the United States, an even more serious bottleneck existed in the AOR, in large part because airlift customers insisted that a disproportionately large number of their missions land at Dhahran Air Base, Saudi Arabia. General Kondra used this analogy to describe the difficulties: "We had a four foot opening trying to push airlift through that 7,000 mile long hose and come out a four inch nozzle at the other end. It didn't work very well. You've got to have the offload bases to handle what you're pushing through the flow.

Subsequent analysis of Operation Desert Storm revealed the need for patience by both the customer and the airlift provider. Pressures to move passengers and cargo too quickly in the operation's first days wasted airlift when C-5s and C-141s were sent to aerial ports of embarkation only to find that the cargoes were not ready to load. Pushing the maximum number of aircraft through the en route system had been done instinctively at first, because neither MAC nor its customers would accept anything less

C-17 first flight on Sept 15, 1991. (Courtesy of Ted Venturini.)

than an intense all-out effort. General Kondra later stated that the haste to demonstrate a maximum effort had "clogged the system and worked it to death for no appreciable gain.

Desert Shield confirmed that airflow planning was an art, not a science. The three factors of changing priorities, en route system variables, and delays in unloading and servicing aircraft in the AOR required great scheduling flexibility. Techniques developed during years of peacetime exercises soon became established procedures. By late August airflow managers in the CATs at Headquarters MAC, 21st Air Force and 322nd Airlift Division, Ramstein Air Base, Germany began an unofficial system of assigning each aircraft a slot time (or time window) during which an aircraft was scheduled to arrive at one of the airfields in Saudi Arabia. Slot times, assigned in 15-minute increments, enabled airflow managers to know consistently about how many aircraft were at each of the offload locations. In late August and September, slot times were assigned two days in advance. When the system became more refined in October 1990, slot times were assigned up to four days prior to an aircraft's scheduled arrival. Slot times remained in use throughout the Gulf War. By Desert Shield's third month some airlift customers were reporting unacceptable delays in MAC's delivery of critical spare parts to the Arabian peninsula. To ensure customer satisfaction, on October 31 MAC initiated Operation Desert Express, which provided for the next-day delivery of "show stopper" spare parts to Dhahran and Riyadh. These highest priority spare parts were transported to the Arabian peninsula on a single C-141 mission that departed Charleston Air Force Base, South Carolina, daily at 1200 local time. Logisticians had only to have their high priority materials at Charleston by 1030 to ensure next-day delivery. Army, Marine, Navy and Air Force customers were each allocated a percentage of the cargo space. Airlift managers adjusted the service allocations as USCENTCOM's requirements dictated. Desert Express resulted in part from a lesson MAC had learned before: If an airlift customer perceives a problem, then a problem exists. As General Kondra put it, "When you provide a service, you have to be 100 percent right every time, because people only remember the bad service. They don't remember the good."

When the deployment of an offensive force began in late November 1990, a large portion of the American forces and equipment sent to Saudi Arabia came from Germany. This created the requirement for an overnight delivery of critical spare parts from Europe as well as from the United States. MAC responded by initiating European Desert Express in December 1990. In this operation, a dedicated C-141 operated daily from Rhein-Main Air Base to Dhahran. When the stateside Desert Express operation began experiencing some cargo backlogs in early 1991, a second dedicated daily flight from Charleston to the Arabia peninsula commenced on February 13, 1991.

Desert Express, however, carried a price. Since airflow planners assigned no slot times to other airlifters within an hour of the dedicated C-141's arrivals at Dhahran and Riyadh, there was some reduced flow in the total airlift system. Planners estimated that this caused the MAC system to lose the equivalent of four C-141 offloads daily. Nevertheless, by the time the redeployment began in early March 1991, the aggregate Desert Express missions had airlifted more than 2,700 short tons of cargo.

The strategic airlift missions to the AOR averaged approximately 100 per day in August and September and 50 missions daily in October and November. The signal for a more intense airlift tempo occurred on November 8, when President Bush ordered an additional 250,000 troops to the Persian Gulf region; thereby, raising the size of the deployed American force to about 430,000 men and women. The reason for this second deployment, the President said, was to create "an offensive military option" if Iraqi forces failed to withdraw from Kuwait by the January 15, 1991, deadline the United Nations had imposed. Many of the Phase II troops came from the United States Army's VII Corps, Stuttgart, Germany. While most of VII Corp's equipment was transported by sea, nearly all its troops were deployed by air. Also sent to Saudi Arabia in Phase II was the 3rd Armored Division, Frankfurt, Germany and several Marine units from the Continental United States.

The Military Airlift Command managed its Phase II airlift drawing upon the lessons it had learned since early August. Most important, the strategic airlift missions were now scheduled based upon the ability to offload and refuel aircraft in the AOR. "How many aircraft can we put through daily to Dhahran, King Fahd, Jubail, and the other Saudi bases?" was the critical question asked. "When we really started becoming efficient was when we worked the requirements backwards." General Kondra explained.

If MAC now recognized the importance of scheduling missions based upon the ability to offload, too many airlift customers had not learned the importance of sending their passengers and cargo to locations other than Dhahran, where 65 percent of the August missions had landed. Dhahran was the best airfield in Saudi Arabia, but there were other bases available to airlift nearly as good. Nevertheless, as the deployment progressed, Army and Marine customers continued to schedule most of their passengers and cargoes for Dhahran in order to marry up their inbound troops with equipment that was arriving by sealift at the nearby Ad Damman and Al Jubayl port facilities. By late September 1990, King Khalid Military City and King Fahd International Airport (the latter only a short distance from Dhahran) had been opened to airlift. Yet not until December 1990 did airlift customers validate more than 50 percent of their monthly cargoes for locations other than Dhahran.

The airlift's second phase began slower than MAC either desired or expected. Although MAC was ready to accelerate the airlift when President Bush ordered the second major deployment, the Phase II airlift did not get fully underway until the first week of December. Airlift planners attributed the delay partially to some European Command organizations being in the mind set of receiving forces rather than of deploying themselves.

Cargo to resupply the deployed forces continued backing up at Dover Air Force Base, Delaware and Tinker Air Force Base, Oklahoma, the two major aerial ports in the continental United States. Partly at MAC's urging, U.S. Central Command established several sustainment cargo diversion teams composed of representatives from all the services. The teams went to Dover, Tinker and the Defense Logistics Agency at Mechanicsburg, Pennsylvania. Here they identified non-priority cargo and arranged for its movement overland to an appropriate seaport for movement by sealift. To illustrate the dimension of the problem, one team diverted more than 1,200 tons of rations for sealift which, if transported by air would have required the equivalent of 60 C-141 missions.

Throughout the Gulf War, MAC relied heavily on the civil airline industry to help satisfy its enormous airlift requirements. On August 7, 1990, the day that President Bush issued the Desert Shield deployment order, two World Airways DC-10 passenger aircraft departed Pope Air Force Base, North Carolina, for Dhahran with more than 500 troops from the 82nd Airborne Division at neighboring Fort Bragg. Volunteered commercial airliners had completed about 30 missions to the Arabian peninsula, when the requirement for airlift became so great that General Johnson activated the CRAF's Stage I on August 17, 1990. His action marked the first time in the CRAF's 39 year history that any of its stages had ever been activated. Although the CRAF Stage I activation put a total of 38 aircraft (17 passenger and 21 cargo aircraft) at MAC's disposal, it actually added only 10 airplanes above those the civil air carriers had already volunteered. The important point, however, is that 12 of the 38 aircraft were the wide bodied Boeing 747s that MAC needed to transport troops.

Since the CRAF had never been activated before Operation Desert Shield, some problems naturally occurred. A troublesome issue became securing Title XIII insurance, which guarantees an airline government-sponsored war risk protection not available from commercial insurance companies. The Federal Aviation Administration (FAA) grants indemnity insurance to the CRAF, and the Department of Defense pays for any aircraft lost. The airlines reported that too much time had been spent in Desert Shield's first days obtaining Title XIII insurance on a mission-by-mission basis. If an aircraft had to be rescheduled, the paperwork began anew. Three weeks after the first CRAF mission arrived on the Arabian peninsula, the FAA authorized blanket approval on a weekly basis for all the commercial flights that were supporting Desert Shield. Still the absence of streamlined procedures for obtaining indemnity insurance during the critical month of August had resulted in duplicated effort, delayed missions and too often an inefficient use of airlift.

The civil air carriers played no less an important role in the Phase II deployment than they had during the Phase I airlift. On January 17, 1991, the day the air war began, the lingering backlog of cargo at Dover Air Force Base was the major reason for activating CRAF's Stage II. Given the urgency of moving more cargo, MAC told the Stage II carriers that only long-range, international (LRI) cargo aircraft need respond, even though the passenger aircraft called up in Stage II later participated in the redeployment airlift. The Stage II activation brought the total number of aircraft activated in both stages to 76 LRI passenger and 40 LRI cargo aircraft.

Meanwhile in the continental United States and Europe, civil airline crews learned from the media of Iraq's Scud missile attacks on Israel and Saudi Arabia and that the missiles might be armed with chemical warheads. Some airlines refused to send missions to the airfields that were within Scud range because their crews did not know how to use chemical warfare protective gear. Airlift managers had planned to distribute chemical warfare equipment to the CRAF aircrews after they landed in the AOR and simply had not anticipated the airlines' concerns.

MAC resolved the problem by sending teams to the airlines' en route stops at Frankfurt, Rome, Brussels, and other locations. Here the CRAF crews were given warfare gear and shown how to use it. The teams also gave the CRAF aircrews intelligence briefings. To address further the concerns of the airline crews, MAC ceased scheduling commercial missions into the AOR between 1600Z and 2100Z, the period when Scud launches were most likely to occur. General Johnson later said that CRAF pilots' concerns should have been anticipated and dealt with sooner.

From August 1900 through May 1991, the civil air carriers completed more than 5,000 missions. Since the CRAF had never been activated before August 1990, the lingering question had always been, "How well will the CRAF operate if called upon to support wartime requirements?" Clearly, the CRAF's most important lesson was a positive one. All in all, the CRAF had worked well, even though some broad questions, such as improving the system of indemnity insurance, remained. So it was, then, that General Johnson described the CRAF carriers as "tremendous heroes." I couldn't be more pleased with the system," he said.

The first major implementation of the total force policy also occurred during the Gulf War and confirmed the critical role of the Air Reserve Component (ARC) to the air mobility mission. In the autumn of 1990, the active strategic airlift force was authorized 76 C-5s and 234 C-141s. Another 28 C-5As and eight C-141s belonged to the Air Force Reserve. The Air National Guard had 11 C-5As and eight C-141s authorized. On Desert Shield's first day, General Johnson directed that all of the ARC's C-5s and C-141s participate in the airlift. He also asked aircrews from the Guard and Reserve to volunteer. Many Guard and Reserve crewmembers did so, but not enough of them volunteered to operate the ARC aircraft for an extended period of time.

In 1990 about half of the C-141 crews and 60 percent of the C-5 crews belonged to the ARC. The Command Analysis Group projected that 44 percent of the C-141 crew force would reach the 150 flying hour ceiling in the airlift's first 30 days. This prompted MAC on August 25, to activate three C-5 and two C-141 Reserve squadrons which equated to about 19 percent of the ARC's strategic airlift aircrew capability. Still the rapid rate at which crewmembers were accumulating flying hours continued. Some crewmembers, especially, C-5 pilots, were exceeding the flying hour limits, but another week passed before MAC decided to activate the ARCs remaining C-5 aircrews. Neither did it activate any additional C-141 squadrons until September 9, 1990. Ultimately, the aircrews of all seven C-5 squadrons and 11 of the 15 C-141 squadrons were activated. General Johnson later said he should have asked for a selected ARC activation sooner.

In the Military Airlift Command, more than 7,600 non-medical personnel were selectively called up before combat operations began in mid-January 1991. When President Bush ordered a partial mobilization of Reservists in all the services on January 18, 1991, 22,000 ARC men and women (60 percent of the Air Force total) belonged to MAC. Commencing with the first selective call-ups in late August 1990, MAC deployed active duty officers and airmen from its east coast bases at McGuire Air Force Base, New Jersey, Charleston and Dover to the European en route locations to support transit operations there. Personnel from the east coast bases were replaced stateside by activated Guardsmen and Reservists. Temporary accommodations had to be found for the ARC men and women who backfilled at the three east coast bases for the ones sent overseas. Military billeting and off-base accommodations at or near the east coast bases were taxed to the limits. Many ARC personnel had to be billeted far away from their TDY bases. Clearly, it would have been better to have kept most of the active duty force at home station and moved the ARC personnel forward.

Air transportation during the Persian Gulf War consisted of much more than thousands of strategic airlift missions. MAC C-130s made up the second component of the total airlift system. By the end of January 1991, approximately 144 MAC C-130s were fulfilling USCENTCOM's requirements for in-theater airlift. These aircraft were assigned to six wings at bases in Oman, the United Arab Emirates, and Saudi Arabia.

Although the beddown of fighter squadrons had been USCENTCOMs top priority in August 1990, theater airlift was also critical to the war effort. By the end of the month, 64 C-130s were operating in the AOR. Brigadier General Fredric N. Buckingham, the first Commander of Airlift Forces (COMALF), managed MAC's C-130s in the AOR on behalf of the Central Air Forces' Commander, Lieutenant General Charles A. Horner, General Buckingham and his staff were among the first United States forces to arrive on the Arabian peninsula. Their responsibilities included setting up the Airlift Control Center, creating a C-130 infrastructure and establishing the 1610th Airlift Division, Provisional, to which the C-130 airlift wings were soon made subordinate. In mid-October, General Buckingham was succeeded by Brigadier General Edwin E. Tenoso, who served as the COMALF until mid-May 1991.

Determining how many C-130s were needed to support General Schwarzkopf proved a challenge for General Tenoso and suggested an issue that might crop up again. Theoretically, MAC's entire C-130 inventory could have been deployed, but the COMALF wanted only enough C-130s to meet USCENTCOM's wartime requirements. During the force build-up from August 1990 to mid-January 1991, General Tenoso kept asking the Army to estimate its airlift requirements once hostilities commenced. Its spokesmen responded that the Army's wartime airlift requirement would be minimal, given its plans to rely principally on ground transportation. General Tenoso, however, believed that once the war started, United States ground forces would want all the airlift they could get. "Airlift will be like free candy," he quipped. "Everybody will want some." The nearly 14,000 C-130 sorties flown from January 17 through the end of February 1991 proved him right.

The Commander of Airlift Forces kept pressing Army and Marine Corps planners to define their airlift requirements for activities such as troop transport, food distribution and aeromedical evacuation. Based upon these discussions and the recommendations of his own staff, General Tenoso proposed 144 C-130s as the number of tactical airlifters needed to support USCENTCOM's wartime theater airlift requirements. USCENTCOM eventually agreed.

That no C-130 aircraft or crews were lost in the AOR confirmed the quality of C-130 flight training. Nevertheless, the Persian Gulf War demonstrated the need for better C-130 aircrews to fly more training missions not just with other airlifters but also with fighters, Airborne Warning and Control System aircraft (E-3s) and other airborne command control and communications systems as they had done during the Persian Gulf War.

Desert Storm also demonstrated the need to add high-altitude airdrop training to C-130s peacetime training curriculum. In January 1991, airlift planners had considered airdropping supplies to Allied forces from high-altitude, given the absence of terrain masking in the desert and the likelihood that the threat at high altitude would be less than from small arms fire at low altitude. The few C-130 crewmembers so trained were mostly Guardsmen and Reservists who had gained their proficiency in high altitude airdrop in Vietnam. Although no high-altitude airdrops were made during the Gulf War, General Tenoso urged that C-130 crewmembers be given this training in preparation for the next contingency.

With the C-130s under USCENTCOM's command and control, the supply system in the AOR required the C-130 wings to obtain spare parts through CENTAF logistical channels. Delays stemming from CENTAF's unfamiliarity with airlift often resulted in the provisional wings circumventing the established supply system and requesting spare parts directly from their home units. The importance of setting up an efficient resupply system quickly when mobility forces "chop" to another command in wartime was demonstrated during Desert Shield/Storm. "If we had to come over...and fight a war in the first two or three weeks, our planes' mission capability could very well have been down...just because our supply system was nonfunctional," General Tenoso cautioned. Yet overall, General Tenoso could rightly state that, "By every measure, we supported our users in an outstanding fashion."

From January 18-28, 1991, C-130s airlifted elements of the XVIII Airborne Corps from King Fahd International Airport to Rafha Airfield, in northern Saudi Arabia, near the Iraqi border. This intense airlift supported General Schwarzkopf's flanking maneuver, which he called the "Hail Mary Pass." C-130s flew the mission corridors at 10-minute intervals in radio silence. During the airlift to Rafha, C-130 sortie rates increased from 200 per day to more than 350 sorties during one 24-hour period. Nearly 14,000 troops and more than 9,300 short tons of cargo were transported in this fast-paced demonstration of air mobility. For all of Operation Desert Shield/Storm, the C-130s flew approximately 13,900 missions, 47,600 sorties and logged 76,000 flying hours to transport more than 242,000 passengers and 174,000 tons of cargo.

MAC shouldered the burden for airlift but did not carry the entire load. Tankers from the Strategic Air Command (SAC) played a significant, role by transporting 74 percent of SAC's passengers an 56 percent of its own cargo. SAC action officers sometimes referred to the tanker airlift as a "Little MAC" operation. Tankers moved a total of 14,208 passengers and 4,817 short tons of cargo. Operating in a purely cargo transport role, the KC-10 transported more than 25,100 tons of cargo and moved 4,185 passengers during the Persian Gulf War. Before the deployment, Air Force planning guidance stated that 23 of SAC's 57 KC-10s (40 percent) would be available for strategic airlift in a major regional contingency. The 23 KC-10s would have provided an additional 2.54 million ton-miles per day. As it happened, an average of only seven KC-10s operated in a dedicated cargo role from mid-August 1990 to the outbreak of hostilities. Not until the war began were 20 KC-10s released and committed exclusively to airlift. A greater use of the KC-10 as an airlifter is anticipated in future contingencies.

Of course air refueling United States and United Nations coalition aircraft was the primary mission of tanker aircraft during Desert Shield/Storm. The Strategic Air Command sent 256 KC-135s and 46 KC-10s to the AOR, and they performed an impressive number of missions. For all of Desert Shield/Storm, USAF tankers flew more than 34,000 sorties. KC-135s and KC-10s made over 85,000 air refuelings and offloaded more than 1.2 billion pounds of fuel. Tankers refueled both American and allied aircraft using the boom/receptacle and probe/drogue configurations.

The Gulf War tanker assessment confirmed much that SAC had been doing correctly in its management of the United States Air Force's air refueling mission. SAC tankers deployed to 21 bases in 21 overseas nations during Desert Shield/Storm. Nearly 100 tankers operating from nine countries maintained the Atlantic and Pacific air refueling bridges that moved more than 1,000 bomber and fighter aircraft to the Persian Gulf theater. As with airlift, ARC volunteers timely and much needed tanker expertise during the deployment's first days. Nearly 5,200 Guard and Reserve men and women were activated to support tanker operations, confirming the critical wartime role of these forces for air refueling as well as for airlift.

A comparatively small number of strategic airlift missions were air refueled. Early in the deployment some missions were air refueled in the AOR because of refueling limitations on the ground in Saudi Arabia. General Johnson pointed out that the long distance from the continental United States to Saudi Arabia negated the advantages of air

C-130H military transport Hercules. (Courtesy of Lockheed Martin.)

refueling on most missions. The en route bases in Europe were roughly equidistant from the continental United States and the Persian Gulf. Each airlift leg from the continental United States to Europe and from Europe to the Arabian peninsula, was about 3,500 nautical miles. Allowable cabin loads were based on these 3,500 mile legs. Little would have been gained from air refueling C-5s and C-141s. Unless it were done at both ends of the mission. It was more practical to launch from the east coast of the United States to a base in Spain or Germany, change crews there, and send the mission on to its final destination.

As Operation Restore Hope in Somalia later demonstrated, air refueling could sustain airlift over these long distances, but only if staging bases were closer to the main area of operations. (Cairo West, Egypt, and first Jeddah New and then Taif Air Bases in Saudi Arabia, were the stage bases for the airlift to Somalia.) Even if there had been stage bases on the Arabian peninsula for air-refueled C-5s and C-141s during the Gulf War, aircrew changes would have still been required at the heavily congested bases. Staging through Europe during Desert Shield had given MAC increased flexibility in managing the crew force since, on most missions, crewmembers did not have to be qualified for air refueling.

Insufficient airspace to support all of the allies' air refueling requirements became the most limiting factor for tanker operations once the war began. Limited airspace was a problem not just over Saudi Arabia but also in the refueling areas along the Turkish/Iraqi border and over the Mediterranean Sea. To make maximum use of the airspace available for air refueling, SAC employed a combination of air refueling tracks and anchors in which tankers remained in orbit to perform short-notice air refuelings. During the most intense periods of flying activity, airspace congestion prevented using additional tankers. Airspace congestion for the tankers was compounded by the language barriers between the tanker crews and many of the air traffic controllers who were coordinating air traffic in the Persian Gulf region.

To some senior military leaders, the Gulf War revealed weaknesses in the organization of the Air Force. Referring specifically to air mobility assets, General Johnson had no doubt that a single manager controlling the majority of airlift and tanker forces would improve future mobility operations. "We found in Desert Shield/Desert Storm that we didn't coordinate very well on bedding down big airplanes," he said. Having most USAF tankers and airlifters assigned to a single command will enable the manager to "better use...limited assets to project forces, to give true Global Reach."

The war, in fact, proved a catalyst for the reorganization of the United States Air Force. A new major command structure soon provided an organizational framework in keeping with the Secretary of the Air Force's landmark policy document, *The Air Force & U.S. National Security: Global Reach-Global Power*, published in June 1990. Activation of the Air Mobility Command and concurrent inactivation of the Military Airlift Command and Strategic Air Command on June 1, 1992, brought into a single organization most of the Air Force's tanker and airlift assets. Air Force planners believed that Global Reach could be accomplished more effectively by integrating the airlift and tanker forces. To further this objective, in early 1992, a new Tanker Airlift Control Center located at Headquarters MAC, began serving as the single planner, scheduler, and executor for all airlift and tanker missions worldwide.

Frequently asked during the Desert Shield/Storm assessment was the question, "Can the Gulf War be seen as a model for future conflicts?" Certain aspects of the conflict set it apart from past contingencies and suggest that a similar set of circumstances is unlikely in the near future. For one thing, Saddam Hussein's hesitation to invade Saudi Arabia gave time to build an international coalition to oppose Iraq's aggression by diplomatic and military means. It enabled forces to train for war in an unfamiliar desert environment and to assemble by airlift and sealift the vast stockpiles of munitions, spare parts, and supplies needed for a campaign against a formidable adversary. The 7,500-mile distance to the Persian Gulf from the United States east coast made Desert Shield a worst case scenario. On the other hand, American forces benefited from an excellent infrastructure on the Arabian peninsula and plentiful supplies of fuel in Saudi Arabia and at the en route bases. And Allied air superiority negated the threat of Iraqi interceptors to the transport aircraft.

In another sense, AMC's leaders viewed Desert Shield/Storm as a model for future military contingencies. Instead of a superpower confrontation, the Persian Gulf War had been a massive, well-orchestrated, multinational response to a regional conflict that threatened the economic stability of many nations. In the future, the Air Mobility Command will likely be called upon to move forces over thousands of miles with a minimum of advance warning. The theaters of operation will probably be places having little or no infrastructure, unlike the Arabian peninsula where American forces benefited from modern airfields, roads, and seaports. The absence of an adequate infrastructure proved a problem in December 1992 with Operation Restore Hope in Somalia.

Restore Hope offers an interesting comparison with air mobility operations in the Persian Gulf War. AMC's Tanker Airlift Control Center, as the single manager for airlift and tanker operations, enabled Restore hope to become a more finely tuned operation than Desert Shield/Storm. The TACC enabled a better tracking of user requirements, aircraft, aircrews, and airfield capabilities. While sophisticated airfield support had existed on the Arabian peninsula, only a rudimentary infrastructure was present in Somalia, where most fuel, water, electricity, and other support assets had to be provided by United States forces.

The Department of Defense drawdown and the reduction of the United States presence overseas has implications for future air mobility operations. In the spring of 1993 the Air Force's entire en route structure came under review. Zaragoza Air Base, Spain, had been closed, and there was no agreement with the Spanish government for the United States to use Zaragoza in wartime. Nearly one-third of Desert Shield's strategic airlift missions had staged through Torrejon Air Base. But plans were underway to place Torrejon in caretaker status in Fiscal Year 1995, while retaining the capability to augment the United States Air Force presence at Torrejon to support another large contingency. Moron Air Base, Spain, used heavily by SAC tankers during the Gulf War, remained in stand-by-ready status. Moron, like Torrejon, could be augmented quickly with mobility forces to support contingency operations as it was during Operation Restore Hope.

The Restore Hope airlift relied only marginally on United States forward basing, in keeping with the post-Gulf War national military strategy that depended less on overseas bases. Most of the strategic airlift missions to Somalia during Restore Hope's surge phase in December 1992 were air refueled. Nevertheless, AMC's tankers had operated extensively from Moron and Lajes Air Base in the Azores. Even if more airlifters are air refueled during future contingencies, forward bases will still be needed to support the tankers.

The changing composition of AMC's airlift fleet also has implications for future air mobility operations. The C-141 is approaching the end of its service life. The nation's future core military airlifter, the C-17 Globemaster III, entered AMCs inventory in June 1993 with delivery of the first aircraft to Charleston Air Force Base, South Carolina. A key player in the Air Force's post-Cold War strategy of "global reach-global power," the versatile C-17 will replace the aging C-141 and deliver twice the cargo for roughly the same operating cost.

The Persian Gulf War made a convincing case for the C-17 and helped clarify its role in future conflicts. C-5 payloads averaged 72 tons in August 1990, while the C-141s average payload was 19 tons. The C-17s advertised average payload is 40 tons over twice the payload of the C-141. The C-17s outsize cargo capability would have been especially useful during the Persian Gulf War, since 25 percent of all United States cargo airlifted in Desert Shield's first two weeks had consisted of outsize cargo. Additionally, the C-17's austere airfield capability would have reduced the need for C-130s and freed up ramp space for additional fighters.

The airline industry is changing, too, in ways

potentially detrimental to air mobility in wartime. Deregulation, consolidation, and the infusion of foreign capital into the American airline industry make the future of several air carriers uncertain. For decades Pan American was one of the world's most prominent airlines, but it is now out of business. This stalwart airline had previously accounted for 11 percent of the CRAF's wartime cargo capability and 10 percent of its wartime passenger commitment. Trans World Airlines and Continental Airlines, two other key CRAF participants in the Persian Gulf War, reorganized under Chapter 11 bankruptcy after the Gulf War. Circumstances such as these prompted General Johnson to describe the future of the CRAF as "very uncertain."

In sum, Persian Gulf operations corroborated the truth of the oft-made statement that airlifters and air refuelers perform their wartime mission in peacetime. When the United States goes to war, the only thing that changes significantly for airlifters and air refuelers is the intensity of the effort. The success of air mobility in the Persian Gulf War was many years in the making. For MAC, the experience garnered in Vietnam, Grenada, Panama, and elsewhere contributed immeasurably to the successes of 1990-1991. This, of course, was by design, for it was by flying hundreds of humanitarian, exercise, and other missions every year for five decades that MAC and its predecessor commands had trained for war. So, too, had the tanker force prepared for the Persian Gulf War by performing tens of thousands of air refueling missions in peacetime and during the Southeast Asia War. As C-5s, C-141s, C-130s, KC-135s, KC-10s, and contracted commercial aircraft streamed toward the Arabian peninsula in August 1990, airlifters and air refuelers drew upon these many years of experience. They emerged from the war with new knowledge which had been garnered from the first major military operation of the post-Cold War era.

Support Hope Ends Chapter

by 1st Lt. Daniel Ortwerth
U.S. Transportation Command Public Affairs

Following more than two months of involvement in Central Africa, aircraft from Air Mobility Command and the air component of the U.S. Transportation Command, airlifted remaining equipment and approximately 478 people from the region during the final well of Operation Support Hope.

By presidential authorization, the Department of Defense officially ended Operation Support Hope, Sept. 30, completing the U.S. military's emergency logistical assistance effort on behalf of millions of Rwandan refugees.

Airport service, logistics base support, water supply and distribution, and the establishment and management of an airlift hub in Uganda were four missions the president assigned to the joint task force, July 22.

AMC played a central role in the accomplishment of those tasks.

In conjunction with the JTF, AMC helped to establish an airlift hub at Entebbe, Uganda, which remained the central hub for relief activities throughout the refugee crisis area.

It also established 24-hour airport operations at Goma, Zaire, and Kigali, Rwanda. At Goma, missions surged from only three to 11 a day, and Kigali went from almost none to a capability of 35 missions a day.

"The Department of Defense, joint task force and U.S. Transportation Command worked together to coordinate the movement of relief supplies from points as distant as Aqaba, Jordan; Brussels, Belgium; and the Netherlands to the crisis area," said Lt. Col. David Phillips of the East Mobility Cell and part of USTRANSCOM's Mobility Control Center.

AMC crews stationed in California, Maine, and England helped provide purified water to more than 1 million refugees at Goma, only 40 hours after being alerted to go to the region.

An AMC C-5 delivered the purification unit by using three in-flight refuelings to fly nonstop from Travis AFB, Calif., to Goma in only 21 hours. JTF water production helped stem the cholera epidemic and reduced he death rate in the refugee camps from 5,000 a day to less than 250.

"The global reach capability of AMC made the movements and our rapid response to the crisis possible," said Phillips.

The magnitude of the Rwandan crisis in early April and the subsequent refugee crisis overwhelmed the United Nations and relief agencies already operating in Rwanda and neighboring countries.

The role of the JTF was to rapidly reinforce relief agencies with its airlift and material assets, until these organizations could recover and resume effective operations.

In support of this effort, as many as 33 different active-duty, National Guard and Air Force Reserve units participated in AMC airlift operations into and throughout the crisis area.

Under the coordination and direction of USTRANSCOM, AMC and civilian aircraft flew 1,138 missions, carrying 10,968 passengers and 23,038 tons of cargo in the initial deployment to the region. Civilian carriers contributed more than 15 percent of the total airlift of cargo for the deployment.

At the direction of the president, military people began deploying to the Rwandan crisis area, July 22.

At their height of operations in early August, DOD had more than 2,350 people assigned to the region.

C-5A in its first flight. Lockheed Georgia factory buildings where it was built can be seen in the background. 30 June 1968. (Courtesy of Jack S. Gibson.)

Charlie Matthews. (Courtesy of Melvin Ledbetter.)

Loading the X 24B at Martin Marietta to be delivered to Edwards AFB for testing. 1972. (Courtesy of Ted Venturini.)

The first YF-16 loading at General Dynamics, Fort Worth, Texas. 1972. (Courtesy of Ted Venturini.)

KC-10 Extender refuels USAF/McDonnell Douglas F-15s. June 1989. (Courtesy of Jim Ramsey.)

Tanker History

Fuel Aloft: A Brief History of Aerial Refueling

Introduction

The idea of refueling aircraft while in flight was born during aviation's early days. In 1918 Lieutenant Godfrey Cabot, an American Navy Reserve pilot, tested the feasibility of aircraft securing gasoline containers from prepositioned floats using a grappling hook. This effort prompted an interest in refueling aircraft short of landing and using conventional methods. Godfrey's idea was proven possible in 1920 when a Navy pilot in the rear cockpit of a Huff-Deland HD-4 aircraft grabbed a five-gallon can of gasoline from a float in the Potomac River. Late in 1921, a stuntman named Wesley May, who had previously bicycled and roller-skated atop the wing of an aircraft, strapped a gas can to his back and walked out to the wing tip of his Lincoln Standard biplane. From there, he stepped onto the wing skid of a Curtis JN-4 and poured five gallons of fuel into the *Jenny*'s tanks. The crowd at Long Beach, California, thought the demonstration was amazing. Viewers proclaimed the dare devil feat the first "air-to-air" refueling.

But the progress made in "aerial refueling" since that November day less than seventy years ago has been even more amazing. Its development has not been steady; rather, it has been characterized by periods of rapid improvement, followed by times of relative stagnation. Through these alternating stages of progress and inactivity aerial refueling has moved from the arena of barnstormers to a setting in which United States Air Force professionals routinely air refuel all types of aircraft, all over the world. The purpose of this brief history is to chronicle that progress.

The plane in the foreground (Question Mark) is a Fokker C-2 trimotor. The tanker (Refueling Airplane #1) is a Douglas C-1 biplane.

Major H.C. Davidson greets Capt. Ross G. Hoyt, pilot of refueling plane, and Lt. Elwood Quesada, Sergeant Roy Hooe, members of Question Mark crew upon their arrival at Bolling Field. 26 Jan 1929.

The Era of the Question Mark

Barnstormers may have been the first "air refuelers," but military aviators became the first to see the practical importance of air refueling. Discussing his World War I experiences in France, Army Air Service Lieutenant John Richter complained that his Spad could stay aloft only 20 to 40 minutes in combat:

I had to fly nine sorties on the day the St. Mihiel offensive started...We all wished we could refuel somehow without having to return to our bases just when the action got interesting.

Years later, Richter did something about the problem; supported by Major Hap Arnold, then commander of Rockwell Field in San Diego, he set out to prove the feasibility of air refueling. The ability to transfer fuel from one aircraft to another using the force of gravity had been demonstrated in April 1923 during a test over San Diego. Two months later the Army Air Service conducted its first successful air refueling. In August 1923 Richter and Lieutenant Lowell Smith kept their Haviland DH-4B aircraft above southern California for more than 37 hours and flew 3,293 miles. During their flight, they received oil, supplies, and 75 gallons of gasoline from another Rockwell Field DH-4B, flown by Lieutenants Virgil Hine and Frank Seifert. The two DH-4B aircraft contacted one another some fifteen times by means of a fuel hose.

The press labeled the feat just another in a long series of barnstorming gimmicks; the airmen were determined to prove them wrong. On October 25, 1923, Richter and Smith flew their DH-4 nonstop from the Canadian border to Tijuana, Mexico. With the help of two refuelings, one over Eugene, Oregon, and another over Sacramento, California, they covered the 1,280 miles in just a little over 12 hours. The mission represented a major success for the two aviators. Tragedy soon diluted their euphoria.

The next month, Lieutenant P.T. Wagner, a Kelly Field pilot, died in an air refueling accident when the fuel hose became entangled in the wings of both aircraft. Experiments ended, and in the wake of this accident little was done to develop further aerial refueling for the next five years. But an incident in April 1928, one which almost ended in tragedy, sparked new interest in air refueling.

The crew of a German Junkers aircraft, the *Bremen* was trying to complete the first west-bound, nonstop trans-Atlantic flight when their objective faded before their eyes. They were forced to land on a remote island off the coast of Labrador. The crew was rescued, but the Germans wanted to get the *Bremen* back, too. They felt if they could get their chief pilot, Frederick Melcheor, to the island, he could fly the plane to safety. The Germans asked for U.S. Army Air Corps help.

The Chief of the Army Air Corps, Major General James Fechet, agreed to help, but only after his aide, Captain Ira Eaker, determined that the job could be done. Soon General Fechet, Captain Ira Eaker, Lieutenant Elwood "Pete" Quesada and Frederick Melcheor were headed north and anticipating positive results.

The mission was successful, but just barely: one aircraft crashed on a mountain, another, an amphibian, nearly sank in a Canadian port, and Melcheor had to parachute down to the island. But the real problems, the ones that renewed interest in air refueling, came on the way home, over Portland, Maine. The pilots aware that fuel supplies were running low, worried whether or not they could find an opening in the solid overcast to land safely. The question now was the closeness of the ceiling to the ground and what they might hit before they broke out of the overcast. The answer could be fatal. They sweated time and weather, hoping they'd have enough of one before they ran out of the other. Their luck held. The cloud cover broke, and down they came with a few drops of gas left in the tanks.

Contemplating their brush with danger, Pete Quesada and Ira Eaker speculated how nice it would have been to have had an aerial gas station from which they could refill their tanks. Further reflection focused on the possibility of a refueling flight using a bigger plane than that used earlier by Smith and Richter. If successful, the usefulness of aircraft would take on a perspective far different than the popularly held view that airplanes were something akin to adult toys and of little practical value.

The idea was born, and General Fechet authorized the attempt. At the Middletown Air Depot in Pennsylvania, two aircraft were specially modified for the flight. A Fokker C-2 trimotor monoplane, picked as the receiver, was fitted with additional tanks and other special equipment. A Douglas C-1 biplane was chosen as the refueling tanker, and configured with two 150-gallon fuel tanks, and a 50-foot length of hose. The hose had a lead weight attached to the end, and was to be lowered through a trap door in the bottom of the C-1s fuselage. Following practice and demonstration flights in the Washington, DC area, the Fokker (by now christened *Question Mark*, because no one knew how long it could remain aloft) and the C-1 called *Refueling Airplane #1* (RP#1) left for California.

The flight to the west coast was routine, at least until RP#1 was ready to depart from Dallas enroute to Midland, Texas. With full throttle the heavily loaded C-1 shuddered and shook, but was unable to taxi on the soft field. The copilot and mechanic climbed out and, finally, with full power and both men pushing RP#1 began to move. The copilot and mechanic scrambled aboard through a trap door, and RP#1 lifted off, just clearing high-tension lines at the end of the field.

In California, the Fokker and the C-1 joined a second C-1 tanker, identical to RP#1. Together, the three crews and their aircraft spent the last week of 1928 practicing for the upcoming endurance flight. On December 31, the *Question Mark* and RP#2 left Rockwell Field in San Diego for Metropolitan Airport in Los Angles; RP#1 remained at Rockwell.

The *Question Mark* took off from Metropolitan Airport early the next morning, January 1, 1929, with five men aboard: Major Carl Spaatz, Captain Ira Eaker, Lieutenants Harry Halverson and Pete Quesada, and Staff Sergeant Roy Hooe. RP#2, flown by Lieutenant Odas Moon and crewed by Lieutenants Joseph Hopkins and Andrew Salter, did the first refueling, just 45 minutes after *Question Mark*'s takeoff. Officials of the Rose Bowl had requested a refueling over the stadium, and the mission commander, Major Spaatz, agreed. While Georgia Tech was defeating California by a score of 8-7, the *Question Mark* and RP#2 refueled overhead. As retired Lieutenant General Eaker recalled 1975:

We went over the Rose Bowl. It was very bumpy, as you can appreciate, as we should have appreciated, up against those mountains in January and the refueling plane and the *Question Mark* were torn apart. I was piloting the *Question Mark* and I realized that General Spaatz had probably been saturated in high octane gasoline.

General (then major) Spaatz indeed had been drenched. The other crew members tore off his clothes, and rubbed down his inflamed skin with lubricating oil. Major Spaatz told his crew that if he needed further treatment he would parachute from the *Question Mark*. Under no condition would the crew of the *Question Mark* abandon the mission for his sake.

The mission was not abandoned, nor did the commander of the *Question Mark* have to use his parachute. Despite ailing engines, darkness of night, fog, wind, constant noise, numbing fatigue, and primitive communications (none of the aircraft had radios) the crew of RP#1 and RP#2 kept the *Question Mark* and its crew aloft for more than six days. But on January 7th, seven days into the flight, its failing engines caught up with the *Question Mark*. The left engine, which had been an intermittent problem since the third day, was failing rapidly, and the aircraft was losing altitude. The crew, reluctant to quit, decided to try to fix the engine. Major Spaatz spelled out details in his log:

Sgt. Hooe taped the lower ends of his trousers, prepared the tool kit. The door to the catwalk was opened. Hooe climbed out and Quesada stood by at the window to hand out the tools. The attempt at repairs was unsuccessful.

With the two good motors wide open... and Hooe on the catwalk, the ship rapidly settled from 5,000 to 2,550 feet and was continuing to settle when the decision was made to land.

Sergeant Hooe was called back in, and Captain Eaker landed the *Question Mark* in Los Angles. The final statistics of the mission are impressive, even by today's standards: RP#1 and RP#2 had off-loaded 40 tons of material, including 5,660 gallons of gasoline, 245 gallons of oil, meals, water, batteries, and other supplies. Interestingly, all four officers of the *Question Mark* eventually became general officers, as did two of the officers aboard RP#1 and RP#2.

While the accomplishments of the *Question Mark* crew are well known, history has rarely given proper credit to the crews of the refuelers. All the crew members of the *Question Mark* were awarded the Distinguished Flying Cross; the crews of the refueling aircraft received only letters of commendation. The contributions of the fliers aboard RP#1 and RP#2 went relatively unnoticed for nearly half a century. But on May 26, 1976, the two surviving members of the refueling crews, retired Brigadiers Generals Ross G. Hoyt and Joseph G. Hopkins, were awarded the Distinguished Flying Cross by Air Force Chief of Staff General David Jones. Generals Hoyt and Hopkins were cited for their exceptional contributions to the development of air-to-air refueling.

The crews of *Question Mark*, RP#1, and RP#2 succeeded in their mission of demonstrating the practical value of aerial refueling and the endurance of man and machine in flight. But the crews also succeeded in another task, thought by many to be equally important. That job, according to General Eaker, was to keep military aviation in the public view, and to let Congress and the nation know that airpower was no longer a dream or a barnstorming sideshow, but a serious, professional effort to forge a strong link in our nation's chain of defense.

Progress After the Flight of the Question Mark

Following the flight of the *Question Mark* in 1929, American air refueling efforts met with only limited success. Later that same year, in 1929, a Boeing mail plane, called the *Hornet Shuttle* and flown by Captain Ira Eaker, set out on a non-stop transcontinental flight. An Army Air Corps Douglas C-1 and specially equipped Boeing model 40B-4s refueled the mail plane. But during a midair transfer of oil from a Boeing refueler, a full bucket was accidentally dropped through the *Hornet Shuttle's* wing, forcing Captain Eaker to land at Salt Lake City. The cross-country attempt was abandoned but illustrated possible military uses of cross-country aerial refueling. A test in August 1929 and another in November of the same year when Bobbi Trout and Eleanor Smith set a women's flight endurance record emphasized possible civilian uses of aerial refueling.

Most of the air refueling progress in the 1920s took place in Britain. Sir Alan Cobham, after years of study, established an aerial refueling corporation, known as Flight Refueling, Limited. The British Air Ministry made obsolete "Harrow" bombers available, and Flight Refueling, Limited began to refuel Imperial Airlines aircraft on their trans-Atlantic runs. A Shell Oil Company executive named Jimmy Doolittle - who, less than three years later, would lead the famed raid on Tokyo - explained the British system to the Air Corps Chief Major General Hap Arnold in a 1939 letter:

The ship to be refueled lets out a couple of hundred feet of cable, on the end of which is lead weight ... the tanker ... below and slightly behind ... shoots a harpoon carrying a cable to intercept the wire trailing from the ship to be refueled. The harpoon line and the cable ... are then pulled into the tanker as it gradually takes position above, behind, and offside of the other ship ... The hose, which is on a reel in the tanker, is then secured to the end of the cable and lowered away as the cable is reeled in by the receiving ship. When the cable is reeled all the way in, an automatic fitting on the end of the hose engages in a socket and, by means of a system of flag signals, fueling is started ... Fueling is accomplished at a maximum rate of about 150 Imperial gallons per minute. As soon as fueling is accomplished, contact is broken, the tanker leaves formation, reels in its hose and lands, and the receiving ships carries on.

Cobham's company used the technique in 16 trans-Atlantic mail service flights. Passenger runs were slated to begin in 1940, but the start of World War II put an abrupt end to further development.

World War II: The Cost of Neglect

When it came to the use of airpower, the American War Department was racked with dissent. The Army General Staff saw the airplane as only good insofar as it supported ground troops; they had no use for Billy Mitchell's ideas on strategic bombardment. The Navy brass were equally backward - they saw the sea, including the air above it, as their realm, and protested any over-water missions by the Army Air Corps. So it happened that there was little activity permitted in the area of extending aircraft range. As late as May 1938, the United States Army Adjutant General told the Assistant Secretary of War:

The Chief of the Air Corps has been informed that experimentation and development for the fiscal years 1939-40 will be restricted to that class of aviation designed for the close-in support of ground troops and for the production of .. medium and light aircraft.

In that same month, the 2nd Bombardment Group at Langley Field tested its navigation ability by sending three new B-17s on what was thought to be a nearly impossible mission - to intercept the Italian liner *Rex,* somewhere hundreds of miles into the Atlantic. The bombers (with Lieutenant Curtis LeMay as the lead navigator) located the *Rex* while it was still 700 miles offshore. The B-17s returned to Langley, their crews elated over their success - they had proven that long-range aviation could intercept an invasion force hundreds of miles out to sea. The Navy was somewhat less elated, however, and the War Department immediately limited all Air Corps activity to within 100 miles of the U.S. Coast.

These developments badly hurt the movement toward long-range military aviation and lessened the need for progress in air refueling. But the extent of the damage was not fully realized until after the United States was drawn into World War II at Pearl Harbor. American bombers of that time were unable to strike enemy targets from U.S. bases. To reach German targets, bombers of the Eighth and Fifteenth Air Forces had to be based in England and the Mediterranean area, well within the range of Luftwaffe retaliatory raids. Additionally, with no air refueling capability, fighter escorts for both Eighth and Fifteenth bombers were extremely limited in range and duration until very late in the war. German fighters simply waited until low fuel forced the American fighter escorts to head for home and then attacked the vulnerable heavy bombers.

But it was the Pacific theater that most clearly demonstrated the inadequacies of long-range American bombing. On January 5, 1942, less than a month after Pearl Harbor, seven Air Corps officers met at Wright Field to discuss a plan to bomb Japan. The purpose of the secret meeting was to examine a proposal which called for an attack on Tokyo by Hawaii based B-24s, which would be refueled enroute by Navy PBY flying boats. After long study, the officers recommended against the attack, primarily because of the lack of crews trained in refueling, and also because of a lack of refueling mechanism. The group concluded, however, that "all schemes for the extension of range of aircraft should be investigated."

Many alternatives were studied, the most promising involving a B-26D tanker and B-17 receiver. Extensive tests, using the British developed techniques, were conducted at Eglin Field, Florida, during the spring of 1943. In one flight a B-24D transferred 1,500 gallons of aviation gas in 18 minutes. The results showed that, although some problems needed to be ironed out, the techniques were practical and deserved further development.

But prevailing opinion on the need for air-to-air refueling continued to vacillate. On September 27, 1943, the Air Staff directed that "further development of the flight refueling project be discontinued." But less than four months later, the Air Staff directed that "this air-to-air refueling project (for fighter aircraft) should be given highest priority." Soon the B-24/B-17 combination was also rekindled; and, by October 1944, studies involving B-24 tankers/B-29 receivers were under way. However, the end of World War II brought to an end the air refueling studies and experimentation.

The American air effort in World War II, however successful, paid a price for the neglect of long-range aviation during the great depression. Repeatedly, bases within range of enemy targets had to be constructed and personnel and supplies had to be relocated. The cost was high, but long-range aircraft and aerial refueling were to become cornerstones of the post-war Strategic Air Command.

Boeing B-50 superfortress Lucky Lady II being refueled by KB-29M tanker during the first non-stop around the world flight in 1949.

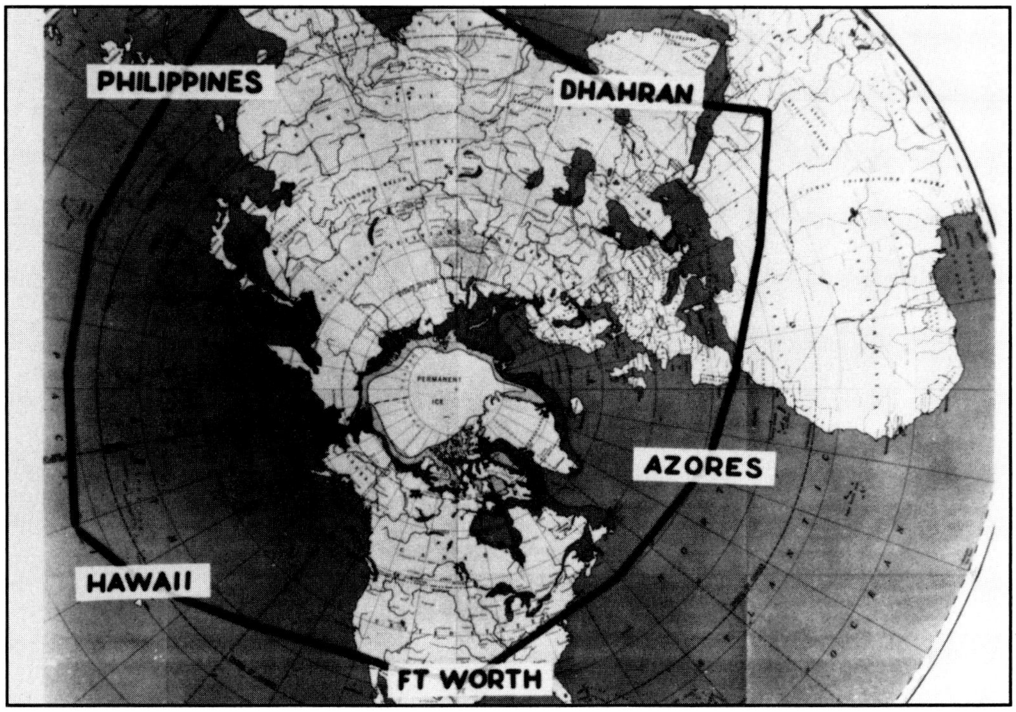

Map showing the flight course of Lucky Lady II.

Refueling in the Early Days of SAC: The KB-29M

On March 21, 1946, the Strategic Air Command was activated at Bolling Air Force Base, Washington, D.C. Its mission was proclaimed by Air Corps Chief General "Tooey" Spaatz:

> To conduct long range offensive operations in any part of the world ... to conduct maximum range reconnaissance over land or sea ... to provide combat units capable of intense and sustained combat operations employing the latest and most advanced weapons; to train units and personnel for the maintenance of the Strategic Forces in all parts of the world; to perform such special missions as the Commanding General, Army Air Force may direct.

While SAC's mission was global, its bomber force was not. Striking worldwide targets was dependent on operating from, or recovering at, foreign bases. At the end of 1946, the SAC bomber force consisted of 148 aircraft, all B-29s. It would not be until 1952, when the B-36 became fully operational, that SAC would have a bomber capable of reaching targets in the Soviet Union without dependence on overseas bases.

In 1947, to bring a truly global capability to the SAC force, planners turned to a concept which had been around for a quarter of a century: inflight refueling. In November 1947, the Air Materiel Command (AMC) asked the Boeing Airplane Company to study air-to-air refueling methods and equipment. A month later Air Force Headquarters planners began reviewing whether B-29s could be used to transfer fuel to P-80 and P-84 jet fighters. In March 1948, Boeing and AMC set out to test the feasibility of transferring fuel, by hose, between two B-29s. Based on the British system of the 1930s, it called for the tanker and bomber to fly in formation, with the tanker above, ahead and to the left; from the tanker trailed a cable (called the "contact line") at the end of which was attached a 50 pound lead weight. The receiving B-29 also trailed a cable (the "hauling line") from its refueling receptacle. A grappling hook at the end of the hauling line caught the tanker's contact line during a crossover maneuver, and a winch in the tanker wound in both lines. The tanker operator separated the lines, then attached the hauling line to the hose; another winch, this one in the receiver, then pulled the hose out of the tanker. When the hose nozzle was seated in the receptacle and locked with hydraulic toggles, a "contact made" signal was sent to the tanker, by now above and behind to the right. Fuel transfer then began.

The first test, set for March 20 in Wichita, was aimed only at making contact between the tanker and bomber. The first four attempts failed, when the contact line just slid over the receiver's hauling line without engaging the grappling hook. But on the fifth attempt, after the hauling line was shortened so it trailed more horizontally, the grappling hook caught.

The test was conducted at an airspeed just under 200 knots, in calm air, with good visibility. The tanker pilot commented that, under those conditions, the procedure was safe, but that operations at night, or in turbulence, would be too risky, since the tanker came to within about 20 feet above, and 75 feet behind, the receiving B-29.

A week later, on the afternoon of March 27, the two B-29s went up for the second test - this time, water was to be transferred from the tanker to the bomber. All went well until the hose had been pulled to within two feet of the bomber. At that point, a hose link failed, and the mission had to be aborted; both aircraft returned to Wichita. Repairs were made overnight and another attempt was made the next day, but it, too, failed. The third try, later that same afternoon, succeeded: 400 gallons of water were transferred between the two B-29s in about 12 minutes.

Based on this success, the Air Materiel Command authorized Boeing to proceed on two fronts: immediate incorporation of hose-type refueling equipment in B-29s and B-50s, and also, for the longer term, development of a system superior to the hose method.

To convert the B-29s - some into hose-type tankers, some into receivers - Boeing reopened its Wichita Plant #2 in 1948. Here, 92 B-29s were converted to hose-equipped tankers, referred to as KB-29Ms. In addition to the pumps, hose, and powered reel in the aft fuselage, each bomb bay was fitted with jettisonable tanks holding about 2,300 gallons of fuel. These tanks were integrated with the normal aircraft fuel system. Seventy-four B-29s were modified to act as receivers for the KB-29M. These receivers, called B-29M Rs, had a 2,500 gallon tank installed in the aft bomb bay.

In June 1948, the 43d and 509th Air Refueling Squadrons became the first air refueling units in the United States Air Force. Based at Davis-Monthan and Roswell Air Force Bases, the 43d and 509th began receiving their KB-29Ms in late 1948. On December 9, 1948, a B-36B and a B-50A completed a nonstop round trip from Carswell Air Force Base, Texas, to Hawaii, and back to Carswell. The B-36 did it unrefueled but the B-50 used three inflight refuelings from KB-29Ms of the 43d and 509th squadrons. By the close of 1948 the Strategic Air Command with its two medium air refueling units of twenty KB-29Ms each, faced a world with a military flexibility heretofore unknown.

First Around the World Non-Stop: The Lucky Lady II

In the midst of the world tension caused by the Soviet blockade of Berlin which began in late June 1948, SAC began to prepare for a non-stop, round the world flight which would demonstrate its newly acquired global capability. Five B-50As were sent to Oklahoma City Air Materiel Area for airframe and engine overhauls, and the addition of auxiliary fuel and oil tanks. After two weeks of intensive aerial refueling practice and final flight planning at Davis-Monthan Air Force Base, the five B-50s flew to Carswell Air Force Base, Texas, on February 22, 1949.

Three days later, the aircraft chosen to make the attempt - the *Global Queen* - taxied out, followed by its backup, *Lucky Lady II*. The *Global Queen* departed to the east on schedule, and *Lucky Lady II* returned to her hangar. But late that night, engine trouble forced *Global Queen* to land in the Azores; the next day, *Lucky Lady II* got her chance.

Shortly before noon on February 26, *Lucky Lady II* headed east on her attempt at the record-breaking flight. *Lucky Lady II* (the crew had named her after the B-29 "Lucky Lady," which the second pilot, Lieutenant Arthur Neal, had flown around the world - stopping for fuel - just the year before) took on her first fuel from KB-29Ms over the Azores. Because the flight was shrouded in secrecy, *Lucky Lady II* confused any flight-following attempts by switching call signs and tail numbers with the KB-29 tankers, which flew from bases in the Azores, Saudi Arabia, the Philippines and Hawaii. At 0930 CST on March 2, just over 94 hours after takeoff, *Lucky Lady II* returned to Carswell after a 23, 452 mile flight. Only after it was safely on the ground was the news of the mission released to the public.

Each member of the crew of *Lucky Lady II* was awarded the Distinguished Flying Cross, and the aircraft commander, Captain James Gallagher also won the year's Mackay Trophy, the first in SAC. The citation accompanying it read, "These inflight refuelings ... resulted in a graphic demonstration that the range of modern military aircraft is unlimited." But while the mission was a success, it had a high, but little-known, price: one of the KB-29 tankers was lost enroute to Clark Air Force Base in the Philippines. The remaining KB-29s searched for the missing tanker for several days, until ordered back to Davis-Monthan. Two weeks after the disappearance, the wreckage was found near the top of a 4600-foot mountain on Luzon Island, not far from Clark. It had crashed during its approach to Clark Air Base, killing all aboard.

The crew of the B-50 taken at Davis-Monthan Air Force Base, Tucson, AZ, prior to the history making round-the-world, nonstop flight. The crew members pictured are: Left to Right, Front Row: 1st Lt. A.M. Neil, pilot; Capt. G.E. Hacker, navigator; 1st Lt. E.L. Rigor, navigator; Capt. D.B. Parmelee, flight engineer; 1st Lt. W.F. Caffrey, radar operator; 1st Lt. R.B. Bonner, radar operator; Capt. James G. Gallagher, aircraft commander on flight; and Capt. J.H. Morris, co-pilot. Back Row: S/Sgt. M.G. Davis, gunner; S/Sgt. R.G. Davis, flight engineer; T/Sgt. V.L. Youny, flight engineer; S/Sgt. D.G. Traugh, gunner; S/Sgt. R.R. McLeRoy, radio operator, and T/Sgt. B.E. Cantrell, radio operator.

Refueling in Combat: KB-29Ms in the Korean War

The outbreak of the Korean War in June 1950 highlighted the need for more sophisticated air refueling procedures that could be used for combat aircraft. A handful of KB-29Ms were modified with the probe-and-drogue system, to test in-flight refueling of combat aircraft during the Korean War. The KB-29Ms, assigned to Detachment 4, FEAF Bomber Command, were home-based at Yokota Air Base Japan, and operated out of both Yokota and Misawa. One of the KB-29Ms trailed refueling hoses from pods on both wing tips, and also one from a centerline position. The remaining tankers had one hose only; it trailed from the centerline position. The fighter aircraft flew into position behind the tanker, so that its probe could be inserted in the drogue at the end of the tanker's hose. Once contact had been made, fuel transfer was accomplished. The modified KB-29s were used to refuel RB-45s of the 91st Strategic Reconnaissance Squadron and F-84s from the 159th Fighter Bomber Squadron.

The First "Flying Boom": The KB-29P

The KB-29M hose refueling system worked, but it had obvious limitations, among them the amount of time needed to make contact, the slow rate of fuel transfer, and the limited airspeed (216 knots) imposed by the hose. Additionally, hose-type receiver equipment was so bulky that it was unsuitable for fighter aircraft. Accordingly, the Air Force pushed Boeing to speed up work on an aerodynamically-controlled, swiveling and telescoping arm known as a "Flying Boom." In a May 26, 1948, letter to Boeing the Air Material Command stressed the importance it placed on the development of an effective air refueling method. Even a single air refueling, according to an Air Force staff report, could increase the range of a B-29 with a 10,000 pound bomb load to 2,700 miles. The new air refueling method could enhance significantly operational capability.

This "flying boom" was controlled by an operator positioned in what was formerly the B-29's tail turret; the operator "flew" the boom nozzle, by means of airfoils called "ruffevators," into a receptacle, usually located atop the receiver's fuselage. The pilot of the receiver had to keep his aircraft within a prescribed envelope to maintain contact, and was assisted in this task by indicator lights along the bottom of the tanker. These indicators, connected to limit switches in the base of the boom, directed the receiver pilot by use of colored lights, and the terms "right," "left," "up," "down," "fwd," and "aft." In 1950-51, Boeing modified 116 B-29s with the "Flying Boom;" these new tankers became known as KB-29Ps. On September 1, 1950, the 97th Air Refueling Squadron, Biggs Air Force Base, Texas, received the first KB-29P.

KB-29s continued to be involved in record-setting, award-winning flights. In early July 1952, KB-29s of the 2nd, 91st, and 93rd Air Refueling Squadrons led 58 F-84Gs of the 31st Fighter-Escort Wing from Turner Air Force Base, Georgia, to bases in Japan. The first refueling rendezvous took place between Georgia and California and the second between the west coast and Hawaii. For this operation, called Fox Peter One, the 31st Fighter-Escort Wing became the first organization to win the Air Force Outstanding Unit Award. A second major fighter deployment to Japan using KB-29s took place on October 3, 1952, when 75 F-84Gs flew from Bergstrom Air Force Base, Texas, to Misawa Air Base Japan with each fighter having received two aerial refuelings.

Later that same month, a 91st Strategic Reconnaissance Wing RB-45C made the first

KB-29P refueling F-84 Thunderjet over South Korea.

nonstop, trans-Pacific jet flight from Elmendorf Air Force Base, Alaska, to Yokota Air Base, Japan. This flight, made possible by two KB-29P refuelings, earned the RB-45 crew the Mackay Trophy for 1952. The next year, KB-29s, assisted by the new KC-97s, made possible yet another award, this time for the 40th Air Division: On August 20, 1953, Operation Longstride, involving the trans-Atlantic movement of F-84s from Turner Air Force Base, Georgia, began. By the time it ended 23 days later, 28 F-84Gs had been ferried to England and Morocco, and then returned to Turner. For this exercise, the 40th Air Division won the Mackay Trophy for 1953.

The end of 1951 found 187 KB-29s - mostly the boom equipped "P" models - in SAC. But the advent of the KC-97, faster than the KB-29s and better able to keep up with the new jet receivers, foretold the decline of the KB-29. By December 1953, KC-97s outnumbered the 143 remaining KB-29s by more than two-to-one. The numbers of KB-29s dwindled steadily until November 25, 1957, when the last two KB-29s flew from Bergstrom Air Force Base, Texas, to the Military Aircraft Storage and Disposition Center at Davis-Monthan Air Force Base, Arizona.

Of the 92 KB-29Ms and 116 KB-29Ps, only two survived past the mid-1950s. Both were KB-29Ms, and were transferred to the United States Navy, where they saw service in towing targets for Navy fighter pilots. By the close of the decade the KB-29 had been replaced by a more modern air refueler.

A unique variant of the KB-29 involved a test in which a hose-equipped KB-29M was reconfigured with a triple probe-and-drogue system. Only one aircraft was modified - it became known as a YKB-29T - and the test program achieved only partial success. The triple-hose arrangement worked well, and the YKB-29T proved that it could successfully refuel three fighters simultaneously. The tanker's relatively poor performance, however, meant that jet fighters had to descend to lower altitudes, and slow to dangerously low airspeeds for the refueling. Planners accordingly urged the development of the triple-hose method, but only on an airframe faster than the old KB-29.

After some debate, test directors selected the aging B-50s as the tankers for the triple-hose modification. They offered improvements in performance and off-load capability over the KB-29s. Of the 136 aircraft modified, 112 were B-50As and RB-50s; these were redesignated KB-50J. The remaining 24, TB-50Hs, had most recently been used as unarmed navigation/bombing trainers; these were designated KB-50Ks. One hose appeared in the tail, and one in a pod under each wing tip. To allow the KB-50s to match better the speeds and altitudes of jet fighter receivers, a 9,200 lb.-thrust J-47 jet engine was mounted under each wing.

The KB-50 made its first flight in December 1955, and it served in TAC, PACAF and USAFE for the next 10 years. Pressed into service in the early stages of the Vietnam War, KB-50s sometimes were called upon to refuel fighters at low altitudes and over enemy territory. But a pair of in-flight fires caused crashes in August and October 1964; accident investigators determined the cause to be severe corrosion. For these reasons and others the KB-50 fleet rapidly phased out, with some of their pumping equipment and jet pods being installed in KC-97L tankers of the Air National Guard. While only briefly on the scene, the KB-50s performed a valuable service in Southeast Asia. Only two KB-50s are known to exist today - both KB-50Js. Serial number 49-389 is on display outside the Air Force Museum, Wright-Patterson Air Force Base, Ohio; serial number 49-372 is at the Pima Air Museum near Tuscon, AZ.

SAC Refueling In The Early Fifties: The KC-97

The KB-29s did not perform well as refuelers for the more powerful B-50, which had begun to enter the SAC inventory in early 1948. Furthermore, new all-jet bombers already under development - the B47 and B-52 - would be even less compatible with the aging KB-29s. With an eye to the future, Boeing announced in late 1950 the development of the KC-97A Stratofreighter, a convertible tanker-transport, designed to refuel the coming generation of jet fighters and bombers. The KC-97As - only three were built - amounted to a modified version of the C-97 transport, already being used by the Military Air Transport Service. The boom on the KC-97A was similar to the one on the KB-29P, but located differently. The KB-29P boom operator sat in what used to be the old B-29 tail turret position; in the new KC-97, the boom had been moved forward, and the operator directed refuelings in a pod located in the aft fuselage. Looking down the boom made it easier for the boom operator to place the nozzle precisely into the receptacle.

Boeing received an Air Force contract for sixty KC-97s in December 1950. These improved Stratofreighters, designated KC-97E, were powered by four Pratt and Whitney R4360-35C engines each generating 3,500 horsepower.

The E-model also featured a stronger cargo floor, more modern instrumentation, and better communication equipment. In addition, four additional fuel tanks holding a total of 7,200 gallons, were fitted in the upper cargo area. The KC-97E, with a maximum gross weight of 175,000 pounds, could be switched to a cargo configuration by removing the upper deck tanks, and by replacing the inflight refueling pod with a set of cargo doors.

KC-97Es began to be delivered to SAC just eight months after the contract was awarded. The 306th Air Refueling Squadron at MacDill Air Force Base, Florida, accepted the first plane on July 14, 1951. The following March, the new KC-97s refueled

KC-97.

a B-47 Stratojet on a nonstop 12,000 mile mission, providing a graphic demonstration of SAC's growing intercontinental attack capability.

The success of the KC-97E led to Boeing's development of two more models. In April 1952 the corporation announced the KC-97F. Featuring improved engines and instrumentation, 159 F-models came off Boeing's Renton, Washington, assembly line. But the biggest contract came in mid-1953, when the Air Force purchased the KC-97G. The -G offered much improved flexibility: a 700-gallon capacity drop tank was added beneath each wing; the upper deck tanks of the -E and -F, used for air refueling, were moved into the lower fuselage, so the -G was immediately usable as a cargo aircraft. The G model soon became the major production version of the KC-97: a total of 592 Gs were built before the KC-135 replaced it on the Renton assembly line in 1956.

Sept. 1953: the first jet tanker wasn't a KC-135, but a KB-47G, a modified B-47B; the receiver was a YB-47F. Only one of each ws ever built: both appeared before 215,000 people at the natinal Aircraft Show in Dayton, OH.

Propeller Problems

While the KC-97 suffered from the usual problems associated with a new aircraft, it was also afflicted with a fatal flaw in the design of its propellers. The unusual structure of the propeller blades - built of lightweight hollow steel - made them especially susceptible to metal fatigue. Further, any dents, bumps, or even scratches could result in almost invisible hairline cracks. At high RPM, these small cracks led to blade failure, which in turn caused a severe vibration resulting in engine separation from the wing. Although crews were warned of the hazard, and emergency procedures were developed - there was even an "Engine Separation" emergency checklist - propeller failures almost always proved fatal.

An interim "fix" involved a careful magnaflux inspection of the propellers before each flight, but the inspection was time-consuming and not always effective. By 1958, SAC had lost at least five aircraft and 39 crewmen because of propeller failure, but a long term remedy was in process and would be adopted pending successful test results. It involved replacement of all KC-97 blades with heavier ones made of "dural," an aluminum alloy. The dural props added 1800 pounds to the 97's gross weight, and cut its climb rate by 33%. The redesigned hub cut airflow over the engine, and for a while, caused engine overheats and failures. But these problems engineers accepted as a small price to pay for the increase in reliability and safety, and the heavy propellers remained on the KC-97s until they left the inventory.

Record-Setting KC-97 Flights

Like their KB-29 predecessors, KC-97s quickly began to enter the record books. In August 1953, just 13 months after the first deployment of fighters across the Pacific using air refuelers they joined KB-29Ps in Operation Longstride, the trans-Atlantic F-84 deployment for the 40th Air Division won SAC's fourth Mackay Trophy. By the end of 1953, B-47 Stratojets had teamed with KC-97s to set two trans-Atlantic speed records, an east-to-west Atlantic crossing in less than nine hours falling to a B-47 with KC-97 refueling support. In June 1954, Major General Walter C. Sweeney, Fifteenth Air Force Commander, led a flight of three 22nd Bomb Wing B-47s on a nonstop flight from March Air Force Base to Yokota Air Base, Japan. This flight supported by KC-97 inflight refuelings, became the longest point-to-point B-47 flight to that date, and marked the first appearance of the B-47 in the Far East. On August 6-7, 1954, KC-97s refueled B-47s during Operation Leap Frog, involving 10,000 mile nonstop flights from Hunter Air Force Base, Georgia, to French Morocco and back to Hunter. By the end of the year, 32 medium air refueling squadrons and 683 assigned tankers attested to the importance of inflight refueling in Strategic Air Command operations.

A KC-97F refuels a B-47B. The high-performance B-47 had to slow to dangerously low airspeeds to refuel from KC-97. As the bomber onloaded fuel and became heavier, both aircraft would toboggan. In 1957, through, the situation started to reverse: then it was the B-47s who were unable to keep up with SAC's new KC-135s.

Saving the B-52: "Quick Kick" and P.D. Eldred

SAC's response to the Suez Crisis in late 1956 resembled the response to the Berlin blockade eight years earlier: bombers and tankers combined to demonstrate to the world the high state of SAC's readiness. But this time a new complication surfaced: while the B-47 had proven itself as a reliable weapon system, SAC's new strategic bomber - the B-52 - was under fire, as a result of several crashes earlier in 1956. A reporter, P.D. Eldred, had interviewed B-52 crews and maintenance people at Castle Air Force Base, and had gathered information for an article highly critical of the B-52. Somehow, SAC learned of Eldred's upcoming article, and began planning a counter-offensive - an impressive demonstration which would show not only the Soviets but also the American people that the B-52 was an effective and safe weapon system.

The result was called "Quick Kick," an endurance flight involving B-52s and KC-97s. On 24 and 25 November, eight Stratofortresses - four from the 93rd Bomb Wing and four from the 42nd Bomber Wing - flew non-stop around the perimeter of North America, supported by KC-97 refuelings. This flight received wide publicity, and for a very short time was thought to have effectively neutralized any controversy which might result from Mr. Eldred's story.

But, just five days after Quick Kick, still another B-52 crashed, shortly after takeoff from Castle Air

Force Base. The cause of the accident, which killed all 10 men aboard, could not be determined, but immediately resurrected SAC's concerns about P.D. Eldred's newspaper article, which by then had been sold to the Associated Press. Two December exercises - Power House and Road Block - would send more than 1,000 B-47s and KC-97s on simulated combat missions over North America and the Arctic. While this demonstrated American resolve to the Soviets - by then threatening to send "volunteers" into the Middle East - it did nothing for the maligned B-52s. So another B-52 exercise was planned for the next month - it was to be called "Power Flite."

Power Flite, and the Flight of the Third Lucky Lady

Just before Christmas in 1956, three operations officers from 93rd Bombardment Wing met with Major General Archie J. Old, Jr., Commander of the Fifteenth Air Force, to plan an around-the-world B-52 flight. As a result of this meeting, Fifteenth Air Force published Operations Order 28-57, which called for "a long range mission involving B-52 aircraft," nicknamed the operation "Power Flite," and tasked the 93rd Bomb Wing to provide five aircraft and crews.

On January 7, 1957, Brigadier General William E. Eubank, Jr., Commander of the 93rd Bomb Wing, selected five aircrews. In addition to the normal crew of six an extra pilot and navigator were included, as well as the aircraft crew chief. General Old would serve as the mission commander, and Lieutenant Colonel James Morris would be aircraft commander of the lead B-52, *Lucky Lady III*. Colonel Morris was no stranger to endurance flights - he had been the copilot on the *Lucky Lady II* in 1949, and had commanded one of the Operation Quick Kick B-52s just a month before.

Maintenance people searched aircraft records to find the best aircraft available, and chose five which happened to have sequential serial numbers: 53-394, -395, -396, -397, and -398. These were modified with the addition of bunks, cooking facilities, 2-man life rafts and other necessities. All aircraft systems were carefully scrutinized: when the five B-52s thundered down Castle's runway just after noon on January 16, they were thought to be in perfect condition.

But such was not the case. One of the three primary aircraft - *La Vittoria,* named after Magellan's ship, the first to circumnavigate the globe - proved to have more than its share of problems. After takeoff, one outrigger gear failed to retract; extensive efforts to raise the gear failed until a little later in the flight, when it retracted on its own accord. But that was just the start of *La Vittoria's* problems; over Newfoundland, ice in the refueling receptacle resisted the crew's best efforts to dislodge it, precluding a successful refueling. According to *La Vittoria's* navigator:

We couldn't onload any fuel. We tried every tanker in the formation and worked a solid hour with them. The anguish of the hour was beyond description ... there were tears in everyone's eyes when we reached the tanker formation's turn around point and the decision had to be made to return to Goose Bay. The Number One Spare was as elated as we were dejected as he replaced us to continue on the historic globe-circling mission.

La Vittoria returned to Goose Bay, and the second spare, Number 53-395, left the group over North Africa and landed in England. The three remaining bombers headed across Africa, Saudi Arabia, and the Persian Gulf. They simulated a bomb run on a railway crossing in Malaya, and then crossed the Pacific on their way to California.

By the time the Stratofortresses finished their last refueling over the western Pacific, nearly 100 KC-97s had taken part in "Power Flite." Twenty tankers from the 98th AREFS had performed the refueling over eastern Canada: 305th, 100th and 98th AREFS, operating from Ben Guerir Air Base in Morocco. Over the Persian Gulf, 18 Dhahran-based KC-97s from the 100th AREFS refueled the three B-52s still on the mission. A brief refueling by 12 tankers of 11th AREFS over the Philippines got the bombers to Guam, where 26 KC-97s from the 11th and 22nd AREFS provided the final air refueling.

All three B-52s landed at March Air Force Base on the morning of January 18. Their flight time - 45 hours, 19 minutes - was less than half that required by Lucky Lady II just eight years before. General Curtis E. LeMay, SAC Commander-in-Chief, greeted the returning aviators and presented each of the 27 crew members with the Distinguished Flying Cross.

Tremendous pubic attention now focused on the three bomber crews. The crew of *Lucky Lady III* rode a float in President Eisenhower's inaugural parade just two days after the mission, and appeared on several nation-wide television programs. The aircraft commander of -398, which comedian George Gobel had christened "Lonesome George" prior to the mission, was interviewed on Gobel's prime-time television show. P.D. Eldred's article about the inadequacies of the B-52 went unnoticed in the excitement.

But, as had been true since the flight of the *Question Mark* 28 years before, little mention was made of the 98 KC-97 tankers and their crews. SAC - for reasons of security - would not even discuss from where the tankers had operated. In an official account published not long after the mission, the KC-97s rated only one mention in the entire 23-page text:

Occasionally - exactly how many times is secret information - KC-97s rose from enroute bases to meet the heavy bombers and replenish their fuel supplies.

In his remarks during the award ceremony for the bomber crews, General LeMay told them that their flight was a "demonstration of SAC's capabilities to strike any target on the face of the earth." Later, the National Aeronautic Association recognized POWER FLITE as the outstanding flight of 1952, and named the 93rd Bombardment Wing as recipient of SAC's sixth Mackay Trophy. *Lucky Lady III* in time would become an eloquent symbol of the accomplishments of SAC air crews.

The KC-97 on Alert

However dramatic the accomplishments of operations like QUICK KICK and POWER FLITE, they represent only a small part of the Strategic Air Command's KC-97 story. The greatest efforts - and perhaps the greatest success - involved not short moments of drama, but long years of routine. They involved hundreds of aircraft and thousands of aircrews, sitting and waiting at bases from Guam to Bermuda to Greenland. These tankers - on constant alert - allowed the relatively short range B-47s to serve as long-range bombers, and constituted an essential part of our strategic deterrent posture for nearly a decade, from 1956 to 1965.

SAC's decision to place its aircrews on ground alert came in response to the rapid development of the Soviet intercontinental attack capability in the 1950s. The first efforts aimed for one-third of the bomber and tanker force on alert, with the first aircraft launching within thirty minutes of notification. By 1959, each KC-97 and KC-135 squadron had seven of its 22 tankers on ground alert, and by 1960 the response time had been cut from 30 to 15 minutes.

Not all ground alerts occurred at CONUS bases. In a long-term operation called "REFLEX ACTION," B-47s and KC-97s went on 7-14 day alert rotations at overseas bases. The KC-97s mostly "reflexed" to Royal Canadian Air Force Bases at Churchill, Cold Lake, Namao and Frobisher; and to two USAF-controlled bases: Earnest Harmon Air Force Base in Newfoundland, and Goose Air Base in Labrador. Keeping this force constantly ready for launch required tremendous efforts, especially in the winter months as ground and flight crews battled freezing rain, snow, ice, fog and sometimes temperatures as low as -55oF. But, as proven in Unit Simulated Combat Missions, the tankers could be relied upon to accomplish the mission.

A warm-weather variation of REFLEX known as AIR MAIL, involved B-47/KC-97 alert at Andersen Air Force Base, Guam. Like REFLEX, AIR MAIL began in the late 1950s, as a response to the growing Soviet threat. The temporary duty assignments were longer than for REFLEX: 30 days was a normal AIR MAIL tour. But unlike the 100% alert REFLEX force, only two-thirds of the AIR MAIL aircraft - 10 B-47s and 2 KC-97s - remained on constant alert.

The REFLEX force declined along with the size of the B-47/KC-97 fleets. By the end of 1963, only four bases - three in Canada, one in Greenland - supported KC-97 REFLEX Operations. On March 31, 1964, Air Mail ended, and the last KC-97 REFLEX alerts followed soon after. On November 10, 1965, the last KC-97 - from the 9th Air Refueling Squadron at Mountain Home Air Force Base, Idaho - was removed from alert. The next month, the last two KC-97s in SAC, one from the 100th and one from the 384th AREFS, flew to the Military Aircraft Storage and Disposition Center at Davis-Monthan Air Force Base, Arizona.

The KC-97L: Refueling in the Air National Guard

In 1957, two separate developments made it clear that the SAC days of the KC-97 were numbered: The command received its first KC-135, and then began to phase out the B-47 force in favor of the B-52, bigger in size but fewer in numbers. SAC's year-end KC-97 strength peaked in 1958, with 780 Stratofreighters flying from bases throughout the world.

After KC-97s left the SAC inventory, some went to the Air National Guard; they were to be used primarily to refuel TAC fighters, which by then had outgrown TAC's modest KB-50 fleet. In April 1961, the 185th Air Refueling Squadron in Oklahoma City received the first Guard KC-97, and F-model. The first Air Guard refueling occurred on September 6, 1961, involving a KC-97F and F84F, both from the Illinois Air National Guard.

The -F and -G models had one major limitation in the Air Guard role: they could not travel fast enough to keep up with their jet fighter receivers. But a solution surfaced: the retirement of KC-50s had made available a great number of General Electric J-47 engines, and it was decided to put those engines to use on some of the KC-97s. In 1964, the Air Force awarded a contract to Hayes Industries - the same company that had added the J-47s to the KB-50s nearly ten years before. Hayes modified more than 70 -G models, adding a J-47 outboard to the numbers one and four engines. The resulting aircraft were called KC-97L.

To the Air National Guard aircrews who flew it, the KC-97 represented a thing of beauty, and the -L was the best of them all. The J-47s drastically reduced take-off roll, and the rate of climb increased fourfold. Cruise airspeed was 30 knots higher than the KC-97F and -G with a refueling airspeed that was even more improved. The Air National Guard finally possessed a sophisticated air refueling capability that would soon be tested.

KC-97Ls went overseas in April 1967, to Rhein

Main Air Base in Germany. The operation, known as CREEK PARTY involved the rotation every two weeks of the entire six aircraft tanker task force, including operations, maintenance, and support personnel. Creek Party lasted ten years, until April 1977. During that time, there were over 250 unit rotations involving over 3,000 trans-Atlantic flights, and more than 10,000 refueling sorties. All this occurred without a major accident.

In 1978, the last Air National Guard KC-97s were deactivated, as part of a program which transferred KC-135s into the Air Guard (and the Air Force Reserve) inventory. On June 23, 1978, the last Guard KC-97 was flown to the Military Aircraft Storage and Disposition Center (MASDC) where many of them remain today. The SAC Chief of Staff acknowledged the departure of the KC-97 in a 1978 message to the Director of the Air National Guard:

It is with mixed emotions that we at SAC acknowledge the end of an era ... Please convey the sincere appreciation of the SAC staff to your personnel for their enthusiasm and dedicated support ... The men and women of the ANG KC-97 units played an important role in this part of Air Force history.

Indeed, an important chapter in the story of air refueling had ended.

All-Jet Refueling: The KC-135

Boeing Development of the KC-135

In the early 1950s, American aircraft manufacturers recognized the potential of jets in the commercial transport field. A British jet transport - the Comet - had first flown in 1949, and was threatening to dominate the world airline market. But while American aircraft builders worked on jet design, both the airlines and the military remained reluctant to support financially the development of a jet transport. The costs could be astronomically high and available aircraft design had not been thoroughly proven.

In 1952, Boeing saw three distinct needs not yet adequately filled by current aircraft: a commercial jet transport, a military jet transport, and a military jet tanker. Unable to gain either government or airline backing, Boeing's board of directors approved, in April 1952, a proposal to spend $16,000,000 of company funds to develop one airplane - a prototype jet tanker/transport, which would be ready to fly in the summer of 1954.

Boeing's gamble paid off. On May 14, 1954, the prototype rolled out of the Renton assembly plant. Boeing called it the Model 367-80, because it was the eightieth design configuration of the Model 367, known to the Air Force as the C/KC-97. But any resemblance to the KC-97 was well-camouflaged. The wings were swept back 35°, giving the *Dash 80* a sleek and rakish look. The basic fuselage structure used the KC-97's figure 8 pattern, but the pronounced midriff crease had been smoothed out; fuselage width - 132 inches - was identical to that of the KC-97. Four Pratt and Whitney JT3 engines provided 10,000 pounds of thrust each. It had an empty weight of 92,120 pounds, and a maximum gross weight of 190,000 pounds. The *Dash 80* - Boeing began to call it the Model 707 in the publicity releases - made its first flight, 1 hour and 24 minutes, on July 15, 1954.

While Boeing developed the *Dash 80*, the Air Force defined its jet tanker requirements. Tests in late 1953 and early 1954, showed that KC-97/B-52 refueling was feasible, but obvious differences in speeds and altitudes, as well as the KC-97's relatively limited offload capability, made the KC-97/B-52 an imperfect combination. In November 1953, General LeMay stated SAC's position: 200 jet tankers should be procured with fiscal year 1954 funds. Air Staff directed an immediate evaluation of tankers to support the B-52 and B-58 programs.

On May 5, 1954, Headquarters USAF announced that a jet tanker would be procured. Six companies were invited to participate in a design competition; five - Boeing, Lockheed, Convair, Douglas and Fairchild - accepted. But in August 1954, while the competition was still under way, and just three weeks after the *Dash 80*'s maiden flight, the Air Force announced it would procure 29 "interim jet tankers" called KC-135s - from Boeing. So, even when Lockheed was judged the winner of the design competition on November 10, 1954, Boeing maintained its strong position: their design was available quickly, and cost less than Lockheed's. Furthermore, SAC emphasized that to support two types of tankers would create too many logistical and crew training problems. The Lockheed design was discarded, and in the spring of 1955, an additional 68 KC-135s were ordered.

The first flight of a KC-135 came on August 31,

This sign was found on the receptacle of a 92d Stragetic Aerospace Wing B-52, when it went into Boeing's Wichita plant for modification in 1964.

KC-135 refueling a B-52.

T/Sgt Simpson, 307 CAMS, inspects a KC-135 boom nozzle.

1956, but it was nearly a year later, on June 28, 1957, that Castle's 93rd Air Refueling Squadron took delivery of SAC's first KC-135A. Not until eight years later - January 12, 1965 - did the last KC-135A appear at the 380th AREFS at Plattsburgh Air Force Base. In all 820 C-135s were built.

The KC-135 development program was not trouble-free. When the aircraft caught on, and orders for its sister 707s began to flow in, Boeing and the Air Force had considerable disagreement over priorities. Unreliable Lear MC-1 autopilots plagued the early production models. Stator vane welds in the early J-57 engines were faulty, and fuel tank contamination because of poor housekeeping at Boeing caused some additional difficulties.

Even more serious problems arose from ice formation in the fuel system: many engine failures developed by aircraft descending from high altitudes, as water first separated from the JP-4 and then froze, blocking the fuel system. On June 3, 1958, a KC-135 had an especially close call: descending from 45,000 feet, two engines suddenly flamed out from fuel starvation: both were restarted, but then three engines quit just as the tanker landed at Westover Air Force Base, Massachusetts. To overcome this problem, Air Materiel Command directed installation of fuel system heaters.

Other developmental problems involved hydraulic system overheats, and a severe flutter in the rudder assembly. These were soon resolved by redesigns and modifications clearing the way for the KC-135 to dominate air refueling for at least the next 25 years, and perhaps for the rest of the twentieth century as well.

The KC-135 in SAC: The Early Record-Breakers

In the record-setting flights of the KB-29s and KC-97s, the tankers enjoyed only supporting roles: it was the bombers - B-50s, B-47s and then B-52s - that always had the center stage. But the KC-135 represented such a major technological advance that, within months after its delivery to SAC, it too began to enter the record books - this time on its own. Just five months after the first -135 was delivered, General Curtis E. LeMay, then Air Force Vice Chief of Staff, set a world record in a KC-135 for non-stop, non-refueled flight. Flying in Operation Long Legs on November 11-12, 1957, General LeMay flew 6,320 miles in a little over 13 hours, from Westover Air Force Base, Massachusetts, to Buenos Aires. The return flight on November 13 set a record as well, flying from Buenos Aires to Washington, D.C., in just over 11 hours. A second phase of Operation Long Legs - which was part of Argentina's Annual Aeronautics Week - involved a non-stop flight of six 42nd Bomb Wing B-52s from Homestead Air Force Base, Florida, to Buenos Aires, then to Plattsburgh Air Force Base, New York. Three refuelings were needed - two by KC-97, and one by the new KC-135.

The next year - 1958 - saw even more record breaking flights by the KC-135. On April 7-8, a 93 AFRES tanker set two world records for speed and distance, Brigadier General William E. Eubank, Jr., the 93rd Bomb Wing Commander, led the operation which was called "Jet Stream." On June 27, three 99 AREFS KC-135s set out from Westover Air Force Base to break the New York-London-New York speed records. One tanker crashed on takeoff, killing all 15 aboard, including Brigadier General Douglas W. Sauders, 57th Air Division Commander, and Lieutenant Colonel George M. Broutsas, 99 AREFS Commander. Six journalists on board also died in the crash. Despite the tragedy, the other two aircraft completed the mission - called TOP SAIL - and the lead -135 set another world record.

The quest for records led the KC-135 in some unusual areas. On September 24, another 99 AREFS KC-135, piloted by Captain William H. Howell, captured the official world weight lifting record by carrying a 78,089 pound payload to an altitude of about 6,000 feet. The effort - which required loading the tanker with nails, concrete block and steel plate - can best be understood by putting it into its Cold War context - a Russian TU-104A had grabbed the world record just 18 days before.

Chrome Dome: Support of Airborne Alert

In 1958, amid growing concern about Soviet Intercontinental Ballistic Missile capability, SAC began to study the feasibility of maintaining a portion of the B-52 force on continuous airborne alert. The first such test - called HEAD START - began in late 1958, and involved B-52s from the 42nd Bomb Wing. Despite the inability to gain Congressional funding, other SAC wings continued unpublicized flights under a variety of nicknames: STEEL TRAP, HIGH TRIP, FRONT RAIL, DOWN FIELD, COVER ALL, and CLEAR ROAD. On January 18, 1961, after more than 6,000 airborne alert sorties had already been flown, General Thomas S. Power, the Commander in Chief of SAC, publicly announced the program, and stated that a portion of the SAC alert force would remain airborne at all times. Later that same year, the B-52 missions were rechristened "CHROME DOME," a name which remained for nearly seven years.

CHROME DOME missions averaged 24 hours in length, and covered two basic routes: the southern route crossed the Atlantic, entered the Mediterranean (where refueling took place) and then returned to the United States. The northern route went up the eastern coast of Canada, then crossed Canada westbound at about 84 degrees north latitude. North of Alaska the route cut south, and continued down the west coast of North America.

There were two refueling areas on the north route - "Black Goat" on the east coast, and "Cold Coffee" over Alaska. Cold Coffee was especially critical: the B-52s reached Alaska many hours into the mission, and the bomber crews were tired. A missed air refueling at this juncture meant an unscheduled landing somewhere in Alaska or the Yukon, where suitable fields were - to say the least - few and far between. The KC-135 crews met the challenge: in the first five years of CHROME DOME operations, tankers of the Eielson Tanker Task Force compiled a 99 percent reliability rate in the Cold Coffee refueling area.

CHROME DOME underwent a number of modifications in its seven year life. Routes were changed, and the Mediterranean portion dropped out completely after a B-52/KC-135 collision off Palomares, Spain, on January 17, 1966. Even the name CHROME DOME was changed to GIANT WHEEL in 1967. But on January 22, 1968, another airborne alert B-52 crashed, scattering nuclear weapons on the Greenland ice cap. The end of B-52 airborne alert followed shortly thereafter thus terminating also an important period in the KC-135's history.

KC-135 Refueling in Southeast Asia

The Early Years

In the spring of 1964, it became clear that the situation in Southeast Asia was becoming increasingly serious. Communist Pathet Lao guerrillas had overrun key Laotian installations on the Plain of Jars. In late May, unarmed U.S. Navy reconnaissance jets avoided ground fire over central Laos, and in early June, two Navy F-8s - one an unarmed reconnaissance version, were shot down.

On June 7, 1964, the Joint Chiefs of Staff told SAC to move six KC-135s from Andersen Air Force Base, Guam, to Clark Air Force Base in the Philippines, to support a retaliatory strike by American fighters. Two days later, four of the six tankers refueled eight F-100s on their way to strike Pathet Lao gun emplacements on the Plain of Jars. The KC-135s, nicknamed the flight Tanker Task Force, landed at Clark, and then, six days later, all six of the flight tankers returned to Guam.

But just two months later, on August 5, 1964, eight *Yankee Team* tankers were back at Clark, following the Gulf of Tonkin incident. On that same day, 48 KC-135s took 84 TAC fighters to the Western Pacific. The *Yankee Team* name was changed first to

A U-2 from the 4080th SRW takes on fuel from a KC-135. U-2s are not normally air refuelable, but a limited number of U-2Cs were converted to U-2Fs, with a refueling receptale directly behind the canpy. In February 1963, three U-2Fs made a 13-hour, 30 minute flight from Hawaii to the Philippines. (Courtesy of Col. Jerry Sinclair.)

(Courtesy of Troy E. Minton.)

Foreign Legion, then, in January 1966, to *Young Tiger*, a nickname that remained throughout the rest of the Southeast Asia War.

In June 1965, the B-52s began the first ARC LIGHT operations. Soon one-third of the flight force was being used to support ARC LIGHT while the remaining two-thirds were refueling Pacific Air Force fighters. By mid-1965, the Southeast Asia tanker force numbered 49 KC-135s; 45 at Kadena, and 4-the Tiger Cub Tanker Task Force - at Don Muang airport near Bangkok.

"Silent Refueling" The Early Years of ARC LIGHT

Kadena-based KC-135s flew most ARC LIGHT missions, refueling B-52s launched from Andersen Air Force Base, Guam. Early ARC LIGHT missions observed total radio silence, since the long B-52 flight times to target - about six hours from Guam - gave the enemy plenty of time to change his location if it was learned the bombers were on the way. Most missions consisted of between 18 and 30 tankers, supporting an equal number of bombers. Each tanker formation consisted of a "three-ship" cell, the leader taxied after getting clearance by a light signal from a ramp vehicle, and the rest of the formation fell in behind. Takeoff clearance came by way of a green light from the tower; the tankers of each formation launched one minute apart, with three-minute spacing between formations. When the 135s broke through the usual overcast, they were greeted by a black sky full of rotating beacons. The task was to find the right beacon, and then form up behind it, one mile back. Timing points were the responsibility of the formation leader, and rendezvous with the receiving B-52s was by electronic beacon, often in bad weather, usually at night, and always - at least in the early years - in complete radio silence. As they proceeded down the refueling track, each tanker would silently offload about 90,000 pounds of fuel to its B-52, and then return to Kadena.

Eventually the North Vietnamese located "fishing trawlers" near Guam; they ensured that the Viet Cong were notified of any B-52 departures, and made the radio silence procedures no longer necessary. The reintroduction of normal radio communications made the refueling missions safer - poor communications had already contributed to the loss of two B-52s in a midair collision in one of the crowded tracks. This brought an end to a unique and harrowing chapter in air refueling.

Tanker Saves in Southeast Asia

Tanker crews continued to build a record of impressive accomplishments throughout Southeast Asia, and were credited with hundreds of fighter saves. On November 22, 1965, a Kadena-based KC-135, commanded by Captain Ross C. Evers, saved an F-105 returning from a bombing mission in North Vietnam. The *Thunderchief* was down to its last 200 pounds of fuel when it made contact with the tanker, well inside North Vietnam. On July 6, 1966, a similar mission - this time by a crew from the 301 AREFS - saved the entire "Peanut Flight," four F-105s which had used nearly all their fuel shaking loose from a MIG attack over the North.

Another save, on May 3, 1967, was an even closer call. Two F-105s had been flying protective cover for a downed airman, and found themselves unable to reach any friendly base. A KC-135, commanded by Major Alvin Lewis, headed for the F-105s, reaching them just as one pilot was preparing to eject. The KC-135 was in a twenty degree descent when contact was made, and when the F-105s engine flamed out seconds later, Major Lewis maintained contact increasing the tanker's dive to thirty degrees. The first Thunderchief pilot restarted his engine, and the second F-105 was refueled, again just in time.

But perhaps the best known save came later that same month, and it was unique in several ways. It was the first - perhaps only - three-level refueling: a KC-135, flown by a crew from the 902 AREFS at Clinton-Sherman Air Force Base, was scheduled to refuel two F-104s over the Gulf of Tonkin. But while refueling the F-104s of "Falstaff Flight," the KC-135 was redirected toward two Navy A-3 tankers from the carrier Hancock. While both A-3s had only minutes of useable fuel left, they had 4,000 pounds of fuel in tanks from which they could transfer, but not use themselves.

In the meantime, two Navy F-8s arrived, so low on gas that they immediately made contact with the two A-3s, which continued to take turns refueling from the KC-135. By then, it was time for the F-104s to refuel again from the tanker: soon, two Navy F-4s from the carrier *Constellation* arrived, both critically low on fuel. The KC-135 refueled these, and then filled up the F-104s again. By this time, the tanker itself was unable to return home, and it diverted into Da Nang, landing with just 10,000 pounds of fuel on board. The mission resulted in eight aircraft saved, with 50,000 pounds of fuel transferred in 14 separate contacts. For the flight, the KC-135 crew won the Mackay Trophy for the most meritorious flight of the year.

While tankers had contributed to so many of SAC's previous eight Mackay-winning missions, this was the first time that a tanker crew had actually won the award. On July 11, 1968, the Mackay Trophy was presented to the entire tanker crew: Major John Casteel, Captain Richard Trail, Captain Dean Hoar, and Master Sergeant Nathan Campbell.

For the rest of the Southeast Asia war, the number of tankers involved fluctuated, serving as a barometer for the level of bombing activity. In January 1968, North Korea seized the USS *Pueblo* and the North Vietnamese started the Tet Offensive; by the next month, the number of KC-135s supporting

A U.S. Air Force HH-53 "Jolly Green Giant" helicopter assigend to the 40th Aerospace Rescue and Recovery Squadron (ARRS) approaches the refueling drogue of an HC-130P as another HH-53 awaits its turn. Nhakhon Phanom AB, Thailand. Jan, 1973.

Steel Tiger - TEXACO in the sky.

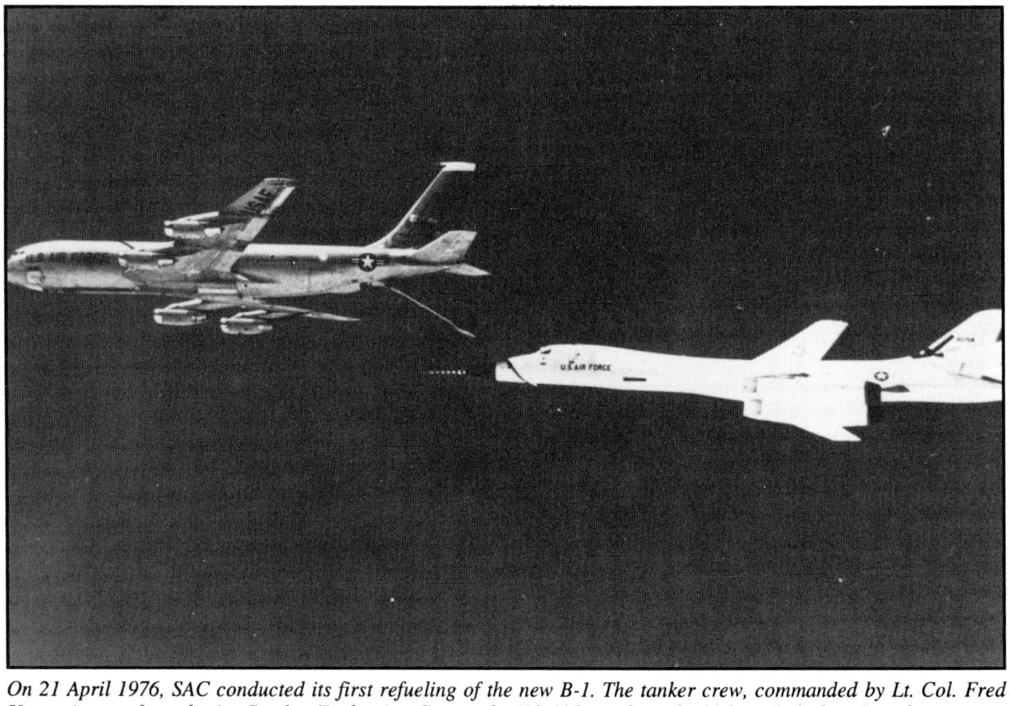

On 21 April 1976, SAC conducted its first refueling of the new B-1. The tanker crew, commanded by Lt. Col. Fred Hartstein, was from the 1st Combat Evaluation Group; the KC-135 was from the 22d Air Refueling Squadron.

the war effort reached a new high - 94 flying from five bases: U-Tapao and Takhli in Thailand, Ching Chuan Kang on Taiwan; Kadena Air Base, Japan; and Andersen Air Force Base, Guam. In 1968, President Johnson's bombing halts reduced the need for air refueling support, and the trend in tanker operations spiraled downward until 1971. By December 1970, tanker numbers had declined to 48; they no longer flew from CCK and Takhli; 31 tankers were at U-Tapao, with another 15 at Kadena; two more sat on strip alert at Guam. The need for air-refueling support had been scaled back.

1972: Bullet Shot, Constant Guard, and Linebacker II

With the dawn of 1972 came an increase in enemy activity throughout South Vietnam, and American air power throughout the area was quickly boosted. Tankers returned to duty in the Pacific. KC-135s ferried whole squadrons of F-4s and F-105Gs to Thai bases in a series of operations called CONSTANT GUARD. Similar movements involved deployment of B-52s - including, for the first time, the more advanced G models - to Guam and U-Tapao; these deployments became know as Bullet Shot.

By June 1972, the number of KC-135s committed to Southeast Asian operations peaked at 172, flying from six bases. At Kadena alone, 58 KC-135s lined the ramp; 46 more were at U-Tapao; 40 more were spread across the flight lines of Korat, Takhli, and Don Muang in Thailand; the remaining 28 were at Clark Air Base in the Philippines, until torrential rains and flooding forced a return to CCK. The number of monthly tanker sorties reached a wartime peak in September, when the KC-135s flew 3,902 sorties in support of fighter and bomber operations.

In the meantime, negotiations in Paris were underway, aimed at ending the decade-long war. When the North Vietnamese reneged on commitments in December President Nixon ordered B-52 bombings in the Hanoi area. From December 18-29, KC-135s refueled B-52s from Guam as they joined B-52s from U-Tapao attacking targets around the capital city of Hanoi and its port of Haiphong. To the world, it was called the "the Christmas bombing" or the "Eleven Day War;" officially, the operations became known as LINEBACKER II. It had its desired effect: the North Vietnamese returned to the negotiating table, and several weeks, later, the peace treaty was signed.

Southeast Asia Epilogue

As American forces returned home, tanker strength declined. By May 1974, just 42 tankers remained in the area, based only at U-Tapao and Kadena. KC-135s provided refueling for the fighters which covered the final American evacuation of Saigon in April 1975. The next month, the final KC-135 combat sorties were flown, this time refueling fighters involved in the successful *Mayaguez* rescue mission on May 12. The fighter cover was maintained for nearly 24 hours, with the help of 20 *Young Tiger* aircrews flying orbits over the Gulf of Siam. Their presence made the mission possible.

On December 21, 1975, 17 KC-135s left U-Tapao, escorting 36 A-7s departing from Korat. They flew together to Andersen Air Force Base, then to Hickam and on to Hill Air Force Base, Utah. Then, the KC-135s returned to their home bases, having just flown the last of the YOUNG TIGER missions. In the years since the first flight tanker crews flew from Clark in 1964, KC-135s had flown 194,687 sorties, providing nearly 9 billion pounds of fuel in 813, 878 refuelings. But perhaps the real measure of their success rested in accomplishments

KC-10 giving a little guy a drink.

not so easily documented: the hundreds of aircraft and aircrews that they saved, and the American mastery of the skies that they made possible.

Upgrading the KC-135

By the late 1970s, with the age of the early KC-135s approaching the quarter century mark, it was clear that the tankers were developing problems that could only be remedied by major modifications. Two especially serious problems - metal fatigue in the wings, and the age and inefficiency of the J-57 jet engines - needed to be addressed.

Wing Strengthening Programs

Although the average KC-135 had accrued far fewer hours than its civilian 707-720 counterpart - 8.151 hours vs. 49,960 hours, as of early 1979 - metal fatigue began to be a serious problem, especially in the center wing structure. A number of modifications were begun in an attempt to avert a catastrophic wing failure. One approach - called "ECP 439" - involved "structural fatigue packages" which replaced fasteners in areas which had proven especially susceptible to fatigue cracks. Four fatigue packages had been prepared for use, and the choice of package depended on the number of flying hours accrued on the particular KC-135. Fasteners were first removed, and the holes inspected for hairline cracks. Technicians then enlarged the holes and installed larger fasteners. Depending upon the aircraft, as many as 8,000 fasteners could be removed and replaced as part of the modification.

Another program centered on the placement of an external splice across a wing production break which had shown considerable fatigue cracking. The splice, designed to be strong enough to hold the wing together if the internal plates failed, was installed on all -135s that reached the 8,500-hour mark, but had not yet undergone a wing reskin program.

Still another modification would include 135s not yet reskinned involved the installation of the Acoustic Crack Detection System (ACDS). Transducers installed in the underside center wing area increased airframe reliability by detecting and reporting small cracks before they could lead to structural failure. The Acoustic Crack Detection System forestalled what many felt would develop into an integrated improvement program.

These programs amounted to just stopgap measures, designed to compensate for the fact that the original 7178 aluminum alloy used under the 135's wing could no longer be trusted to perform safely. The ultimate fix technicians called "ECP 405," the C/KC-135 wing reskin program. In this effort, 22 panels running from just outboard of number one to outboard of number four engines - covering about 130 feet of the tanker's wingspan - were replaced with 13 larger panels, made of a stronger and more crack-resistant 2024 aluminum alloy. The job was extensive - just the panel installation took 32 days to complete - but it eliminated the need for the other interim measures, like ACDS and structural fatigue packages. Original schedules called for modification of three aircraft per month, but this rate was doubled in 1979. The entire retrograde was scheduled for completion by 1988, making the tanker a much safer and more reliable aircraft.

KC-135 Re-Engining

In the late 1970s, a project called "Pacer Angler" - the transfer of KC-135As to the Air Force Reserve and the Air National Guard - quickly made a long-standing problem critical: with KC-135As beginning to operate from commercial airports across the country, their noisy, smoky engines caused a great deal of criticism and controversy. Loud objections from nearby communities brought a new impetus to an old idea: could not these veterans be given new life with new engines?

Three engines emerged as prime contenders for the KC-135 re-engine program: the GE/Snecma CFM56, with 24,000 pound thrust (compared to 13,370 pounds thrust for the 1950 vintage J-57 engines), was eventually chosen over two Pratt and Whitney entries. But better community relations - the noise footprint dropped with re-engining - were only some of the benefits the Air Force hoped to gain from the re-engine program. A single KC-135 with new engines installed could handle more receivers and transfer often double the amount of fuel expected from a KC-135A. As planners estimated the number of potential customers to increase to 14,000 by the mid-1980s, interest was high at the Air Staff and in NATO and elsewhere. The new engines used far less fuel than their predecessors and planners estimated at least a 25 percent cut in fuel usage or a savings thought to exceed 150 million dollars a year. The new CFM56 jet engines required far less maintenance than the older J-57 engines. For these reasons the Strategic Air Command began pushing to have all 615 KC-135As re-engined as soon as possible. The realities of tight budgets and other priorities made SAC's goal rather unlikely but by 1990 a substantial portion of the fleet had been updated.

ATCA: The Development and Deployment of the KC-10

Changing world situations - pointing toward a need to be able to rapidly project American tactical air power into critical areas like the Persian Gulf - led in the early mid-1970s to the concept of "ATCA," the Advanced Tanker/Cargo Aircraft. ATCA would combine the cargo capability of a strategic airlifter with the refueling capability of an oversize tanker. Unlike previous generations of tankers, ATCA's primary mission would not be to support SAC bombers, but rather to enhance the mobility of the Tactical Airlift Command's fighters and Military Airlift Command's C-5s and C-141Bs.

The demand for large cargo capacity, as well as great fuel transfer capability, naturally led Air Force planners in the direction of the American "wide body" jets. Boeing had a 747 prototype equipped with a boom and an operator compartment as early as 1973. McDonnell-Douglas had a DC-10 in convertible freighter version. The 747 had the advantage of already being a proven tanker - the Iranian Air Force was already using 747s to refuel their fighter fleet; but the KC-10 with its three-engine design, offered better fuel economy and maintainability.

A version of the McDonnell Douglas entry - the DC-10-30CF - was ultimately selected as the ATCA. The program began in January 1978, and two months later Air Force officials agreed upon the designation KC-10A. Work began on the first KC-10 in June 1979, and the first aircraft - serial number 79-0433 - rolled out of the Long Beach plant on April 16, 1980.

Early artists' conceptions had shown the KC-10 painted in much the same was as a KC-135 - basic gray, with standard USAF and SAC markings. So the fashionable appearance of the first KC-10 - by now tagged with the unromantic name of *Extender* - came as a surprise to many. Its glossy white top and gray bottom were separated by a band of dark blue. Engine nacelles were a shiny silver and a stylized American flag graced the 58-foot high tail. (McDonnell Douglas had even offered to add red pinstripes that would run the length of the fuselage, but at that point, the Air Force opted for simpler designs.) The capabilities of the KC-10 far exceeded its already attractive appearance. Despite problems highlighted by an American Airline DC-10 crash in 1979 - later attributed to faulty maintenance practices - the KC-10 offered the benefits of a proven design: the first of its more than 300 DC-10 counterparts had been flying in commercial service since 1971. Forty-five airlines had flown their DC-10s over 5 million hours and high time aircraft have already logged 32,000 hours, and 12,000 flights.

The KC-10 also offered an important logistical benefit. Eighty-eight percent of the plane's components could be used on DC-10s, allowing the Air Force to make use of a world wide support network and supply inventory already in existence.

As a tanker, the KC-10 offered substantial improvements over the KC-135. Unlike standard KC-135As, the KC-10s could be refueled using a universal receptacle located atop the fuselage, just aft of the cockpit. Also unlike the KC-135, the KC-10 could refuel receivers using both the boom and probe-and-drogue methods on the same mission, since the KC-10 housed a separate hose and reel arrangement in the aft fuselage tanks. These tanks with a capacity of nearly 118,000 pounds, when added to the normal four wing tanks of the DC-10, raised the KC-10s fuel capacity to 356,065 pounds or nearly twice that of the KC-135.

Refueling a F-116 from a KC-135 tanker. April 24, 1984. (Courtesy of James Pargoe.)

The KC-10 also featured a radically different boom area. Like the old KB-29P, the boom operator performed refueling tasks seated and not from a prone position as in KC-97s and KC-135s. Three seats were installed (one each for an instructor and an observer) and the operator viewed the receiver through a large 25 x 54 inch window. A system of mirrors allowed him to monitor other aircraft in formation off the KC-10 wing tips. An advanced boom, ten feet longer than the KC-135s featured "fly by wire" flight controls and increased fuel transfer rates (up to 1,500 gallons per minute). Its rolling, rather than yawing, motion permitted receiver contact to be maintained more securely and under wider range of flight conditions.

As a cargo carrier, the KC-10 was less revolutionary, but still impressive. It could hold 27 standard 463L pallets, more than twice the 13 pallets that could be handled by even the stretched C-141B. With a 102x140 inch access door, and a payload of nearly 170,000 pounds, the KC-10 could be loaded with nearly all the equipment needed to support the deployment of a tactical fighter squadron. Future missions would verify the importance of this attribute.

KC-10 Program Status

In 1979, 12 F-15 Eagles, 115 tons of support equipment, and 209 people deployed to Saudi Arabia. The move required two days, using 16 KC-135s, 3 C-141, and 2 C-5s, and two overseas bases - Lajes in the Azores and Torrejon in Spain. It is estimated that, had KC-10s been available, only six of them would have been needed - no C-141s, no C-5s, and no forward bases. The move could have been completed in half the time, with a savings of 600,000 gallons of fuel. But as capable as the KC-10 would appear, the program in 1980 faced an uncertain future. At one time, planners envisioned a buy of as many as 59 aircraft; by 1976, that number had dropped to 41; when the first KC-10 (the second production model) was delivered to SAC's 4200th Test and Evaluation Squadron in March 1981, the planned buy stood at 20, with firm, funded commitments for only 12.

Meanwhile, aircrews demonstrated the impressive range and fuel-transfer capabilities of the KC-10. In October 1982, KC-10s supported the longest non-stop F-15 deployment on record. The tankers made possible the movements of the fighters from Yokota Air Base, Japan to Eglin Air Force Base, Florida. The tankers transferred 500,000 pounds of fuel before the fighters touched down in Florida. During operation EARLY CALL to Egypt in February, 1983, the KC-10s proved themselves useful as both refuelers and cargo carriers. The following year the KC-10 won honors in refueling U.S. Navy aircraft over the Pacific when fog prevented A-6s and A-7s from making carrier landings. The naval aircraft flew safely to alternate sites thanks to fuel received from the KC-10s. By 1985 the Strategic Air Command was rapidly integrating the KC-10 into all phases of its refueling mission which seemed to grow and expand monthly. As the 1980s came to a close, the KC-10s, some of which would be equipped with wing-mounted drogue refueling pods, had proven invaluable in supporting the air refueling needs of the United States Armed Forces and its allies.

Epilogue

Early on the wintry morning of March 12, 1980, a pair of B-52As lifted of the runway at K.I. Sawyer Air Force Base, Michigan. They flew over Europe, and on to the Indian Ocean, where they photographed portions of the Soviet fleet. After a low pass over the carrier *Nimitz*, both bombers headed for home, by the way of the South China Sea and the North Pacific. They landed at K.I. Sawyer on the morning of March 14, having flown around the world, nonstop, in less than 43 hours. The flight generated little of the excitement and media attention which typified POWER FLITE twenty-three years before, but that came as no surprise. The B-52 had been proven over and over again for a quarter of a century. Equally well-proven in this mission and literally thousands to follow were the crews and aircraft of SAC's air refueling force. The tanker force - often the unsung heros of Air Force flight operations - had become an essential ingredient in the use and application of American air power. The fundamentally important role of air refueling surfaced in the American strike in Grenada, the series of crises in the Middle East, the United States' air raid against targets in Libya in 1986, and more recently in American actions in Panama, in Operations DESERT SHIELD/STORM and in Operation RESTORE HOPE. And now, as a component of the Air Mobility Command, it is a role that is unlikely to change in future air operations.

A Brief History of the Air Mobility Command

By Dr. John W. Leland HQ AMC/HO

As part of the largest restructuring of the United States Air Force since its creation as a separate service in 1947, a new era in airpower history began on 1 June 1992 when the Military Airlift Command (MAC) and the Strategic Air Command were deactivated and the Air Mobility Command (AMC) formed from elements of these two historic organizations. Among the reasons for the reorganization were the end of the Cold War, the blurring of strategic and tactical requirements after the American victory in the Persian Gulf War, and the Air Force's new strategic vision of Global Reach...Global Power.

AMC melded a worldwide airlift system with a tanker force that had been freed from its strategic commitments by the collapse of the Soviet Union. General Hansford T. Johnson, dual-hatted as Commander in Chief of the United States Transportation Command and MAC, became AMC's first commander. On 25 August 1992, he was succeeded by General Ronald R. Fogleman. Approximately 181,000 active duty, Air National Guard, and Air Force Reserve officers, airmen, and civilians performed AMC's mission of global reach.

In AMC's first year, one of Gen. Fogleman's principal objectives was to focus on the core mission of air mobility and to divest the command of activities not directly related to global reach. Thus, AMC relinquished ownership of Hurlburt Field FL, and Kirtland AFB NM, and transferred the Air Rescue Service, intratheater aeromedical airlift forces based overseas, and much of the operational support airlift fleet to other commands. Plans were also underway to reassign the command's C-130s to the Air Combat Command.

New ways of doing business characterized the Air Mobility Command. The Tanker Airlift Control Center (TACC), located with the command headquarters at Scott AFB IL provided centralized scheduling and execution of all AMC airlift and air refueling missions. Centralization simplified customer access to mobility forces and streamlined the chain of command, eliminating bureaucratic delays and increasing responsiveness.

The Air Mobility Command's ability to provide global reach was tested daily during its first year. In the 30-day period from 25 August to 24 September 1992, three weather disasters combined with heavy routine commitments to produce a degree of activity that approached wartime intensity. After Hurricane Andrew lay waste to much of southern Florida on 24 August 1992, the Air Mobility Command began what Gen. Fogleman described as "the biggest and fastest domestic relief effort done by air in the history of the Air Force." By the time AMC's support of the relief effort ended in late September 1992, the command had airlifted nearly 21,500 tons of relief supplies and more than 13,500 relief workers to the disaster area using 724 missions. Typhoon Omar struck Guam just four days after Hurricane Andrew hit Florida, and in a four-week period AMC aircraft delivered nearly 2,000 tons of relief cargo and 750 passengers to the western Pacific island aboard 59 missions. Hurricane Iniki, the third weather disaster, ravaged the Hawaiian Islands on 11 September 1992. This time between 12 September and 18 October 1992, AMC active duty C-130s flew 257 missions to move disaster assistance cargo and personnel to the Aloha State.

Although the Cold War was over, religious and ethnic hatreds, poverty, and ideological differences still threatened world peace and stability. Throughout the year, AMC was employed to implement national policy objectives around the world. Command C-130s augmented the United States Air Forces in Europe to airlift and airdrop humanitarian relief supplies to the beleaguered people of war-torn Bosnia-Herzegovina in Operation Provide Promise. The Soviet Union's disbanding on Christmas Day 1991 signaled the end of the Cold War. In the short-term struggle to establish democratic governments and market economies, however, the former Soviet republics needed humanitarian assistance. Consequently, AMC continued supporting Operation Provide Hope, which the Military Airlift Command had begun in February 1992. By May 1992, the command had delivered more than 2,400 tons of cargo to the former Soviet states using 109 missions. The Air Mobility Command also continued assisting the Kurds living in the United Nations guaranteed safety zone of northern Iraq by flying Provide Comfort supply missions to Turkey.

From August to October 1992, AMC C-130s deployed to Angola for Operation Provide Transition and furnished transportation for soldiers demobilized after 16 years of civil war, in preparation for national elections. When the United Nations banned Iraqi flights below the 32d parallel, AMC transported U.S. forces and deployed tankers in support of Operation Southern Watch I and II.

Clearly, the year's greatest test of air mobility came in Somalia, devastated by anarchy and mass starvation. In August 1992, AMC C-130s operating from Kenya began airlift of food to Somalian towns. By the time Operation Provide Relief ended on 28 February 1993, nearly 3,100 missions had delivered 34,000 tons of cargo from Kenya. Air mobility operations in Somalia grew dramatically in December 1992 when President Bush sent American troops to bring order and protect international relief agencies distributing food and medicine in Operation Restore Hope. By 10 May 1993, AMC had flown 1,182 strategic airlift missions to transport more than 51,431 people and 41,243 tons of cargo. Operating from bases in the United States and deployed locations overseas, command tankers supported the air bridge, completing more than 1,170 refueling missions and offloading 82.4 million pounds of fuel. Gen. Fogleman described Restore Hope as "the first time our air mobility forces have been engaged in a major exercise in their post-Cold War configuration."

AMC stands ready to meet whatever new challenges arise in the command's second year. As Gen. Fogleman has said: "While others train for the next battle, every day members of this command...provide services that save lives and relieve suffering. We are a total mobility team engaged in peacetime pursuits, yet ready if called to battle."

Air Mobility Command Bases

Altus, OK
Andrews, MD
Charleston, SC
Dover, DE
Grissom, IN
Kurtland, NM
Little Rock, AK
Malmstrom, MT
March, CA
McChord, WA
McGuire, NJ
Norton, CA
Plattsburgh, NY
Scott, IL
Travis, CA

The first flight formation of four C-17 on the anniversary of the first FLT. Flight passing the 500 hour mark of testing. Sept. 15, 1992. (Courtesy of Ted Venturini.)

A KC-10 Extender refuels USAF/McDonnell Douglas F-15s. (Courtesy of Jim Ramsey.)

Corporate Sponsors

AAR Cadillac Manufacturing

ISUTM Container with 5" and 8" trays.

ISU-90-I Container.

Brooks & Perkins (B&P) was incorporated in May of 1948 and headquartered in Southfield, Michigan. Manufacturing plants and offices were in several areas of Michigan, New York and California. One of the Company's facilities, operated under the name Brownline, was located at Heathrow Airport, London, England.

The Company was the successor to two corporations which were merged in August 1948: Brooks & Perkins, Incorporated and Vanette, Incorporated, both Michigan companies.

In 1948, Brooks & Perkins was involved primarily in the magnesium parts fabricating business with products manufactured for the defense and aerospace industries. Following World War II, B&P was a supplier of the major magnesium components for the B-36, B-47 and B-52 aircraft. Up to and including the Korean Conflict, the corporation prospered in this product area.

However, like so many other defense-oriented companies, after the Korean Conflict, B&P's business declined. In 1953, the Company began a commercial division to apply magnesium to other uses and, with the advent of the space age, its advanced technology in the forming and assembly of magnesium components and products was further developed and expanded to include design, development and manufacture of products and components in aluminum, titanium, zirconium, beryllium, columbium, exotic steels and other metals.

During the 1960s, the corporation diversified into manufacturing air cargo handling equipment and industrial materials handling products made out of lightweight, durable metals.

The list of Brooks & Perkins' early accomplishments is impressive:

The first nonmetal producer to commercially deep draw and severely hot form magnesium, titanium, zirconium and beryllium sheet.

Pioneered magnesium-lithium product development.

The first company to have electroplating facilities for magnesium and other exotic metals.

Installed the first commercial HAE anodize facility in the world and had one of the earliest installations for Dow 17.

Designed and built the first vertical plotting board system for radar and air defense.

Has been a supplier of major components for such aircraft as the B-36, B-47, B-52 as well as most aircraft in use today.

Produced the Vanguard satellites (payload structures and instrument cans, exclusive of instrumentation), America's first space venture ... as well as structures and components for other satellite systems.

Produced assemblies such as nose cones and missile body sections for numerous missiles.

Designed, developed and produced aerial delivery systems for the U.S. Air Force and developed and produced the aircraft conveyor/restraint rail systems standardized by the Air Force for aerial delivery and cargo handling. Has designed and manufactured cargo handling and delivery systems for many different types of military and commercial aircraft.

Was one of the initial developers and producers of air cargo containers for the airline industry...progressing to the world's largest producer.

Producer of precision cabinets, rack and transit cases for electronics packaging.

In 1977, Brooks & Perkins' business involved the design, development, manufacture and marketing of components and systems for materials movement, containerization and storage, serving a wide variety of industries and markets. In addition, B&P excelled in the area of advanced technologies in providing proprietary production of specialized precision parts, components and structures.

Substantial effort was put into the development of manufacturing processes and techniques - some applying to metals, some to other materials such as plastics, balsa and other woods. Diversification in its product lines and in its capabilities enabled the company to build a reputation as a worldwide leader in a number of areas.

In 1981, Brooks & Perkins, with its divisions Advanced Structures, Brownline, Brownline Ltd., Cadillac Manufacturing, Materials Handling and Skydyne, was acquired by AAR CORP., a global aviation support and services conglomerate. The name Brooks & Perkins was subsequently phased out and operating units became AAR Advanced Structure (which absorbed Brownline), AAR Cadillac Manufacturing, AAR Handling Systems, AAR Skydyne and AAR PowerBoss.

Today, AAR Cadillac and AAR Advanced Structures are companies which, utilizing state-of-the-art technology and computer design and engineering, manufacture innovative products in support of the U.S. government's rapid deployment program. Air cargo handling and logistics systems, specialized containers and multi-purpose shelter systems are just a few of these products.

With a rich history of manufacturing top quality, durable products, AAR is well positioned for future demands. Our responsive network of product support and customer service is in place to provide technical assistance, field engineering, system installation and operational training services. As it was in 1948, our mission remains today, "We will be the best at what we do in each of the respective markets for the products and services we provide."

Armed Services Mutual Benefit Association

ASMBA's Headquarters in Nashville, TN.

The time was 1963, the event was the Vietnam War. The situation was that many military people needed life insurance coverage. Many service personnel wanted, in the event of their death, to provide protection for their families.

That's when the founders of ASMBA (Armed Services Mutual Benefit Association) - active duty Servicemen and Airlifters - discovered there was a real need for low cost, high value term coverage protection. Protection that came without exclusions, restrictions, or war clauses.

As you know, during the Vietnam War most insurance companies, and even some military associations, had several restrictions on coverage.

There were rank restrictions, usually offering coverage for officers only. There were time-in-Service restrictions, requiring a long period of Service (usually two years) before coverage. The young draftee or combat arms officer who came into the Service couldn't get coverage until completion of two years of Service! This meant that coverage would be made available to him only after he returned from the war zone! In addition, there were geographical restrictions. Coverage from some outfits was good anywhere in the world, except Southeast Asia!

Also, some commercial companies simply inserted a war clause for military people, meaning they would not pay if the insured's death was combat-related.

We didn't think that was fair then, and it's not fair today. Military personnel need worldwide coverage for all possible situations, and that coverage is needed now more than ever before. That's why today, just as in 1963, ASMBA continues to meet the needs of all ranks of military people - with honor, without assignment restrictions.

It's true that during the Vietnam War we sustained high mortality rates. Yes, we paid many death claims because we didn't refuse coverage to anyone who was in good health and needed life insurance.

We don't intend to ever exclude anyone because of rank, hazardous duties, location, or combat. That's what ASMBA is all about - military people caring for other military people. And we'll keep it that way.

But that's not the whole ASMBA story. Since 1963, we have come a long, long way. We now have thousands of military members - from privates to four-star generals. Thanks to our members, and avid supporters like the Airlift/Tanker Association, ASMBA is a true success story. We now offer the military member twelve different coverages - coverage for themselves and their family members, ranging from plans of Permanent Life or High Value Level Term life to supplemental plans of coverage for spouses and children. And, there are more plans on the horizon! All the coverage a military family will ever need is available at ASMBA - without any restrictions on members.

We are proud of each and every ASMBA member and vow to continue our service to them. Likewise, we are proud to be associated with the Airlift/Tanker Association. We respect the contributions and sacrifices our members have made in the defense of this great nation and our way of life. The early founders of ASMBA made a tough and courageous decision to organize ASMBA - and we would like to take this opportunity to salute you and them for a job well done!

Beech Aircraft Corporation

U.S. Air Force T-1A Jayhawk.

Beech Staggerwing.

America was close to rock bottom in 1932. Crude oil sold for 10 cents a barrel, wheat for 25 cents a bushel, and General Motors stock was quoted at less than $8 a share, with no buyers. Bread lines and soup kitchens kept the unemployed from starving. The entire aviation industry produced a total of only 549 commercial planes valued at little more than $2 million.

If there was ever a time for a man who had any money to hold on to it tightly, that was it. Walter H. Beech and his wife, Olive Ann, had accumulated a tidy sum from the sale of their holdings in Travel Air to the Curtiss-Wright Airplane Company. They could have played safe and waited out the depression, making it an occasion for a long vacation. They did not. The Beeches wanted to build the finest airplanes in the world, and the most dramatic economic slump in history would not stop them.

So, in spring of 1932, Walter and Olive Ann turned away from New York City and the circles of high finance to come home to Wichita, Kansas. There they organized the Beech Aircraft Company, predecessor to the present corporation. Walter was president of the new company and Olive Ann, who had demonstrated extraordinary business ability as Walter's assistant and office manager, was named secretary-treasurer and director.

In small quarters rented in a depression-closed factory, the Beech team went to work to design and build a five-place biplane having the interior luxury and passenger comfort of a fine sedan, a top speed of 200 mph or better, non-stop range of close to 1,000 miles, easy controllability and sound aerodynamic characteristics. The result: the Beech Staggerwing, now considered a classic among civil aircraft of the world.

The Beeches followed this successful product with the Model 18 Twin Beech. For the military, the Model 18 served as a excellent platform for bombardier and navigation training. During World War II, Beech built the SNB, a gunnery and bombardier trainer for the Navy; AT-7, a navigation trainer for the Army Air Force; and AT-11, a bomber trainer for the U.S. Army. Additionally, the company's Model 25 answered the U.S. Army's need for an advanced multi-engine pilot trainer that could be built with non-strategic materials, primarily wood.

After the war, Beech continued to build military training aircraft. In 1953, the U.S. Air Force placed its first order for Beech T-34As, which were used as basic trainers. The following year, the Navy ordered T-34B pilot trainers. This workhorse offspring of an early model piston-powered Bonanza served as the Navy's primary trainer until the 1970s, when an upgraded turbine-powered version, the T-34C, replaced it. The T-34C is still in use today.

In the 1970s, the Navy selected a King Air derivative, designated the T-44A, as its multi-engine trainer for maritime pilots. It is also still flown today.

More recently, the U.S. Air Force selected the Beechjet as winner of a hotly contested competition for an advanced multi-engine trainer. It was renamed the T-1A Jayhawk, and is used to train pilots who will be assigned to either airlift or tanker missions. The T-1A is a militarized version of Beech's top-of-the-line aircraft, the Beechjet 400A. When combined with the T-1A, Beechjet production accounts for the second largest light jet manufacturing operation in the world.

In addition to the Beechjet, Beech builds a full line of jetprop aircraft, including the Super King Air 350, the King Air C90B, and the King Air C90SE, as well as the Super King Air B200, which has been produced for all U.S. military services as the C-12 Huron, a personnel and cargo transport aircraft.

Other Beech-built jetprops include the all-composite Starship 2000A and the Beech 1900 Airliner, which is produced for regional/commuter markets and flies throughout the world in the colors of many major airlines. Beech's piston-powered aircraft line includes the famous single-engine Bonanzas, produced in three models, and the twin-engine Baron 58.

Besides its own products, Beech performs completion work for Raytheon Corporate Jets Hawker 1000 and Hawker 800 medium jets, installing interiors and avionics, and provides subcontracting support for companies such as Boeing and McDonnell Douglas.

Beech also produces two missile target systems, the AQM-37C and the MQM-107, for use by the U.S. armed forces and for international export to allied nations.

In addition, Beech operates Beech Aerospace Services, Inc. (BASI), which provides contractor support to U.S. military aircraft around the world. BASI recently won a new U.S. Air Force contract for all management, equipment, personnel and services necessary to support the 15th Flying Training Wing mission at Columbus AFB, Mississippi.

Beech Aircraft is a subsidiary of the Raytheon Company and a leading manufacturer of business and military aircraft. With headquarters and major facilities in Wichita, and major manufacturing operations in Andover and Salina, Kansas, the company has produced more than 50,000 aircraft since its founding. Beech has major facilities in 11 states.

Beech and Raytheon Corporate Jets constitute the Raytheon Aircraft Products Group, an organization with the broadest and strongest product line in business aviation.

Above: Beech Staggerwing.

Below: Support crew members prepare a Beech T-1A Jayhawk for flight at Reese AFB.

The Boeing Company

In 1949, Boeing developed the "flying boom," an aerodynamically controlled swiveling and telescoping arm that a crew member, using cockpit-like flight controls, guided into a receptacle on top of the receiving aircraft's fuselage. The boom, first used on B-29s, quickly became the aerial transfer system of choice and was subsequently used in the KC-97 tankers and, later, in the KC-135 jet tankers.

On July 15, 1916, William E. Boeing incorporated what is now The Boeing Company under the name of Pacific Aero Products. Its first product was a single-engine seaplane with a top speed of 75 mph. By the 1920s, Boeing had pioneered early airmail routes and established the company as a major manufacturer of military trainers and fighters.

During the 1930s, the company built its first all-metal transports, which included the high-speed Model 247 and the transoceanic Model 314 Clipper.

Boeing played a key part in World War II with its famous B-17 and B-29 bombers. After the war, the company built the world's first swept-wing jet bomber, the B-47, and later developed the B-52, which still is relied upon by the U.S. Air Force today.

In 1954, Boeing developed the Dash 80, prototype for the 707, which was destined to revolutionize world air travel. The 707 was followed by the 727, 737, 747, 757, 767, and 777, establishing the company as the world's leading manufacturer of commercial jet transports.

Since the late 1950s, Boeing has been involved in a wide range of aerospace programs, including the development of helicopters, missiles, electronic systems and aircraft for the military, as well as numerous space projects. Boeing built the first stage of the Apollo moon rockets and remains a major NASA contractor.

Boeing Contributions to the Airlift/Tanker Industry

C-97 Stratofreighter: The C-97 Stratofreighter was the high-flying cargo and transport plane that evolved from the B-29 bomber during World War II.

The pressurized C-97s had a double-lobe fuselage consisting of two intersecting circular sections. The lower deck had the same diameter as the B-29. The 74-foot-long upper deck had a greater diameter. Cargo was loaded through large clamshell doors in the belly using a built-in ramp and hoist.

To save production and maintenance costs, the prototype XC-97s shared wing design, engines, landing gear, and other parts with the B-29. In January 1944, an XC-97 set a transport speed record by flying 383 mph between Seattle and Washington, D.C., with 20,000 pounds of cargo.

The next variant of the cargo plane, the YC-97, first flew in 1947, two years after the war ended. By that time the B-50 bomber had been launched, and the YC-97 incorporated elements of the B-50 rather than those used in the earlier B-29.

The Army ordered 10 YC-97s as service test models. Similar in appearance to the XC-97, the YC-97 had a taller vertical tail, greater fuel capacity, and improved engines, nacelles, and electrical systems. It used new aluminum alloys in its construction.

A production version, the first of 50 C-97As ordered by the U.S. Military Transport Service, flew in June 1949. Its range was increased to 4,683 miles by using outboard wing tanks. Other improvements included a chin-mounted radar and new Hamilton hydromatic propellers.

During the Korean War, the interiors of 14 C-97As were modified for the evacuation of casualties, and the airplanes were redesignated as C-97Cs. When the cargo doors of three C-97A airframes were replaced with smaller passenger-entry type doors and the interiors completed with office and living quarters, the Stratofreighters became flying command posts (VC-97D) for Strategic Air Command personnel.

KC-97 aerial tanker: The tanker version of the C-97 was the prime refueling aircraft of its day and set the stage for future development of the KC-135 jet tanker.

Because early jet bombers were limited in range, the Strategic Air Command needed an in-flight refueling system that would give them the ability to perform intercontinental missions. However, modified B-29s, designated KB-29Ps, were not able to match the minimum speed of the B-47s and had a hard time catching up with the more powerful B-50 propeller-powered bombers, when loaded with fuel and slowed by the drag of the flying boom.

Boeing installed its flying boom on three C-97A Stratofreighters, redesignated KC-97As. After proving that the system worked, and that the aircraft were fast enough, the company built KC-97Es specifically as flying-boom tankers.

SAC accepted its first KC-97E on July 14, 1951. The KC-97E's internal tanks held an additional 7,200 gallons of fuel, and it was able to match the minimum speed of the new jets. Sixty KC-97Es were delivered, followed by 159 more powerful KC-97Fs. The last flying-boom equipped KC-97s were G models, with additional 700-gallon B-50-type external fuel tanks under each wing. The versatile G model had space for cargo and troops as well as fuel. It could dispense 8,513 gallons of fuel while carrying 96 troops or heavy equipment.

Until it was replaced by the KC-135, the KC-97 was integral to SAC operations. It was an important link between the B-29 tanker conversions and the next generation KC-135s, and performed admirably during its time. The KC-97s served until 1977, when the last one left the Air National Guard inventory.

In all, Boeing produced 888 C/KC-97s between 1947 and 1958.

KC-135 Stratotanker: The KC-135 tanker/transport was the first offspring of the Dash 80. It was the only jet airplane designed specifically for aerial refueling, and more than 550 of the 732 tankers built—all by Boeing at its Renton, Washington, plant—remain in service.

The KC-135's high speed and high transfer rate of 1,000 gallons per minute, using the Boeing-developed flying boom, made aerial refueling faster and easier than with the venerable KC-97.

The torch was passed on July 18, 1956, when the first KC-135 jet tanker rolled out right behind the last KC-97. KC-135s were the Air Force's only tanker for the next 15 years and are still the major part of the tanker fleet, although the last was delivered in 1965. The company also delivered 12 KC-135 tankers to the French Air Force.

The KC-135s aerial refueling equipment is in the lower deck, leaving the upper deck clear for cargo or troops. The tanker can accommodate 80 passengers or 25 tons of cargo, or any combination of both. In later versions, a special hose and drogue could be attached to the boom to refuel probe-equipped aircraft.

The KC-135s design was extremely adaptable, allowing Boeing to build 88 C-135 aircraft for other Air Force uses. Seventeen KC-135Bs, powered with turbofans, served as SAC Airborne Command Posts. Five became VC-135Bs equipped as transports for high-ranking government officials.

The cargo-passenger version, the C-135, was capable of carrying up to 126 troops, and, for increased airlift capacity, did not include a refueling boom. Boeing built 15 C-135As and 30 C-135Bs, the latter having turbofan engines.

Four RC-135As carried camera equipment instead of the refueling system for aerial reconnaissance. Later reconnaissance versions used turbofans and updated electronic equipment. In addition, airplanes were converted for so many special purposes that type came to be known as the "plane of a thousand faces."

The final special-purpose version of the tanker-transport was delivered during late 1966. Three years later, responsibility for the KC-135 was transferred to Boeing in Wichita, where modification and upgrading of the tankers continues.

KC-135s first saw military action in June 1964 in Southeast Asia. Over a nine-year period, KC-135s performed more than 813,000 aerial refuelings of combat aircraft and transferred nearly 9 billion pounds of fuel.

Modification programs at Boeing have contributed to the life and performance of the airplane. A wing re-skin program, started in 1976 and completed in 1988, extended the service capability of the KC-135 well into the next century.

Because original J-57 engines in the KC-135s were of 1950s technology, they did not meet modern standards for increased fuel efficiency, reduced pollution, and reduced noise levels. Therefore, starting in 1979, Boeing began installing new CFM56 engines on many of the KC-135As, giving rise to the new designation KC-135R—over 350 have been converted, with 400 currently under contract. In addition, as a means of more rapidly improving Air Guard KC-135A operational capabilities, over 150 have been converted to "KC-135Es" by re-engining with JT3D engines taken from retiring commercial 707s.

The KC-135s remain the Air Force's prime aerial refueling airplanes. During the Persian Gulf War, the tankers engaged in some 18,700 separate hookups for the transfer of 278 million pounds of fuel.

Future Boeing Airlift/Tanker Aircraft

In 1993, Boeing began providing information to the Air Force, at the service's request, about the company's 747-400 and 767-300 commercial freighters, and how they could meet a predicted shortfall in the country's airlift capability. As much as 70 percent of the military's airlift requirements are not military-unique. This includes transporting bulk cargo on pallets and oversize cargo such as trucks and towed howitzers between main operating bases around the world.

While Boeing commercial derivative airlifters cannot do the military-unique missions envisioned for the C-17—such as operations on unimproved airstrips, roll-on/roll-off cargo and airdrop—they can do the part of the general airlift job now done by aging C-141s. Boeing 747 and 767 freighters use contemporary technology, meet or exceed noise and pollution requirements, are fuel efficient and operate with two-person crews.

A 767-derivative tanker, with an advanced refueling system, is being developed as a possible future replacement for the U.S. Air Force KC-135 and for international customers.

Ever since the introduction of the 707 Freighter 30 years ago, Boeing has been the world leader in civilian air cargo. And Boeing airplanes have been a major factor in defense as well, with decades of U.S. Air Force airlift service, including bulk and oversized cargo delivery during Desert Storm and Operation Restore Hope. The 747 Freighter provides 42 percent of the world's freighter fleet capability. The freighters are fuel efficient, operate with two-person crews, use new technology and meet or exceed noise and pollution requirements. In 1993, Boeing began providing information to the U.S. Air Force about the company's freighters and how they could meet the service's general airlift requirements. It appears that as much as 70 percent of the military's airlift needs fall in this general category, which includes transporting bulk cargo on pallets and oversized cargo such as trucks and towed howitzers between main operating bases around the world. Boeing commercial airplanes were used in Desert Storm and have proven their capability to meet these kinds of airlift demands. The 747-400 Freighter, shown here in U.S. Air Force livery, provides maximum cargo handling flexibility with its large side and nose doors, and an automated handling system.

GE Aircraft Engines

The year 1876 marked America's centennial. While many reflected quietly on a century of notable industrial and scientific achievement, a small, select group was ambitiously seeking new targets for invention. Increasingly, electricity became the common denominator for those efforts.

One of these pioneers, Thomas Edison, moved to a laboratory in Menlo Park, New Jersey, where he could more effectively explore and develop many of his new inventions. The merger of Edison General Electric Company and the Thomson-Houston Electric Company approximately 15 years later marked the official beginning of the General Electric Company's historical lineage.

By 1939, GE propulsion innovations were in wide use, and helped a Boeing B-17 establish the cross-country flight record of nine hours, 14 minutes.

Two years later, GE was awarded a government contract to produce America's first jet engine - which powered the United States' first jet aircraft (the Bell XP-59A) in October 1942.

Lockheed C-5A.

GE Aircraft Engines traces its tanker/transport roots to the 1960s, when J47 engines were introduced on KC-97L tankers. Before their introduction, KC-97Ls were unable to climb above high cloud ceilings or keep pace with a faster generation of "Century Series" fighters (F100s, F104s, etc.). J47s helped improve time-to-climb ratios by a significant factor and added more speed to the mission, providing straight, level flight and eliminating the slowdowns and aircraft control problems that plagued earlier refueling efforts.

In 1965 GE was awarded the contract to develop TF39 engines for the USAF's C-5A "Galaxy," ushering in the day of the supersized transport and marking the introduction of the high-bypass turbofan. Prior to the TF39, bypass ratios of turbofan engines were less than two-to-one. The TF39 boosted those ratios to eight-to-one, resulting in a 25% fuel consumption edge compared to other engines available in the mid-'60s. Rated at 41,000 lbs, thrust, the TF39s also enabled the C-5 to carry up to 132 tons of oversized cargo - and materials that could never before be airlifted.

Deploying forces rapidly and providing timely material support to friendly nations have always been essential factors in U.S. policy. The F103 engine capitalizes on more than 75-million CF6 commercial flying hours in supporting these challenging objectives, and in adding global mobility to today's Air Force. During Operation Desert Storm, the F103-powered KC-10s to near perfection, providing rapid deployment of tactical aircraft, support equipment and personnel.

Boeing KC-135R with B-2 Bomber.

The USAF has also chosen a version of the CFM56 (known as the F108) to re-engine its existing Boeing KC-135 tankers. These engines offer added range, increased fuel offload, low noise and emission levels to refueling missions. Entering its second decade of operation, the CFM56 is the most reliable engine in the USAF inventory.

T64 engines, continuously improved as a result of eight-million flight hours and the application of modern technologies, new materials and improved manufacturing techniques, have been selected to power the USAF's C-27 utility transport. Designed to meet the requirements of rugged transport for short to medium ranges as an intra-theater airlifter, this twin-engine aircraft is particularly well suited for tactical operations into semi-prepared airfields of less than 1,800 feet.

Above: McDonnel Douglas KC-10.

Below: Alenia/Chrysler Technologies C-27.

CAE-Link Corporation

The first Link flying school.

C-130 Aircrew training system part-task trainer - Little Rock Air Force Base, AZ.

Establishment

The Link name has been synonymous with simulation since Edwin A. Link patented the first ground-based pilot training device in 1929. The "Pilotmaker" consisted of an enclosed cockpit with typical aircraft instruments; the cockpit was mounted on pneumatic bellows to provide motion effects and instrument readouts.

The aviation trainer business took off in 1934 when Link piloted his one plane from Binghamton, New York, through foul weather to a meeting with U.S. Army Air Corps officers in Philadelphia. The Air Corps was experiencing a high aircraft accident rate with inexperienced pilots flying mail deliveries through inclement weather. Inventor Link's ability to navigate despite the day's low clouds and heavy fog - a skill self-taught in his trainer - so impressed the officers that the Air Corps ordered six instrument flight trainers.

With the onset of World War II, the demand for military pilot training became critical and sales of the Link "Blue Boxes," as the trainers commonly became known, increased dramatically. More than half a million Allied airmen trained in 10,000 Blue Boxes before entering the air war over Europe and the Pacific.

The continued requirement for flight training in the 50s, 60s, and 70s swelled the roster of Link-developed simulator systems to read like an aerospace history. Training systems were designed and produced for almost every type of aircraft and application - primary trainers, jet fighters, bombers, transports, tankers, strategic reconnaissance, helicopters, commercial jets, corporate aircraft, and even manned spacecraft.

Development

Training systems technology was also extended to numerous non-flight areas including naval ship simulators, tank driver and gunnery trainers, command-and-control trainers, power plant simulators, and railroad simulators.

The hallmark of Link's philosophy has been continuous technological innovation. A sampling of the company's numerous milestones:

1929	Ed Link patents first instrument flight trainer "The Blue Box"
	Link Flying School Established
1930	Amelia Earhart Trains at Link
1934	First Significant Sale, U.S. Army Air Corps
1941	35 countries using Link trainers: 10,000 "Blue Boxes" used to train 500,000 airmen during WWII
1943	First radar simulator
1950	First electronic trainers
	First jet trainer
1955	First realistic cockpit motion system
1960	Navigation trainer used by Mercury astronaut Alan Shephard
1963	Apollo Mission Simulator
	First digital flight simulator, B-727
1964	SR-71 'Blackbird' simulator delivered
1968	Link acquired by The Singer Company
1969	Locomotive simulator
	Nuclear power plan simulator
1970	U.S. Army Aviation Synthetic Flight Training System
1973	Digital Radar Landmass Simulator
1976	Shuttle Mission Simulator
1977	F-16 Tactical Flight Simulator
1981	First microprocessor-based simulator
1985	First fighting vehicle driving simulator
	First helicopter combat mission simulator, AH-64 Apache
1986	First eye-slaved, area-of-interest, full field-of-view, motion compatible dome visual display, ESPRIT F-117A stealth fighter/bomber simulator
1987	Largest contract-managed aircrew training system, C-130 ATS
1988	Link acquired by CAE Inc.
1989	NASA Training Systems Contract
1991	Project Desert STAARS force level simulation
	RAH-66 Comanche training contract
1992	Virtual Controller air traffic control simulator TeleMedicine, Medical College of Georgia
1993	B-2 aircrew training devices delivered Virtual Anesthesiology training simulation system
1994	MH-60 and MH-47 Aircrew Training Devices delivered

CAE-Link's focus is *Advancing Human Performance* by applying innovative technologies to satisfy the training and productivity needs of military, government, and commercial customers worldwide.

Today/Future

CAE-Link is a member of the largest military aircrew training-focused organization in the world, the *Aerospace & Electronics Group* of CAE, Inc. Under one company umbrella, CAE-Link provides the total spectrum of training systems analysis, design, development, integration, and management.

Flight Simulators have steadily grown more complex. Today their realism is no less than stunning. CAE-Link has successfully developed/operated many of the most sophisticated Air Force training systems ever fielded, among these - the C-130/HC-130 Aircrew Training System, the KC-10 Aircrew Training Program, the F-117A Pilot Training System, the E-3 (AWACS)/E-6 (TACAMO) Flight Crew Training System, as well as the B-2, F-4, F-15, F-16, and F-111. CAE Electronics has also provided the C-5 (Galaxy) WST for the C-5 ATS.

CAE-Link is proud of its heritage and service to the airlift/tanker community, and feel an important part of not only the past but also of its future mission.

HFSI

HFSI Corporate Headquarters in McLean, VA.

HFSI's roots in the computer industry stretch back 35 years, when it was part of the Honeywell corporation and was introducing automated data processing to federal customers for the first time. Over the years, HFSI's installed base of computer solutions spread across the federal community to include sites in the United States, Europe, and the Far East.

The company's long association with the Military Airlift Command (MAC) began during the 60s, when HFSI first supported data processing requirements of the U.S. Air Force major command headquarters. In the early 70s, with the Worldwide Military Command and Control Systems (WWMCCS) computer systems modernization contract, HFSI served not only MAC, but 30 other commands within the Department of Defense. During the 80s, HFSI's solution for the Consolidated Aerial Port Subsystem (CAPS) brought minicomputers and local information processing to MAC sites around the world. Later, through the Deployment Flow Computer Systems, HFSI provided MAC advanced systems to support crisis/exercise airlift planning, peacetime airlift development, and airlift simulation modeling.

In the 90s, as an independent corporation and leading systems integrator to the federal government, HFSI is applying its strengths in integration, networking, and trusted systems to help customers move to open, distributed environments while protecting the security of missions, as well as their investment in existing systems. A pioneer in trusted systems, HFSI provided MAC the first secure high speed electronic connection between a top-secret WWMCCS host and secret level military Crisis Action Teams. This multi-level security system ushered in a new era of trusted computing for the DoD command and control environment.

Looking ahead, HFSI hopes to continue serving the Air Mobility Command (formerly MAC) and Air Force community, helping them leverage current information technology while working toward the integrated enterprise of the future.

Lockheed Martin Aeronautical Systems

Advanced Flight Station - The heart of the C-130J's advanced technology is its modern flight station with Liquid Crystal Display (LCD) instrumentation readouts for aircraft flight control, operating systems and navigation. In addition to its two head-up displays, the "J" has four multifunctional head-down LCD displays. The displays are Night Vision Imaging System compatible, enabling the crew to operate the transport into and out of areas of total darkness with special night vision devices, where special operations dictate.

The 2,000th Hercules transport built by Lockheed Aeronautical Systems Company (LASC) comes in for a landing at Dobbins Air Force Base, Georgia, following an evaluation flight by the U.S. Air Force. The C-130H military transport sports several new factory installed systems, including a self contained navigation system and an electronic countermeasures defensive system. It was delivered May 15, 1992, to the Kentucky Air National Guard's 123rd Airlift Wing at Standiford Field in Louisville.

The aviation roots of Lockheed Martin Aeronautical Systems began over San Francisco Bay in 1913 when two brothers, Alan and Malcolm Loughhead (pronounced Lockheed) successfully flew their first hand-built three passenger seaplane. Three years later the two founded a company in Santa Barbara, CA to manufacture a twin engine 10 passenger flying boat, two seaplanes for the Navy and a small biplane.

Alan Loughhead helped form the Lockheed Aircraft Company in 1926 and set up operations in Hollywood. Thriving airplane sales encouraged the company to expand and move to Burbank, CA. But in 1929 the stock market crashed and the company was forced into receivership. At the time the company's assets were valued at $129,961. A group of young men, headed by Robert E. Gross, bought the company assets for $40,000.

The first model 10 Electra took off in 1934, and so did the company. For the next 15 years civilian and military aircraft poured off the assembly line with more than 20,000 Lockheed planes built for World War II.

In the early 50s, Lockheed was contracted to operate Air Force Plant #6, Marietta, GA, which Bell Aircraft had operated during the war to build B-29 bombers. Following refurbishment of the facility, then called the Lockheed-Georgia Company, its initial contract with the government called for modifying B-29s for use in the Korean War. The company also modified and built B-47 bombers. Between 1952 and 1957, 120 B-29s and 394 B-47s were completed.

A most significant accomplishment of the corporation was the winning of the competition to build a new medium-sized logistic and tactical military airlifter, capable of operation from short, unprepared airstrips. Following a successful prototype program in California, the production program was moved to the Marietta facility. The new transport was designated the C-130 Hercules and by the end of 1957, 150 of the new transports were in the U.S. Air Force inventory.

The JetStar production program, like the C-130, was assigned to the Georgia plant after a successful prototype program in Burbank. It was the world's first executive jet.

In 1961 the company won a contract to develop and build the C-141 high speed jet cargo and troop carrier for the Air Force, and in 1965 the Air Force selected the company to build what was at the time the world's largest airplane, the C-5A Galaxy. Twenty years after the C-5A contract award, the first of 50 new C-5Bs was delivered.

Other major work for the U.S. Government included the C-141B stretch mod program that added a third more cube cargo capacity and in-flight refueling, and the C-5A wing modification program. During the 70s and 80s, Lockheed's Marietta, GA facility was known as the "Airlift Capital of the World."

The popular C-130, on-going during the C-141A, C-141B, C-5A, C-5B and JetStar I and II programs, enjoys the longest uninterrupted production run of any other U.S. military transport. Over the past 40 years, Lockheed Martin has delivered more than 2,100 Hercules. They are operated by 64 countries and exist in more than 70 versions.

In 1992, Lockheed Martin Aeronautical Systems began developing the C-130J, the most advanced, high technology derivative in the Hercules family of airlifters. With roll-out and first flight in the third quarter of 1995, deliveries will start in late 1996. The C-130J features a two pilot flight station and state-of-the-art digital avionics. Incorporating modern fighter technology, the C-130J has mission computers for more accurate airdrops and is more economical to operate and maintain than its predecessors. Currently Aeronautical Systems is under contract to build 25 C-130Js for the UK Royal Air Force.

In addition to cargo aircraft, Lockheed Martin Aeronautical Systems designs and builds antisubmarine warfare/maritime patrol and fighter aircraft. Engineering and manufacturing operations are active for development of the F-22, latest generation air superiority fighter. The F-22 will replace F-15s currently used by the U.S. Air Force.

Lockheed Martin Aeronautical Systems is a unit of the Lockheed Martin Aeronautics Sector, Lockheed Martin Corporation. Headquartered in Bethesda, Maryland, the Sector employs about 38,000 people in six operating units nationwide. The sector was established on March 15, 1995, with the merger of Lockheed Corporation and Martin Marietta Corporation. The Aeronautics Sector is headed by James A. (Micky) Blackwell, President and Chief Operating Officer.

Above: *The first C-103J for the U.S. Air Force is getting its outer left wing installed on the final assembly line at Lockheed Martin Aeronautical Systems in Marietta, GA. It is the second of five C-130Js in final assembly to get its wings installed.* ***Below:*** *The C-130J, scheduled for roll-out at LASC in Marietta, GA in October 1995, is the most dramatically improved version of the popular Hercules since Lockheed began its production in 1954. The new "J" model; will be equipped with an advanced technology flight station, more powerful engines and propellers and mission computers to operate its advanced avionics.*

McDonnell Douglas Corporation

C-17 Globemaster III.

The history of the airlift/tanker industry is inextricably intertwined with the history of the McDonnell Douglas Corporation.

Douglas Aircraft company - which joined to form McDonnell Douglas in 1967—designed, developed and produced the industry's first military transports and the first tanker aircraft. Today, the company is responsible for the newest aircraft in the nation's airlift/tanker inventory — the C-17 Globemaster III and the KC-10 Extender.

The company's heritage began in 1925 with delivery of the C-1, the first U.S. aircraft designed as a logistic transport, to the Army Air Service. This versatile bi-wing carried spare aircraft engines to users and was also used for aeromedical evacuation, as an in-flight refueling tanker, and to transport personnel.

Then followed the C-33 and the C-39, military derivatives of the DC-2. Next came the famed C-47 Skytrain — the "workhorse" transport of World War II — flying supplies from India into China over the "Hump" and dropping paratroopers and supplies in the Allied invasion of Normandy in 1944. Many venerable Douglas "Gooney Birds" are still flying today, some 50 years later.

After the war, the four-engine C-54 Skymaster performed yeoman service in helping circumvent the Russian land blockade of West Berlin. Ninety percent of U.S. aircraft airlifting more than 2.3 million tons of needed supplies in the Berlin Airlift were built by Douglas.

The first of the "Globemaster" transports — the C-74 — became the first airplane carrying more than 100 persons to fly the Atlantic nonstop. It was followed by the C-124 Globemaster II, which proved the viability of large military transports. Lovingly called "Old Shakey," it was the mainstay of Korean War airlift. Some 448 C-124s were delivered to the Air Force from 1949-1955.

The C-133 Cargomaster, a four-engine turboprop, was developed as the Air Force's largest long-range heavy transport. The Cargomaster set a world cargo airlift record of 117, 900 pounds in 1959. The twinjet C-9A Nightingale, aeromedical airlift version of Douglas' popular DC-9 commercial jet, is used as a flying hospital ward and another version is used by the Navy and Marine Corps in support missions.

The YC-15 AMST (advanced medium-sized short takeoff and landing) transport flew in 1975 and was the pint-sized forerunner of the new C-17 Globemaster III. Only two YC-15 prototypes were built, but the technology learned from this vehicle's design and flight testing was transferred to the C-17, now in production by McDonnell Douglas.

The C-17 is the versatile new transport designed to be the U.S. Air Force's "core airlifter of the future." It can carry outsized equipment, such as the Army's M-1 tank, supplies and troops directly to forward austere airfields. Combining the capabilities of three transport currently in the airlift fleet, the C-17 can also airdrop paratroopers and supplies.

In the tanker arena, that same C-1 that pioneered carrying military cargo also accomplished the world's initial operational inflight refueling of an Army Air Corps tri-motor during a six-day, 11,000-mile flight. It was an ancestor of the McDonnell Douglas KC-10 Extender cargo/tanker, currently a critical element of the Air Force's aerial refueling fleet.

Designed to refuel tactical fighters, bombers, and strategic airlifters — as well as carry maintenance crew and cargo — KC-10s support Air Force, Navy and NATO missions around the globe. Between 1981 and 1990, the USAF took delivery of 60 KC-10s.

With continued production of the C-17 envisioned well into the next century and growing demand for KC-10 type aircraft, the company has solidly positioned itself as a leader in the future of the airlift/tanker industry.

Above: C-124 Globemaster II. First flight: Nov. 1949.

Below: C-17 Globemaster III. First Flight: Sept. 1991

Southern Air Transport

Southern Air Transport is a U.S. based charter airline with over 500 employees stationed around the globe. They operate the world's largest commercial fleet of L-100 Hercules (18), four DC8-71/73s and six 747-100/200s. The Company's aircraft are available for wet-lease, long term contract programs and adhoc charters to other airlines, integrated carriers, freight forwarders and shippers worldwide.

Southern Air Transport began service in 1947 in Miami, Florida, with a single C-46 cargo aircraft. Since then, Southern Air has owned and operated virtually every aircraft from DC3s to large jet transports, moving all types of cargo throughout the world. In nearly 50 years of continuous operation, Southern Air has amassed a formidable record of air cargo achievements and is now recognized as a leader in the industry.

B 747 - 200.

L-100 Hercules.

DC-8.

Airlift/Tanker Association

Board of Officers

Chairman	General Alfred G. Hansen, USAF, (Ret.)
Chairman Emeritus	General Robert E. (Dutch) Huyser, USAF, (Ret.)
President	Mr. Joel Storch
Senior Vice-President	Colonel George E. Dockery, USAF, (Ret.)
Vice-President Industry Affairs	Mr. Charles (Chuck) Kline Jr.
Vice-President Programs	Colonel Dennis L. Murphy, USAF, (Ret.)
Secretary	Mr. Lawrence B. Ryan
Treasurer	Colonel Thomas P. (TP) Williams, USAF, (Ret.)

The Early Years, 1969-1979

In 1969 former members of the 834th Air Division in Vietnam, held their first reunion in Las Vegas. Their primary purpose was to maintain the brotherhood developed during combat in Vietnam. They decided to get together on an annual basis and to increase attendance as well as conduct formal and educational programs at each reunion under auspices of "ad hoc' volunteers with no designated board of directors.

During the latter part of this period, the Air Force merged the strategic and tactical airlift mission, and reassigned Tactical Airlift Command C-130 assets to Military Airlift Command. This had the effect of increased membership in the Airlift Association and increased association support from the aerospace industry.

At the 1975 reunion in New Orleans, attending members voted to incorporate the association as a nonprofit entity to promote the enhancement of aviation. Among the stated goals of the association was the sponsoring of seminars and symposiums on the advancement of airlift. Thus was formally formed the Airlift Association, Inc. with its management under an elected board of officers.

In 1978 based on guidance from the chairman of the board, Lieutenant General Bob Coverdale, USAF (RET), the association decided to publish a newsletter and hold annual election of officers at the reunions. In 1979 elections were held and by-laws were approved in Nashville, and a revitalization was effected.

During those early years, a great deal of credit goes to the pioneers and visionaries of the Airlift Association. They include General Bill Moore, Lieutenant General Bob Coverdale, Lieutenant General "Moose" Hardin, Major General Jim Chatfield, Jim Downing, Colonel Bill Morley, Major General Ralph Saunders, Mr. Bob Eisenhart, Chief Master Sergeant Dave Pelletier and others. They helped to form and nurture an organization of 157 members as of September 1979.

Years of Growth 1980-1990

During this period the Airlift Association continued to grow as it achieved its objective of providing a forum for people interested in improving airlift capability. During this decade the fledgling association newsletter evolved into the *Airlift Quarterly*, a 24 page plus magazine that has been recognized as one of the best among all associations. Corporate membership continued to grow as industry was attached to this growing organization. The Association sponsored an Airlift Hall of Fame program that selects outstanding airlift leaders from the past and present whose bronze busts are installed in the Hall of Fame at Scott Air Force Base, Illinois. The association also developed a scholarship program awarding $1,000 to needy, high potential students involved in Air Force ROTC college programs. Additional programs included recognition at the annual convention of outstanding young leaders with the Air Military Airline Commands as well as outstanding MAC air crew members.

Credit for much of the progress and growth of the Airlift Association during these years can be attributed to the leadership, vision and wisdom of General "Dutch" Huyser who served as the Chairman of the Board of Officers for five years. He was later induced into the Hall of fame in 1995.

1991-1995

The Present Years - A New Direction

In 1991 the Air Force went through a major reorganization. This resulted in both a name change and new mission direction. The Air Mobility Command replaced the Military Airlift Command and with the expanded mission of Air Mobility and the Tanker forces were added to the Command. Also Airlift forces were assigned direct to Theater Commanders and the C-130s returned to the Tactical Air Command now called Air Combat Command. This moved the Airlift forces away from AMC and presented a major decision to the Airlift association.

During the 1992 National Convention Member Business Meeting, General Al Hansen who replaced General Huyser as chairman presented a proposal to the membership to change the name of the Airlift Association to the Airlift Tanker Association. This change would align the association to the new mission of the Air Mobility Command by including the Tanker forces. In a near unanimous vote the membership approved the proposal. At this now historic meeting General Hansen also proposed turning the direction of the convention to include professional education seminars. Over the next three years mainly due to the dedication of Colonel Dennis Murphy and Colonel Dave Patterson, these seminars have achieved an outstanding level of professionalism and are recognized as formal education for active duty personnel. As a result, the membership has grown and now includes many young officers and enlisted personnel - a long term objective of the Association.

Today the membership has grown to over 3600 members and 40 corporate members. Recognized as a premier professional organization the association meets the needs of mobility personnel throughout the world. It provides a forum for retired and active duty people, aerospace industry people and others who are concerned about maintaining a strong national mobility force.

At the 1992 Association Convention, held in Atlanta, Georgia, General Alfred G. Hanson, Chairman of the Board of Officers of the newly named Airlift/Tanker Association and General Ronald R. Fogleman, then USCINCTRANS and Commander AMC, signed a new charter re-establishing the Airlift Hall of Fame as the "Airlift/Tanker Hall of Fame."

Chapters

Adirondack Chapter, Plattsburgh AFB, NY
Airlifter Country Chapter, Dobbins AFB, GA
Alamo Chapter, Kelly AFB, TX
Big Country Chapter, Dyess AFB, TX
Capital Chapter, Washington DC
Denali Chapter, Elmendorf AFB, AK
Diamond Head Chapter, Hickman AFB, HI
Eagle Chapter, Dover AFB, DE
Golden Bear Chapter, Travis AFB, CA
Huyser Chapter, Scott AFB, IL
Keeper of the Plains, McConnell AFB, KS
Low Country Chapter, Charleston AFB, SC
Luftbrucke Chapter, Rhein Main AB, Germany
Tommy B. McGuire Chapter, McGuire AFB, NJ
March Field Chapter, March AFB, CA
Pacific Northwest Chapter, McChord AFB, WA
Razorback Chapter, Little Rock AFB, AR
Red River Chapter, Altus AFB, OK
See-Seventeen Chapter, Long Beach, CA
Shogun Chapter, Yokota AB, Japan
Tarheel Chapter, Pope AFB, NC
Team Robins Chapter, Robins AFB, GA
Tony Jannus Chapter, Tampa, FL
Warriors of the North Chapter, Grandforks AFB, ND
William H. Tunner Chapter, South Vienna, OH
Wright Chapter, Wright-Patterson AFB, OH
Yankee Clipper Chapter, Hanscom AFB, MA
Young-Tiger Chapter, Malstrom AFB, MT

Col (retired) Gail Havorsen "The Candy Man" at the Berlin Airlift.

Special Stories

Operation Desert Shield-Desert Storm

The Vernon J. Kondra Notes

Transcript prepared by Clayton H. Snedeker
21st Air Force Historian

Closing Comments:

Why "Doing It All Again Tomorrow" Would Be Difficult, Due To The Reduced Availability Of Airframes, Downsizing Of The Force.

Comment/Question: Not, not really...that's all I've got. The war is over, we won, the troops are home, the parade is on...and you're tired. But let's tie this all up with some of your own personal comments, as a sort of a "wrap" here. You said, at one point in one of these interviews, that you didn't think this could be done again. Why do you say that? Suppose tomorrow we all of a sudden had to jump into another contingency?

LtGen Kondra: If we had to do it again tomorrow? Well, there are a number of reasons why I say I don't think it could be done again. The 141s, in the next couple of years, are going to start dropping out of the inventory, so you're going to have less airplanes available - even with the C-17s coming on board. Eventually, you're going to end up losing 234 PAA C-141s, and you're going to replace them with 120 PAA C-17s. Yes, the C-17 will haul more than the 141, but it isn't going to haul that much more...so you're going to lose containers. The makeup of the commercial carriers is changing. FedEx is getting rid of some of their 747s. UPS will get rid of some of their 747s. They're going to smaller airplanes, because the 747 doesn't meet their requirements for Stateside deliveries. They don't need those bigger airplanes, so you're not going to have those 747s available. Pan Am has gone out of business. So the airplanes available, to do the things we did, with the types of airplanes we used...are going to be fewer in number. Not to mention the fact that the airlift force itself, while basically staying the same, is downsizing. So I think it would be very difficult to do it all over again.

Comment: Well, I think all of this is "good stuff," and I hope you agree that it hasn't been a waste of time.

Communicating with People: A Problem And A Leadership Essential

LtGen Kondra: Oh, no, not at all. But if I had to go back and summarize what sticks out in my mind about this whole thing, I think the number one thing that comes out is how difficult it is, and how poor a job we did, to communicate to the aircrews and to the people of MAC what they were doing, and why. We did not do that very well. It goes without saying; if people understand why they're doing something, they'll go to hell and back. But a lot of the complaints, gripes and concerns were because people didn't understand. I mean, when I went out there in December...out flying around Christmastime...and we've got people who still don't understand why they're out there, "we thought we were going to surge," they didn't know they've already surged. They didn't know that we were running at the peak; they don't see it. It's painfully obvious, when you're looking at if from a macro view, but when you're down amongst it and caught in the quagmire you don't see that. All you see it is getting jerked around.

Question: How would you propose resolving that?

LtGen Kondra: I don't know how you do that. I've said that many times.

Comment: Unless you could have some kind of a "town meeting," with everybody in attendance...

C-141.

LtGen Kondra: But the problem is, your "town meeting" is 80,000 people spread all over the world, and it's very difficult to get the word to everybody. But I think that's one of the things that we -I- did not do very well. As hard as we tried, we didn't do it very well. And I don't know how to solve it for the next time...because it "ain't going to go away."

Comment: How about...and I'm just throwing out a thought that came to me, now - how about CNN? Everybody watched CNN...it was covering the war, "live." Wouldn't it be neat if MAC - or AMC, as it is going to become - could get themselves either hooked into CNN, or have its own broadcasting method, where you could say, "OK, crewmembers, here it is." I don't mean just a "rah rah" pitch from the CINC...

LtGen Kondra: Oh, I think it's necessary, to be from the CINC on down. It doesn't have to be "rah rah," it has to be informative.

Comment: Well, I mean...take advantage of the medium of television.

LtGen Kondra: The problem is, of course, that you don't have the infrastructure to do that. CNN certainly isn't going to do it for us. And the problem is, people don't want to watch videotape...but when you go over and visit with them and make yourself available, I'll guarantee you they'll sit down and listen. As I did on that trip at Christmastime...in the lobby of the billeting office at Torrejon...I bet I sat out there for a couple of hours as hundreds of crewmembers filed by. And we just had open sessions. I did the same thing at the 24-hour feeding facility at the Officer's Club at Torrejon, from about 4 in the morning until 6 in the morning; I did that two mornings in a row. But that is such a small number of people you're reaching...

Comment: And you're only one man...

LtGen Kondra: Yes...and I did the same thing at Rhein Main. But it's amazing how you could start off with a group of people who were really hostile, and as you sit and talk to them and explain to them what you're doing and why, and answer questions...you end up with people doing a 180. But I don't know how you'd do that on a mass-media basis.

Comment: I also have to point out here...and I don't mean to be "buttering you up," or anything...but you are a personable man who can get out there and talk man-to-man with anybody on the planet...

LtGen Kondra: Well, you have to do that...

Comment: I know you do, but not every man in your position, or was in your position, or will be in your position, might be able to do that as well. Of course, I think that it's part of being a General, if you want to know my feelings on it.

LtGen Kondra: I think it's a basic thing called "leadership."

Comment: Yes, it certainly is.

LtGen Kondra: But I don't know how you do that, Clay; I don't know how you get that word out to everybody. And that's something that the person who does this the next time needs to be very concerned about.

Comment: Well, as far as I'm concerned...when we finish up here, the product that comes out of this, should be "required reading" for the next AMC/DO...because many of his problems are going to be the same.

The Importance of Patience

LtGen Kondra (continuing): Well, the situations will be different. But there's another problem. If communication was a severe problem, patience is an absolute virtue that you must have when you start something of this magnitude. And it makes no difference whether it's of this magnitude...or if it's Patriot moves, the move of peacekeeping forces, it doesn't make any difference if it's Provide Hope to the Soviet Union. Patience is an absolute must. But patience is the last thing that decision makers, and the people who push these things, have - because they want it all to happen "now." And lack of patience wastes airlift.

The One Constant That Can Be Anticipated: Dynamic, Changing Requirements

LtGen Kondra (continuing): I don't remember that we've talked about it, we may have, earlier...but to summarize what I'm talking about: somebody saying they "need eight C-5s," and then getting them...and then not being able to have loads to put on them - is a waste of airlift. Wanting a 141 "every 30 minutes" and getting them, and not being able to mobilize and process loads, and having airplanes sit - is a waste of airlift. Wanting to move C-5s out of Germany to Tel Aviv with Patriot missiles on board faster than they can be accepted at the other end - is a waste of airlift. People have to understand that trying to push the maximum number of airplanes through an en route system and to an offload point that can't handle them is an inefficient way of doing business. You work much harder than you need to work, and you accomplish less, because you clog the system up and work it to death for no more appreciable gain. Now, there is a fine line between where do you push it at the maximum it can handle and where you step across that line and put too much into it, so there are things that you are not doing that you should be doing. That's a balancing act. I would say that in this (Desert Shield) we erred on the wrong side. We tried to push too much and we ended up clogging the system and it became inefficient. Nobody will stand for not doing everything you can do...and consequently you tend to overdo; but that's a fact of life, and it's not going to go away. I'll guarantee you that 20 years from now, 50 years from now, with all the command and control, all the computers, all the things you'll have, you are still going to have some of these problems. And whoever works this kind of an operation next time will be faced with these same issues. The other thing that will have to be dealt with is the lack of requirements and the changing of known requirements. In other words, those things like people, equipment, stuff, that you're gonna take from Point A to Point B, is going to be an unknown. It is going to be an evolving, changing, dynamic process that will never be identified, and I don't care how good your TPFDDLs are. You can have the TPFDDLs, you can have the plans, you can have it all there...but when you get ready to execute, the theater commander is going to adjust things as he sees fit. And once that happens, now it becomes a dynamic, evolving, changing thing. You'd better understand it's gonna happen, no matter how hard you try, and you're gonna have to deal with it, in my opinion. It absolutely will not go away. And for the next time...the potential exists for having non-experts pushing a system that they don't understand.

Comment: Yes...you've had lots of that happen to you - "Airlift 101," you've mentioned it a number of times.

LtGen Kondra: Yes, That doesn't mean that airlift can't get better - it will. But in my opinion there will be certain things that will have to be dealt with, and some of them are going to be costly.

Closing Comment: Thanks for all your time. It's certainly been interesting, and I've got to tell you...in all my history writing career, I think this is one of the best things I've ever had the privilege of working on, and I think we've done something really worthwhile here. I deeply appreciate your candidness and your taking me into your confidence; it's been a real pleasure working for and with you.

'Grandpa' is Still Playing War Games with the U.S. Air Force

submitted by James J. Pargoe

Norfolk City Councilman Claude J. Staylor dropped by the other day with a clipping from a Daytona Beach newspaper about an old friend of his, a man who is more or less ignoring the advances of time.

The man is James Pargoe, a Norfolk native now living in Daytona Beach. He and Staylor joined the Norfolk police department at the same time, back in the mid-1930s. Staylor worked his way up the ladder to become chief of police.

Pargoe was a good enough cop. But he was a better electrical engineer, so he stayed with the police department only a couple of years before taking a job at the Naval Shipyard in Portsmouth.

"The pay was better," he said in a telephone interview from his Florida home.

Pargoe's uncle, Linwood "Piggy" Pargoe, was chief of detectives in Norfolk for many years.

What is unusual about Jim Pargoe isn't that he was a policeman and an electrical engineer. The important thing is the fact that, at 77, he is still playing war games with the U.S. Air Force.

And, creeping toward his octogenarian years, Pargoe still flies his own plane, a single-engine, four-passenger Mooney, all over the country. Not a lot of 77-year-old men are doing that, either.

Pargoe is a retired Air Force major who unretired for a brief, shining moment back in April for a Strategic Air Command war games exercise in the Caribbean.

The younger pilots called him "Grandpa," but that didn't matter. Pargoe was flying and that's what he likes to do.

Joe Wisniewski, a commander assigned to Warner Robins Air Force Base in Georgia, is a good friend of Pargoe's. It was Wisniewski who suggested to the SAC people that Pargoe be allowed to fly in the exercise, which involved refueling F-16 jets in flight.

Pargoe learned to fly a plane years ago at Langley Field, known now as Langley Air Force Base. He wasn't about to refuse the SAC offer. So, at age 77, he flew his own plane to Georgia. He attended a briefing on a Monday afternoon, then took off the next day in a KC-135 tanker for the war games.

"A perfect flight,: he called it.

He didn't actually fly the big tanker. But he was there as an "observer." Refueling the jets while traveling 400 miles an hour through the air brought a glint to the old major's eyes.

"This is all new stuff to me," he told the Daytona newspaper.

But he has never been one to turn down a challenge, this chipper graduate of Maury High School and Virginia Military Institute. A year before the United States declared war on Japan after the bombing of Pearl Harbor, Pargoe already was doing his part to aid the allied cause.

Canada, already in the war because England was under attack by the Germans, desperately needed trained people to instruct pilots for the Royal Canadian Air Force. Pargoe was asked to lend a hand and he did.

By 1944, he was flying bombing missions for the U.S. Air Force in Asia.

Later, he would return to the U.S. and be assigned to the Special Air Mission Squadron in Washington. His job there was to serve as pilot for VIPs - including Gen. Mark Clark.

After the war, Pargoe never returned to Norfolk to live. H retired from the Air Force in 1951 and went to work for the U.S. Public Health Service in Washington. He retired from that job in 1969, and moved to Daytona Beach.

Flying airplanes, however, still runs in his blood. He has flown his private plane all over the country, visiting friends and relatives. He has a sister-in-law, Marguerite, in Norfolk. He flies to Pennsylvania to see his son and to California to visit his daughter.

"I've flown my plane across the country three times," he said in the telephone interview.

And, come September, he and his wife, Martha, will climb into the Mooney and fly out to California again.

Pargoe may be getting on in years, but in his heart he'll always be a flyboy.

LtCol James Pargoe at age 77, Oct. 1994. (Courtesy of James Pargoe.)

C-130 Crews Appear in John Wayne Film

submitted by Gerald J. Holland

Three McGuire C-130 crews co-star with John Wayne in his film, "The Green Berets."

While participating in a joint Airborne training mission at Lawson Army Air Field, Ft. Benning, Georgia, the crews, two from the 45th Military Airlift Squadron and one from the 29th, were asked to take part in the movie, which was being filmed at Ft. Benning.

"The Green Berets," an adaptation of Robin Moore's best-selling novel, depicts the adventures of the Army's Special Forces in Southeast Asia. It also stars David Janssen and Aldo Ray.

Batjac Productions chose the Georgia Army post because it is similar in appearance to the Southeast Asian area where the film's action takes place.

The McGuire crews appear in an off-loading sequence with the "Berets: de-planing the C-130s after a mission.

While most of the men and aircraft remain in the background to give the sequence authenticity, A1C Roderick Delia, a flight mechanic with the 438th Organizational Maintenance Squadron, is shown in a close-up scene with John Wayne.

"Very exciting, and interesting," was the comment Maj. Charles J. Smith Jr., mission commander, made about the group's movie-making experience.

Capt. Arthur E. Koshak, assistant mission commander, added, "We were impressed and surprised by the long working hours the production crew kept, and the hard work a movie takes."

John Wayne was found to be "extremely personable," according to Major Smith. "He's forceful, but sincere — just as he appears in his movies.

The Biggest Airlift in History

The Lockheed C-141B Starlifter looks rugged, durable, and just plain tough. Combat requires it to be. After flying one throughout the Middle East between August 1990 and April 1991, I'm convinced that it is highly survivable and the best strategic airlifter for the job.

But before I continue, let me state that I will not be revealing any sensitive material and that Operations Security (OPSEC) has been strictly followed in this article. In addition, now that the Gulf war is over I can discuss some other information I could not previously.

The first noticeable difference in these Desert Shield/Storm missions compared to missions during peacetime was how aircrews were notified. On Monday, August 6, everyone was put on Bravo alert, which means that when called we had one hour to be on base and two hours later had to be airborne. During that two hour period we did the usual for any mission - brief the crew, file a flight plan, compute our fuel requirements, check the weather, etc.

The first of literally thousands of McGuire Air Force Base missions began departing on August 7, and the biggest airlift in U.S. history didn't slow until March 1991. My first of many calls came at 10:30 p.m. on Saturday, August 11.

We always get an intelligence briefing prior to departing on any overseas mission, but briefings for Desert Shield/Storm missions were especially important. First Lieutenant Al Wright, the 438 MAW intelligence officer, would brief us on the latest situation, give us a thick classified crew brochure describing numerous Desert Shield/Storm procedures, and take questions. Prior to walking out to our 325,000 lb, four engine jet transport we picked up additional "secrets," flak jackets for all crew members (just like in the movies), survival vests, and mobility "D" bags which contain all our chemical warfare gear.

Home station departure is one of the toughest parts of any mission, and for these missions it was especially a big relief to get airborne. The ramp at McGuire was extremely busy twenty-four hours a day. Even to inexperienced personnel it was obvious that this activity was far from normal. Dozens of maintenance, aerial port, fleet service, and other support troops swarmed everywhere with their trucks, pickups, and specialized material handling equipment. The coordination for all this was immense, but the high departure reliability rate proved that the system worked well - although severely strained and not without occasional problems.

Once enroute, the workload didn't slow. Even though it was a fifteen hour trip to most forward operating bases (FOB) in Southwest Asia, broken up only once for a fuel and crew change stop in Europe, we had lots to do.

After the cruise check was done, I would begin reviewing the secret crew brochure. I had three pockets of information which covered communication procedures, chemical warfare, contingency guidelines, and many other topics. When I finished studying these I would then ensure that all my crew members read and understood them thoroughly.

I'm a firm believer in the five "P's" - prior preparation prevents poor performance, so I always placed a high priority on these reviews. When I read enough, I would turn the aircraft over to the first pilot and hop downstairs to met the troop commander.

The C-141 is the mainstay of the military airlift command. With over 250 in the Air Force inventory it was one of the two workhorses for Desert Shield/Storm; the Lockheed C-5B Galaxy was the other. The C-141 has a crew of between four and eight, depending on the cargo. On these missions we normally had three pilots, two flight engineers, and one loadmaster. We can carry 200 passengers, 70,000 lbs of cargo, or any combination. With our air-refueling capability our range is unlimited.

After a brief introduction, I welcomed him aboard and briefed the itinerary, forecast weather, and the threats. We invited his troops up to the flight deck and told them not to hesitate to ask for anything. Most were usually asleep. I didn't blame them. They had a long day.

By the way - one look at these troops and their equipment and I was very happy that they were on our side. During this airlift - the biggest in history - I carried a variety of cargo, including Patriot missiles, portable operating rooms, Maverick missiles, meals-ready-to-eat (MREs), medical supplies, Special Forces troops, and many other items of war. The entire operation has clearly been an enormous logistics undertaking, one for the record books.

The work didn't stop even four hours after departure. Although most Desert Shield/Storm missions were long days - almost 24 hours - we all managed to rotate some time in the three crew bunks. Fatigue was our biggest enemy until we got into the AOR (Area of Responsibility), so it was critical to get rest when we could. Compared to modern commercial aircraft, our 1960s vintage jet is not very comfortable, but we do have box lunches, coffee, and a few other comforts to make our long fights bearable.

Strict COMSEC (Communications Security) was critical. We never discussed classified information over the radios or interphone system, and were vague about our cargo when calling our command and control center on the high frequency radios. The Iraqis often monitored our radio transmissions.

The eight hour flight over the "pond," as MAC aircrews affectionately call the Atlantic Ocean, seemed to go faster than normal. At cruise altitude we double checked our navigation waypoints and continued to review our classified brochures. When things settled down, we would simply monitor the auto-pilot, flight instruments, and simply talk - especially about how unreal this situation seemed.

With the end of the cold war, who would have believed we would be involved in another conflict? It's incredible - like a well prepared mini-series similar to the recent PBS civil war documentary.

It seemed like yesterday I was a young kid hanging out on a street corner in New York drinking cheap beer and smoking Camel's. All I wanted to be was an Air Force pilot, not a warrior. and now, as an experienced C-141 instructor pilot with over 3,000 hours of flying time, I was in charge of an $80 million aircraft, lots of lives, in the middle of a war. It really seemed unbelievable.

Before we knew it, we were on final approach at our intermediate staging base in Europe. Here we got twelve hours of crew rest before departing on the "downrange" portion of our mission. The bases bursted with activity. Dozens of C-141Bs, C-5s, fighters, and other aircraft packed the ramps. The crew rest facilities were just as crowded - two hundred enlisted per hangar - cots were 18" apart - and three officers to a room was the standard. This setup made Motel Six look like the Hyatt Regency.

Considering this was the biggest airlift in U.S. history, I was impressed things went as well as they normally did.

Just like Operation Just Cause in Panama, everyone pulled together with a patriotic fervor that made hiding one's emotions as tough sometimes as dealing with the stress and fatigue.

The Desert Shield/Storm airlift was one constant, massive flow of aircraft. While aircrews were limited to a 24 hour day, the aircraft were not. For example, one aircrew would ferry a C-141 or C-5 from, let's say, a base in California to Europe, where another crew would take it to the Middle East and back in the same day. That way, cargo and troops made the 15 hour plus trip on the same plane but with different aircrews, allowing the airlift to roll around the clock.

This was demanding on aircrews, support personnel, and aircraft alike. But this was it, the Big One, the worst case scenario that we prepared for, for years, so we pushed the system to its limits.

One of the limiting factors was maximum allowed flying time. Aircrews are permitted to fly no more than 125 hours in any 30 day period, 275 hours within 60 days, and 300 within 90 days. These limits are set by the Federal Aviation Administration (FAA). However, we were waived for the war to 150 hours within 30 days. These maximum flying time limits are one of the reasons why the Reserves and Air National Guard were activated so quickly; we active duty crews simply flew too much too quickly and "burned out" with flying time.

The following day our flight into the AOR was even busier. Everyone on the crew ensured that their chemical warfare mask was within arm's reach, tested, and ready to go. Our Flak jackets were also by our sides with survival vests. The minute the MOPP (Mission Oriented Protection Posture) changed at our destination, we would put all this equipment on in heartbeat. We constantly monitored command and control and our destination on the radios. In addition, we talked with the destination itself for additional threat advisories.

Fatigue was less of a problem while flying in the AOR because our adrenaline was up. Also, everyone was busy reviewing checklists, our classified instructions, the instrument approach procedures, and "what ifs." One important "what if" included an attack on our destination. Our plan was to divert to an alternate while we put on our chem warfare equipment. Since someone has to fly the plane, the aircraft commander - me - would be the last one to don CW protective gear.

Our most prized possession on these missions was our duffel-bag sized mobility "D" bag. Each contained a chemical warfare protective mask, filter pack, full body suit, and poison gas antidotes. Our biggest concern was not with Iraqi fighter aircraft but with their Scud B surface-to surface ballistic missiles. We had plenty of friendly fighter aircraft in theater, in fact so many that another major concern was of near misses. At that time everyone believed that the Scudel Bs could carry chemical munitions. If our destination was hit, we would have just seconds to don our masks.

We were told that our survivability rate was at least 80% if we just got our masks on in time. If we got them on late and began twitching, experiencing dizziness and nausea - sure signs of nerve gas poisoning - then we had to give ourselves two injections of antidote - six inch long tubes which projected syringes deep into the meaty portions of our thighs. Stabbing yourself isn't pleasant, but it's a small price to pay for survival.

As we entered Saudi airspace and established radio contact with Jedda control, we would make our final preparations for landing. Often it was nighttime, the weather was normally fine, and from 35,000 feet clusters of lights clearly marked the Saudi cities in the distance.

The scene was almost surreal - from our altitude all would look so beautiful and calm yet, far below, war raged in scores of killing fields across the once calm desert. Just months before peace was breaking out as the cold war finally appeared to come to a belated close. How quickly things can change.

Like most Americans, our worlds were suddenly turned upside down. But because of thousands of dedicated men and women of the Military Airlift command - the airlifters of the U.S. Air Force - the biggest airlift in history never stopped once, despite tremendous obstacles and challenges.

Finally, on February 28, 1991, as I flew into Spanish airspace enroute to Torrejon Air Base, Lisbon air traffic control said, "MAC Victor 80119, contact Madrid on 125.7. By the way, congratulations!" God, victory felt so good, and always will!

Desert Shield: The Increasing Importance of Strategic Mobility

by Capt. Phil Bossert, Combat Plans Officer, 438th Military Airlift Wing, McGuire AFB, NJ

With the tremendous movement of troops and equipment to the Middle East in Operation Desert Shield, a military term has acquired renewed importance in our nation's defense policy: strategic mobility.

To define it simply, strategic mobility is the ability to move massive amounts of soldiers, supplies, and equipment anywhere, anytime, as quickly as possible. Strategic mobility is essential to any military operation, because a nation can have the best fighting force in the world, but if it can't bring it to bear where and when needed, then it is useless.

Military planners, our national leaders, and the world at large have been amazed how the U.S. Transportation Command, in general, and the Military Airlift command, in particular, were able to move a troop contingent the size of a "small midwestern city" 8,000 miles so quickly. In its September 3 issue, *Business Week* summarized MAC's contribution: "The heroes of Desert Shield may prove to be the pilots that in two weeks ferried to the Middle East some 35,000 troops and 1 billion pounds of cargo—the largest U.S. airlift ever."

I believe that strategic airlift will continue to increase in importance in our nation's defense policy for the following reasons:

1. Conventional forces are increasing in importance relative to nuclear weapons. Two facts show this. One is the number of arms control treaties and the increasing willingness of the Soviet Union to compromise with the United States. The intermediate nuclear force and strategic arms reduction treaties illustrate this. Additional limitations on the nuclear stockpiles can be expected in the future, as both sides realize that the cost of maintaining these stockpiles and their delivery systems is too high relative to their uses, which, except for deterrence, are minimal. The second fact is the realization that the actual use of these weapons, be it tactically or strategically, is militarily, politically, and rationally suicidal.

2. The proliferation of threats will continue. Panama, Liberia, and now the Persian Gulf crisis prove this. To the dismay of the liberals who wanted to hack the defense budget and use the "peace dividend," the United States will continue to require a large, strong military to maintain its super power status. Anyone with thoughts that the United States is declining as a major power should ask him or herself this: "What have Germany and Japan been doing to resolve the current crisis?"

3. Reduced overseas bases. Even with Operation Desert Shield proving the need for a strong military, we will be reducing our overseas commitments in Europe and Asia. In addition, many nations have and probably will continue to deny us landing or overflight rights, as occurred in the 1973 Arab-Israeli conflict. The air-refueling capability of the C-141s and C-5s makes their use even more essential in the strategic mobility equation.

4. Conventional arms reductions in Europe and Asia. With many of our troops being pulled from overseas posts, the ability to transport them to a global hot spot becomes increasingly important. Also, with more personnel entering the Guard and Reserve, the need for strategic mobility intensifies further.

5. The rapid development of conflicts, and the resulting required rapid resolution of conflicts. With the huge technical advances of the last 20 years which have brought the world together as closely politically and economically, the speed of conflict escalation has accelerated. The current Persian Gulf crisis is a perfect example.

To prevent events from escalating, rapid projection of U.S. military power is essential, and strategic airlift provides this rapid response. The quick deployment of troops and equipment to the Saudi Arabian-Kuwaiti frontier prevented an Iraqi advance into the Saudi oil fields, and the continued huge airlift of forces will certainly help resolve this crisis quickly.

With the American public anxious for a quick resolution to the invasion of Kuwait, MAC becomes more essential. Vietnam, the 1988 Kuwaiti tanker escort operations, and Panama proved that the public is intolerant of prolonged military operations.

As Operation Desert Shield is proving once again, USTRANSCOM and MAC have become more important in U.S. defense strategy because of their exceptional display of the power of strategic mobility.

There is no doubt that airlift, AMC, and USTRANSCOM will continue to increase in importance in our nation's defense.

C-130 Ski Operations. Airlifters fly regulary to the North and South Pole regions. (Courtesy of New York Air Guard.)

Rainbows Above the Clouds

A Woman's Views of Life as a Military Pilot

by A.M. Fletcher

In 1981, I joined a small number of women (approximately 240) who had become active duty Air Force pilots. I was the first female pilot assigned to a particular squadron of Starlifter crewmembers. We flew Lockheed C-141 cargo/passenger jets to Europe, Africa, and South America. I soon discovered that the glamorous mystique perpetuated in the men's war-stories was not suitable for continuation from a gentlewoman. Over the next five years, I developed a uniquely feminine set of "war-stories," free to describe experiences which the men would not.

The real challenges of being a military pilot faced me when I became an aircraft commander. As a lieutenant I found myself in command of men ten years my senior, often in countries that expected me to be subservient. In Jordan, for instance, some very cordial female airport clerks could not believe that I had flown the aircraft. They warned my crew to take "very good care of your woman." I reminded my crew of this promise when they later tried to give me a hard time. The Jordanian Air Force pilots, who had trained in America, believed that I flew, but not nearly as well as they could. One of their new pilots took me aside to tell me of his various accomplishments during American pilot training. Of course, much to his consternation, I had accomplished the same feats during my own training. At last he told me a feat that I was happy to accede I had never done. "I stalled and spun my aircraft five times — solo!" he proudly proclaimed.

I felt no need to advocate women's liberation in countries whose cultures prohibited it; my mere presence was proof of my capabilities and I was a guest in their land. Saudi Arabia ten years before the Persian Gulf War, however, was a particular challenge. One day I accidentally stumbled on to the solution for functioning effectively there. I tried to file a flight plan with a Saudi man, signing the aircraft commander's block in front of him. He pointedly turned and asked my co-pilot a question. I answered, but was ignored until my copilot repeated my answer. From that day forward, I always took a male crewmember along to "translate" when I needed to talk with an Arab.

People ask me if I ever felt in danger while I was flying. Flying was the fun part of my job; it was dealing with the bureaucracies, an occasionally resentful peer, or making a mistake on paperwork that seemed most dangerous. In 1983, for example, I was in crew rest in Spain, expecting to fly to Beirut to pick-up the bodies of Marines killed during the terrorist attack on their barracks. I was excited about flying to Beirut, but terrified when I was told to "dead-head" (fly in passenger status) back home. I dismissed the confusing stories from an arriving crew about the U.S. invading a tiny Caribbean island as a joke. Instead I worried about what I had done wrong to be sent home in disgrace when my service was so obviously needed in Europe. Twelve hours later I was solely in command of my first crew, headed toward Grenada. The scariest part of the mission was issuing a gun to my trigger-happy co-pilot. Fortunately, our regulations gave me the latitude to insist he keep his bullets in his pocket. The "fog of war" was so dense that my command and control facility could not decide which island my cargo should go to until I had been airborne for two hours. At last I received a radio call telling me to land at Grenada. No air traffic control facility had been set up, so I joined a stack of American aircraft circling above the island, directing our own landing sequence based on priority and fuel quantity remaining. I landed after reaching my own "bingo" fuel, shutting down my engines to save the precious fuel left. Immediately I was set upon by a full Army Colonel who was obviously having a bad day.

"What's wrong, why did you shut down your engines? What are you doing bringing us this junk? It's supposed to go to St. John's! You have it here, we might as well unload it. Then I want you out of here post-haste! Your crew is an accident waiting to happen!" he bellowed almost in a single breath.

My "accident-waiting-to-happen" crew unloaded the plane and got us airborne within twenty minutes — definitely a record turn-around! A few months later we each received two medals for our actions. I learned then that military medals do not reflect the full story.

Another scary incident occurred when the paperwork announcing an Egyptian policy change did not make it through the Egyptian and American State Departments and Air Forces before the Egyptian Air Traffic Control decided to enforce it. I was en route to Cairo International when I was told that I would be impounded there. I tried to verify my diplomatic clearance number, which permits a military aircraft to pass through another nation's airspace, then radioed my military authorities to warn them of the problem. They arranged for the American Air Attaché to meet me after I landed. I was not terribly worried until after we parked the airplane. Instead of the thirteen year-old boy with a machine gun who normally guards the airplane, a twenty year-old soldier was standing outside our door. The Air Attaché, a brusque Navy Captain, was more intimidating than any of the Egyptian authorities. He stood me aside while he verified my story with my crew. Satisfied, he accompanied me up some rickety stairs (everything in Cairo was rickety) to meet with Egyptian Air Traffic Control. They told me that I had violated their airspace without giving my diplomatic clearance number. I disputed this, asking them to check any recordings of my radio transmissions. Everyone's faces turned ashen, and I could see that this was the wrong answer. At that moment, I felt sorry for the Egyptian controllers, who were obviously more nervous than I was to be found at fault. Under the Captain's tutelage, I signed a handwritten document which basically said I had been a bad girl, would not repeat the mistake, and would inform my superiors of it. The Captain said we handled the incident very well, and the alternative would have been to delay our mission significantly and strain international relations. After he left us, I cried some very bitter tears over having to sign a false confession.

Overall, I have wonderful memories as a cargo pilot. Two of my everyday favorites held joys that the men either never experienced or were too self-conscious to describe. The first happened whenever some young soldier, climbing the aircraft steps, would see me on the flight deck and proclaim, "Whoa, a girl pilot! I'm not going on this airplane!"

"Fine." I always answered sweetly, preferably while located in a remote desert. The soldiers always replied by moving meekly to the back of the jet.

My next favorite memory is of all the times I took off to fly during gloomy, wet weather. When the jet burst out of the top layer of clouds the sunshine dazzled our eyes. Instead of the one or two rainbows that earth-bound women must be content with, the refracting sunlight created dozens of rainbows just above the bright clouds. While the men never mentioned it directly, I think this experience is what they think of when they say, "what a great day to fly!"

15,000 Hours in a C-141

by Lewis H. Fountain

Lewis H. Fountain. (Courtesy of Lewis H. Fountain.)

Another flying record has been established by a New Jerseyan. CMSGT Lewis H. Fountain of Maple Shade has flown more than 15,000 hours in the cockpit of a C-141 "Starlifter" and another 1,000 hours in C-47s in Europe and C-7s in Vietnam.

"They never had anyone fly 15,000 hours in that aircraft. The probability of anyone doing it again is next to impossible," said Fountain, a flight engineer.

Fountain's flying career spanned 31 years. In 1953, he dropped out of high school in his junior year and joined the Air Force. While serving in Turkey in 1961, he flew as a flight mechanic. When he returned to the U.S. to attend flight engineer school, his record-breaking flying career began in earnest.

He requested training in C-141s because he knew there was a need for flight engineers aboard those aircraft and that the giant cargo carriers flew out of McGuire AFB where he preferred to be stationed.

In 1973, Fountain retired from active duty and continued to fly with the reserves. He garnered most of his 15,000 hours in the cockpit of a C-141 as a reserve.

"It's an awesome feeling. I never thought I would achieve this amount of flying time," the modest airman said. "It is possible because of all the crews I have flown with. One is not an island. You don't achieve these things by yourself."

Then the 56-year-old Air Force veteran became nostalgic. "The best part of flying is the serenity, the beauty of flight. You get in touch with a different world, different bodily senses associated with being airborne."

You can take the aviator out of the plane, but you can't take the thrill of flying out of the man.

Excerpted from Propwash, Aviation Hall of Fame of New Jersey, Summer 1992.

First Flight of the C-130 Hercules

by John S. (Jack) Gibson

The first flight of the YC-130A prototype in Burbank, California took place before I arrived at Lockheed. In 1959, I came to work at the Lockheed Georgia Company, Marietta, Georgia where all production C-130s are built. My initial engineering job was helping to refine the improved C-130B version.

First flight witnesses state the afternoon of Monday, August 23, 1954, at the Lockheed Air Terminal in Burbank was warm and very hazy. The number 1 YC-130A made several high speed taxi runs as USAF and Lockheed officials watched from the top of the Terminal Building. Many other Lockheedians watched from nearby buildings and vantage points on the ground adjacent to the runway. At 2:45 p.m. the long awaited moment arrived. The big turboprop airplane moved rapidly down the runway for about 10 seconds, then rose majestically at a very steep angle. Someone said that it took off like a meteor. Spectators as well as the flight crew were truly amazed at its climbout performance. The YC-130A continued to climb in a wide sweeping turn and soon disappeared into the thick overhead haze.

After a successful inaugural flight of 61 minutes, the plane landed at Edwards Air Force Base where it would later accomplish its basic flight test program. Consequently, the first test flight was also the delivery flight to the USAF customer.

The two YC-130A prototypes completed their test programs in California. Then the first Georgia built production C-130A made its premier flight on the morning of April 7, 1955, at the Lockheed-Georgia/Dobbins Air Force Base complex. Thus began a long an proud tradition. The Hercules evolved over the years in many version and varieties for numerous worldwide customers. It has remained in continuous production to this day, a Legend in its own time!

First Flight of the C-141 Starlifter

by John S. (Jack) Gibson

The number one C-141 Starlifter rolled out of the factory for the first time in late August 1963. During the fall months, the aircraft underwent many static and taxi test, as well as last minute adjustments to various systems. By the middle of December all ground testing was complete and the airplane was ready to fly. All of Lockheed was eagerly awaiting the first flight.

On a cool December afternoon a friend in the flight test organization called to say that within the next few minutes the big event could happen. Several of us, who had worked on the C-141 engineering program from its inception, hurried out of our building and were walking to an open area north of the west end of the runway at Lockheed Georgia Co./Dobbins Air Force Base. Before we got there, we heard the C-141 engines go to full power. We could tell by the engine sounds that the plane was moving, but our view was obstructed by a row of pine trees. It was 12:45 p.m. Suddenly the big silver airplane popped up from behind the trees after a very short takeoff run and climbed dramatically upward in an easterly direction. Some of the guys cheered, others coaxed it on with "climb baby, climb!" What a beautiful sight it was!

The Starlifter leveled out after a while, then climbed more slowly and flew out of our view. We felt great pleasure and pride in being witnesses to the first flight and to having been involved in the airplane's design. As we walked back to our building, we all shared a great sense of accomplishment that day, December 17, 1963, which just happened to be the sixtieth anniversary of the Wright brother's first powered aircraft flight!

First Flight of the C-5 Galaxy

by John S. (Jack) Gibson,
Lockheed Aeronautical Systems Company

Rumor had it that Saturday, June 29, 1968, was to be the day for the first flight of the Galaxy. We waited for several hours at the edge of a cemetery which has a high overlook of the southeast end of the runway at Lockheed Georgia Company/Dobbins Air Force Base, Marietta, Georgia. My wife Diane and eight year old daughter Beth were also there, as were numerous Lockheed employees, families and friends. However, Saturday was not to be the day, although high speed taxi-tests were periodically run and everyone cheered each time the C-5 appeared at our end of the runway.

On Sunday morning, June 30, we arose early to get out to our same vantage point. We just barely made it in time because at 7:47 a.m. the first C-5 Galaxy, then the largest airplane in the world, rolled rapidly in an easterly direction - our direction! As it lifted gently from the runway everyone at first gasped, then cheered mightily as it flew upward in front of us into the early morning sun. I had a small Midland aircraft radio and we listened intently to the flight and ground crews conversations during the takeoff. What a thrill to see and hear it fly after working on the C-5 project since the conceptual design stage four-and-a-half years earlier. I saw the first flight of the C-141 on December 17, 1963, and it was very impressive - but the C-5 was something else! Shivers literally ran down my spine!

Swiftly, the emotional and suspenseful moment passed and the airplane climbed majestically and turned northward. Diane, Beth and I hugged each other in exultation and relief, and headed for our car. We knew we had witnessed aviation history that day. We also knew that the next day would usher in another six day, overtime work week for me as Structures Group Engineer on the Galaxy Program. No rest for the Engineering team as a new phase of flight testing was just getting underway.

C-130A first flight, on way from Burbank, CA to Edwards AFB. (Courtesy of John S. Gibson.)

C-141 first flight, Dec. 17, 1963. Lockheed Georgia/Dobbins AFB. (Courtesy of John S. Gibson.)

The Gooney

by Captain Charles W. Gray,
United States Air Force

Although I've never sat down to a sumptuous meal of roast Gooney bird (or Diomedeidae Nigripes- as you Audubon fans call it), I have a feeling that it just wouldn't compare too favorably with Long Island Duckling L'Orange. All bone and gristle I'll bet, with tough meat and nearly impenetrable skin.

How fortunate we have been that the airplane so appropriately tagged "The Goon," shares this additional characteristic with its venerable namesake. The absolute refusal of the C-47 to be eliminated from the active United States Air Force inventory is a wonder of wonders.

The Ole Bird, over 30 years after its introduction to aviation, is now enjoying some of its "finest hours" in the air resupply operation being conducted by the 1st Air Commando Squadron in the Republic of Vietnam.

The statistical record is impressive: 5,361,109 pounds of cargo delivered to U.S. Army Special Forces in the field in 1964. Most of that year, only two C-47s were committed to that operation. On several occasions only one was in the business and toward the end of the year, the operation was completely suspended for a time to complete a higher priority test program.

How then, such an amazing record? Astoundingly simple: A fantastically low abort rate, and the thorough professional competence of the flight crews who deliver the goods under conditions and weather that require the utmost application of the techniques evolved and refined for this type of operation.

Evolution is the key here. For indeed "The Goon" has been this way before; flown over and between the same mountains, battled the same monsoon weather and tested the mettle of its crew with its neurotic behavior in a cross-wind on wet pierced steel planking (PSP).

The 1st Air Commandos have advanced and perfected the techniques originated over 20 years ago by the original Air Commandos.

The basic tenet is unchanged: Deliver precious cargo to a virtually inaccessible piece of jungle geography regardless of obstacles, and do it reliably and safely.

Precise low level navigation techniques are absolutely essential. Straight line navigation is very rarely possible. Most often you follow a valley or slip through a saddle back in mountain ranges between rain showers. Intimate knowledge of the terrain is a literal lifesaver in the mountainous central and northern portions of the country.

In the flat delta regions, the torrential rains obscure drop zones or short, unprepared landing strips to a degree that would close down many a stateside air base. Finding a pinpoint open field in an area of one-fourth mile visibility is tricky business, to say the least.

Once arrived at the point for dispensing the goodies the real work begins.

Is the Drop Zone secure? Are the "friendlies"

awaiting? Or is there a "black pajama: (Viet Cong) reception committee poised in anticipation?

Verification is usually accomplished by a smoke flare, which also aids in surface wind determination. Now the skill and training pay off. Making a maximum gross minimum run landing into a 2,500-foot strip bordering with 150-foot trees all around calls for precise airplane driving.

Crosswinds, seemingly ridiculous runway slope angles and the ever present rain grinds complications into the basic problem. Using low altitude delivery techniques (250 feet absolute minimum) airdrops can be made with amazing accuracy and speed.

A 5,000 pound, 10 bundle load is usually delivered in about seven minutes. Accuracy, speed and a tight drop pattern are all important. Fifty yards off the pinpoint drop zone (DZ) could very well mean a few boxes of cheer for the bad guys.

Driving all over the countryside at 250 feet is decidedly unhealthy. There is no margin for error or the slightest hint of complacency. It is precise, demanding work, and the penalty for failure is high. Yet, while delivering a life-saving load of supplies and ammunition over a DZ, it's a great thrill to sight a harassed and weary group of GI's cheering and waving on every pass made.

So there it is. Mission accomplished—the third this day—and the indestructible old bird is on its way back home for a few hours rest. Dirty and worn; scarred and patched.

The piece of concertina wire holding the cowling together (compliments of a U.S. Army fort commander at a jungle patch) will be replaced while "The Goon" is being gassed up. The dents made by the water buffalo when they pulled you out of the mud will be smoothed.

An airman with a repair kit will crawl all over the still-warm frame and put little round patches here and there where an errant Viet Cong rifleman lucked into a hit through sky-high jungle foliage.

Meanwhile, the loadmaster will stuff another hold full of goodies into the sturdy bird.

These amenities finished in short order, the revived "Goon" will be pulled up to the starting line, ready to go again.

And the same crew, or another whose day also began long before sun-up, will crawl into the cockpit, coax life and power into the still-warm engines, rein the faithful craft to the end of the runway, and head for a clearing somewhere out beyond the murky horizon.

Pride and sense of accomplishment? Indeed we Air Force jockeys have it in overwhelming proportions.

Does the Old Bird share it with us? I hope so. She deserves it!

C-5A first flight June 30, 1968. Lockheed Georgia/Dobbins AFB. (Courtesy of John S. Gibson.)

The Berlin Airlift Candy Bomber

by Gail S. Halvorsen, Colonel USAF (Ret.)

A hazy afternoon sun shown through a window into a packed Airlift Operations School class room at Scott AFB. What the instructor was saying made me temporarily forget that I was about to replace him and talk about the "old days" of the Berlin Airlift. He went on, "Deterrence is the cornerstone of national security strategy. Airlift is the backbone of deterrence. Readiness is the keynote of a nation's military capability. Airlift is the key to the rapid projection of our conventional forces all over the world."

My thoughts went out the window with that last statement to a certain day in early July 1948. I was to learn about rapid projection. Aircrew and ground crews were gathered under the wings of C-54 and C-74 aircraft on the apron at Brookley AFB Mobile, Alabama. We heard Lt. Col. Hahn say, "They want four of our C-54s in Rhein-Main as soon as we can get them there. No crew rest en-route, four crews per aircraft. The Soviets have cut off all ground access to West Berlin. Be prepared to leave in two hours!"

I was bunked at Rhein-Main, Frankfurt, flying to Berlin and soon mixing coal and flour dust in the brow band of my 50 mission hat.

One day through the barbed wire fence on the approach end of the runway at Tempelhof in Berlin I met about 30 children ranging in age from five to 14 years. Many of their clothes were patched. They had seen rough times. The Soviets had them blockaded, tight. No food or supplies for over two million West Berliners, mostly women and children. Airlift was their only hope.

The kids warmly accepted me even though I was a pilot wearing the uniform of those who rained bombs on their heads but a few months before. That uniform had become a symbol of freedom, representing all the air and ground crews dedicated to preserving their right to choose. They didn't want a Hitler and certainly not a Stalin. The feelings of compassion and being united in purpose with those with whom we had so recently been mortal enemies were strange but immensely rewarding.

After a few questions about our airlift capability the children proceeded to lecture me on the importance of freedom. A little girl with wispy hair and blue eyes said, "Please tell the others how we pray for their safety and success. We don't care if we get enough to eat, just enough to stay alive. Someday we will have enough food but if we lose our freedom we may never get it back."

I suddenly realized my transportation was about to leave, said a hasty farewell and headed across the field. A very small prompting stopped me short. "Why were these kids so different?" Then it came to me. I had flown foreign transport operations in many places of the world. Especially during the war, when you walked down a street in those places little groups of kids would chase you demanding gum and candy.

These Berlin kids hadn't had any of that stuff for many months. I had been isolated with them for over an hour yet not one was so ungrateful as to beg for such an extravagance. They were simply so thankful for the dried food. The magnitude of their self control overwhelmed me.

My hand went immediately to my pocket. All that was there were two sticks of gum. I turned back to the fence. Every eye was on my hand and what was in it. Two sticks-thirty kids. There may be a fight. My hesitation was short lived. The two sticks were broken in two and the pieces passed through the barbed wire. No fight but a sharing of the outer then the tin foil wrapper with those who didn't get gum. Smiles were everywhere. Their reaction was beyond anything that could have been imagined.

A landing C-54 flew low over our heads and little puffs of white smoke came off the tires as they squeaked down. In a flash I thought how those who only received a piece of wrapper could have gum. To my surprise I found myself saying, "If you promise to share I will drop enough gum and chocolate for everyone to have a taste when I fly over your heads on my landing approach tomorrow."

Their response was slow until the translation was verified and then it was a thing to behold. Then they got quiet as the little girl with blue eyes gently demanded to know which aircraft I would be in. An aircraft landed every three minutes. They had to know which one had the goods. A signal was arranged - I would wiggle the wings of my plane on the initial overhead approach to Tempelhof.

The next day at Rhein-Main the 30 of us brought on board all of our candy and gum rations firmly attached to handkerchief parachutes. We were apprehensive that we may be caught. There had been no time to gain approval. The weather en-route and at the terminal was good. As we came over Tempelhof we would see the little circle of about 30 waiting kids. They hadn't told another soul. We wiggled the wings of the C-54, heavily laden with 10 tons of flour, and dropped the little parachutes right over their heads.

Taxiing to take off we saw them along the fence. They had shared. Every mouth was active, every face radiant, and every hand waving at each taxiing aircraft. "I wish they wouldn't do that," observed Captain John Pickering who, with Sgt. Elkins and I, were worried about keeping it a secret.

We repeated the operation for three more weeks, the crowd grew, then it happened. On a return flight to Rhein-Main we were met with a request for the pilot to report to Col. Hahn. "What have you been doing Halvorsen?" the good Colonel demanded. "You almost hit a German newspaper man in the head with a candy bar in Berlin yesterday. He's got the story all over the world. The General called to find out the details and I didn't know anything about it!"

For a few minutes I thought I had flown my last airlift mission. "The general thinks it is a good idea. Continue to drop but keep me informed," the Colonel concluded with a half grin.

I would often return to my bunk in the loft of the old barn by the autobahn in Zeppelinheim to find the bunk covered with enough boxes of gum and candy to thrill the lives of hundreds of blockaded Berlin children. My buddies were giving their rations for the cause. Soon supplies came from the USA. Twenty-two schools near Chicopee, Massachusetts used an old fire station to assemble about 800 pounds of goodies distributed on many small sail cloth parachutes. Almost every other day they had that amount on an aircraft departing Westover AFB for Operation Vittles. Our whole squadron got into the operation. We dropped all over West Berlin.

The response from the Berlin children was overwhelming. Thousands of letters came addressed to Uncle Wiggley Wings and the Schokoladen Flieger. Colonel Walter S. Lee, Base Commander of Rhein-Main, gave Operation Little Vittles two secretaries to answer the mail. When the Berlin children heard we had temporarily run out of little parachutes they sent their used ones back for refills.

One nine year old boy, Peter Zimmerman, wrote that he couldn't run very fast and wasn't getting anything. He included a self made parachute and a map. "When you take off from Tempelhof fly along the big canal to the second highway bridge, turn right one block. I live in the bombed out house on the corner. I'll be in the back yard every day at 2 p.m., drop it there."

I couldn't seem to hit Peter. After several tries he wrote, "You are a pilot? I gave you a map. How did you guys win the war anyway?" I took a generous package of gum and candy and mailed it to Peter through the Berlin Post.

A letter from eight year old Mercedes Simon in effect said that we were scaring their white chickens as we swooped over her house to land at Tempelhof. The result was reduced egg production, a most serious event given the circumstances. The last part of her letter was the most important, "When you see the white chickens, drop it there. All will be O.K." We had no better luck with Mercedes than we had with Peter. I wrote her a letter and sent a package as I had done for Peter.

My time came to rotate back to the States. My friend, Captain Eugene Williams took over the project.

Twenty-one years later (1970) I was returned to Tempelhof as the Commander partly because of those two sticks of gum. Many of the Berliners who had caught a parachute in 1948 and 1949, invited us to their homes for dinner and conversation about their particular experience in catching their prizes. Our busy schedule day and night precluded accepting those generous offers. One family didn't give up. An invitation kept coming until in 1972 we relented and accepted that one request.

Alta and I were greeted by a young German couple at the front door of an old Berlin apartment building near Tempelhof Airport. The lady of the house took us upstairs to their apartment and to a position in front of a china cabinet in the front room The lady opened the cabinet door and removed an old letter. "Read this please," she said with a sight quiver in her voice.

The letter was dated November 4, 1948. It said, "I am sorry I can't find your white chickens but I hope this package of gum and candy will be alright. Your chocolate uncle; Gail Halvorsen."

"You silly pilot," the lady said with emphasis, "I am Mercedes and if you will take five steps over to the window I will show you where the chickens were."

After my retirement Mercedes and Peter Wild, her husband and children visited us in Provo, Utah. Peter is a science teacher in the Berlin school system. We contacted Brent Chambers a German teacher in Provo High School and established a new airlift of friendship between Berlin and Provo. Hundreds of students from Berlin and Provo have participated in this on-going exchange program, an airlift of friendship, because of a older airlift and two sticks of gum!

On January 29, 1993, Mercedes, Peter, Alta, and I stood together in Hangar 3 on the flight line at Tempelhof to witness an historic event, the transfer of control of that airport from the United States of America to Germany. Tempelhof, Gatow, and Tegel had witnessed one of the greatest uses of humanitarian airlift in the history of flight. Airlift here protected the nest of freedom that later hatched the force that brought down the Wall and opened the closed borders of Eastern Europe.

The Berlin Airlift is far from being an isolated case. Airlift has provided the wings of freedom for the oppressed on many occasions, in many different ways, and in many different lands.

This personal experience, far from being an isolated case, has had similar parallels involving children of different skin colors and other airlift air and ground crew and other airlift aircraft in other far away places with strange sounding names. It will ever be so.

Given the will we can fly, "anything, anywhere, anytime." In these operations there will always be a human ingredient extended beyond the official duty day the compassion offered to the defenseless, especially the children, by airlift air and ground crews.

I came back to the Airlift Operations School class room with a jolt in time to hear a quote from "An Illustrated History of the Military Airlift Command," May 1991 page 72, quote, "The Berlin Airlift proved what has been confirmed many times since: airlift is a more flexible tool for executing national policy than either fighter or bomber aircraft."

In the same breath the instructor announced that I would be the next speaker, subject: The Berlin Airlift!

Lt Gail Halvorsen prepares candy parachutes on his bunk. (Courtesy of Gail S. Halvorsen.)

Free Enterprise

by Ronald K. Sable

It was an early summer morning in 1969. We were attending the morning "OPS" briefing for the 315th Air Commando Wing at Phan Rang RVN, just prior to launching for the days missions in our C123K "Bookey Bombers."

Every Aircraft Commander in the 310th ACS had their assignment but one, me. It looked like the briefing was over which only complicated my "comfort level" since my name had not been listed on the flying schedule posted the night before at the Happy Valley Officers Club.

The group assembled began to disperse when the Wing Commander came forward. Everyone froze in position. He walked to the podium and asked us to sit down. He then asked that anyone not having a mission that day standup.

As sweat beads began to materialize on the back of my neck, I stood up. In a cursory survey of the room, I noted that I was the only crew member standing. "Captain Sable" he barked "don't you have a mission?" "No Sir" I said, unsure of what was to come. "You do now" he said. That jeep you've been driving for the last three months - return it to its rightful owner in Saigon and pick up the rest of your FRAG there. I expect you'll have a welcoming committee." "Room Dismissed."

Obviously, we returned the jeep to Saigon. As I taxied my aircraft "Ronna Kay" down Charlie Row, it was easy to find our revetment - four Air Force Security Policemen were waiting for us. We backed the jeep out of the aircraft, handed over the keys and our entire crew saluted smartly.

Well, it was fun while it lasted. You see, only Squadron Commanders and above had their own transportation. At Phan Rang, the rest of us rode the shuttle bus from our "hooch" to the Squadron on the other side of the base. Being innovative and recognizing a target of opportunity at Tan Son Nhut early that spring, our crew had simply borrowed one of the Security Police jeeps parking in a revetment and airlifted it to Phan Rang. After a quick paint job and the stencil "Free Enterprise" on the hood, it became a squadron mascot of sorts. We provided no cost assistance to the Squadron Commander (LTC Saunders) when his jeep broke down, carried pilots to the hospital when they were injured (not in the line of duty) and used the jeep to move the 200 lbs. of shrimp cocktail and roast turkey from the airplane that we had taken in trade from the Army in exchange for a pallet of Egg Nog (but that's another story). General (then Colonel) Don Brown was on the Wing Staff at the time. He even rode in the jeep a time or two. If he knew, he never let on. Oh well, he always was a champion of Free Enterprise.

(Courtesy of Gail S. Halvorsen.)

On Wings of Air Mobility

An Ode to Airlift & Tankers

Rev. Kenneth W. Schamens, Colonel, USAF (Ret.)
National Airlift Association Convention
(October 1992)

On wings of Air Mobility, freedom comes!
As a Total-Force family, we have come.
We are here on the wings of those who have sacrificed their lives,
So that in freedom we can live under sunlit skies.

From the "Hump of Burma" airlifters came,
Giving up their lives without gaining fame.
From the "clouds of heaven" the tankers came,
in service to others they made their name.
An Airlift-Tanker marriage was soon arranged
so that America's global reach would be maintained.
Together their one response is "let us serve,"
so that with air mobility, America's freedom is preserved.

From steamy valleys in India to rice paddies in China,
From Berlin to Korea,
From Operation Dew Line to Operation Deep Freeze,
From Hungary to the Suez,
From Beirut to Matsu,
From the Congo to Vietnam,
From Operation Baby-Lift to Operation Homecoming,
From Israel to Grenada to Just Cause in Panama,
From Hurricane Hugo to Earthquake San Francisco,
From refueling operations to support worldwide missions,
to supplies airlifted to Desert Storm positions.
From humanitarian aid for Kurdish and Somalian refugees,
to the support for Hurricane Andrew's devastation and pleas.
Hope as lifted worldwide like a flame,
"Bob Hope" too became an airlifter of fame.

Airlift and Tanker Forces live their commission
on the wings of service to the global-reach mission.
Nothing spectacular—only fuel and medical supplies are downloaded,
While the critically wounded are onloaded.
Supplies and ammunition are airdropped,
And the enemy is soon stopped.

With airlift and tanker craft they are always there,
Lifting equipment and fuel through corridors of air.
In silence they serve, in honor they die.
In the sanctity of time and space,
These men and women have touched our face —
And on wings of Air Mobility, our nation has been touched by God's grace!

Orders, Egos, Weather and Unknowns

by David J. Orth

The air supply for the eviction of the Japanese from Burma was a very costly, pioneering effort as no previous war had been totally supplied by air. It was an epic in jungle canopy flying in dangerous terrain and violent monsoon weather. It had its price but the appalling cost of the hump operation has never left me, especially the needless parts of it. Prior to enlisting in the Air Force I'd been raised on long, hard hours necessary in growing, harvesting and transporting large tonnages of crops where no waste of any kind was tolerated. However, I've never lost the pride of being in on the cutting edge of the hump supply effort and the following record that was set. A diary entry of August 1, 1945, show what the 443rd Troop Carrier Group, flying from Dinjan, India, did that day. However, there has been a lot left out of the official records.

If one looks at ATC's tonnage graphs for the hump during WWII and especially 1945, it's abundantly clear that in June and July there was a dramatic increase in the airlift across the hump to China. Aside from other inputs previously, there was no mention of Troop Carrier Command, Air Commandos, Combat Cargo, the Air Transport Squadrons and bombers which also added to the tonnage landed in China.

As the war in Burma wound down, whole squadrons and groups were fed into the hump effort until there was the equivalent of a least a wing of transports, some 400 aircraft, plus B-24s and B-29s. None of this included ATC's 100% utilization of all C-87s, C-109s and the new C-54s, along with their fleet of C-46s and C-47s. Also, by this time, C-46Ds were fast displacing the C-47s of Troop Carrier and Combat Cargo. January's figures show 44,098 tons; February 40,677 tons; March 46,545 tons; April 44,254 tons and then May's increase back up to 46,393 tons. First Troop's first C-46 trip to Luliang, China, was on May 5, 1945. June finally jumped to 55,386 tons but not without ATC's own increase using C-54s on the lower, southern route. July jumped to 71,042 tons. August first's whopping 5327 tons were airlifted in 24 hours suggesting a potential of 100,000 - 150,000 tons per month was possible.

If one uses the first five month's average of 44.393 and adds circa 60% to it, you come up to the July mark on ATC's graph. This is why, for the record, that this is being written - before all our memories and notes fade away and a wrong has not been set right. The intent here is not to throw gasoline on a long smoldering fire and certainly not to take any tonnage away from ATC's, rightfully and honestly delivered by their aircraft. The main point here is that the so-called "sub airlifters" actually outperformed and airlifted well over half of the July tonnage. No one wanting to win the war thought much about the tonnage being humped upon other carriers' backs. But, herein surfaced a tender nerve and, to my knowledge, there was nary a word of thanks or credit given for the new infusions of airlift that added another wing to a very large, zealous bird - a bird that had started the New Year out with unnecessary and unacceptable losses which somehow got covered over.

But, that's another, unfinished story. Or is it? Along with its records, which were outstanding to say the least, ATC also didn't take a back seat when it came to warts, visible or invisible. Probably the cruelest order ever given in the theater was that there was to be "no more weather over the hump." The terrible waste of men and aircraft and those out looking for them flew in the face of common sense. There had been plenty of warnings such as 25-50 mph ground speeds and some transports even recorded flying backwards. But it would take a couple of squadrons of planes lost on the India and China side (January 1945) to prove that Mother Nature had accepted puny mans' unwise challenge. She provided a plane breaking violence that was to take exceptional discipline and some mighty brave crews to prove that dictum terribly wrong. Also wrong was the gallon wise gamble and plane foolish practice of taking tail wind generated extra fuel saved on the China side, pumping it out and trying for a 20-30 minute reserve on the India side. The odds of expecting two tail winds on a round trip across the hump were about as nil as the intelligence that placed the bet. But, of course, the odds maker was safely behind a desk or flew in a C-54 with a longer range. The gallons of sweat and blood lost far and away outweighed any of the extra reserve fuel left in China. Especially from the shorter range C-47s and C-46s. Just a few years ago C-87 was found 10 minutes short of the Bramaputra valley. It hadn't burned. Like so many others, it had run out of fuel. So, it is to this day, that there has never been a way to calculate accurately or humanly the cost of a gallon of gas delivered in the CBI. There have been estimates made that took circa 25 cents hi-octane and regular at 10-12 cents and turned it into $2.40-$5.00 per gallon landed in China. But the figures are so incomplete when all the time, maintenance, fuel and broken aircraft as well as the bodies, many of which haven't been found as yet, are added up. The cost is one thing that is firmly known and it's embedded in ice, rock or jungle, even most probably the South China Sea. It was astronomical and fits right in with the common monument that was placed at Wright Patterson Museum by the H.P.A. in 1984. Then there's British General Orde Wingate's gravesite in Arlington National Cemetery that holds the unidentified Americans buried with him and another grave not far from them that holds America's unknown soldier. So it is with the final cost per gallon of gasoline that kept China in the war. It too will always be unknown. It had all started with fuel waiting for Doolittle's Raiders B-25s which didn't get vectored to it, and was followed by the 14th and 20th A.F. that always had hungry tanks.

Operation Vittles

by Russ Crane

Having obtained a C-47 for proficiency flying one fine morning in June 1948, I was asked it I would mind making a flight to Berlin. And so it was that I began my participation in Operation Vittles (this was a few days before it had a name or public identification). I was assigned to HQ USAFE, at Wiesbaden, flying out of nearby Y-80.

During those first few days, navigation aids were very limited and the weather very bad (as it was to be all summer). At Tempelhof, the only navigation aid for a time was a very low power homing beacon located nearby. It was very erratic (some said because of Russian electronic activities) and therefore not trusted by many pilots. This made an instrument approach a chancy affair at times. As a result, a good many approaches were aborted requiring the aircraft to return to home base along with its load for Berlin.

This caused a few of us to work out a bootleg beacon approach using Radio RIAS (Radio In American Sector) which was a quite powerful public broadcasting station nearby. We found we could easily hit Tempelhof, on the nose and felt quite comfortable using a 500 minimum and, on occasion, even a little less. The bad news was that Radio RIAS shut down from 2 a.m. to 6 a.m. The good news was that Russian Radio Berlin was even more powerful, not much farther away, and broadcast 24 hours a day. So you can guess what came next, a beacon approach using it between 2 a.m. and 6 a.m. Only a very few, "in the know" pilots, used this approach, and, needless to say, told no one. It was actually quite safe as the stream was time separated enough to not delay a following aircraft and therefore precluded any risk of mid air.

Only when the weather was really terrible did I have occasion to resort to one of these approaches. In each instance the tower was astounded when we reported field in sight and requested landing instructions as immediately preceding aircraft had usually not been able to make it in. Our typical response was that a small hole had opened up momentarily allowing us to get on down. I have often wondered if we would have been suckered had the Russians caught on and moved the transmitter. I would like to believe not.

So far as I know there was never a complaint from them about over flying their sector in these instances. It occurred infrequently, always late at night, and always in very bad weather. Since they did not have an all weather capability, and therefore weren't flying, they were powerless to do anything about it, and apparently did not wish to admit it. That, of course, is how we hoped it would be.

To the best of my knowledge, this is the first time this story has been told and I don't know if anyone else is still around who participated in it. We were pretty fired up with the notion of getting the stuff to Berlin and, because of the terrible weather, weren't doing a very good job of it. So we felt pretty good every time we got in this way even if it was unconventional and unauthorized. Soon, of course, all the problems were resolved, I went back to my regular USAFE, HQ, job, and the Berlin Airlift went on to make history.

Too Far Down the Runway

by Robert M. Murdock

It was very dark and things seemed almost routine as the crew aboard the C-130 Hercules went quickly and efficiently through preparations for landing their aircraft. In the cargo compartment four very tired and anxious Air Force men and their Vietnamese interpreter were just awakening after a one and one-half hour flight from Tan Son Nhut.

On June 8, this plane, its passengers and its crew from the 374th Tactical Airlift Wing, would be the first aircraft to land at the besieged city of Kontum in more than two weeks. The crew's task was to get the Hercules on the ground and unload the five men. These men were the Mobility Team (MOB), and with their giant fork-lift they would stay on the ground at Kontum and unload following aircraft coming in with supplies and equipment.

The aircraft was completely dark inside except for a few dim lights on the pilot's instrument panel and in the cargo area. Looking out the window, the ground appeared just as dark. Lightning cracked the night for a brief second and the area was briefly visible but it was still hard to find the short landing strip. With the aid of the ground controller, four dim lights were finally spotted which marked the beginning of the strip.

The Pilot, Captain Mike McWhirter, 31, from Omaha, Nebraska, gave the word they were going in. The MOB team checked their equipment and got ready. The plane made its approach.

"Too far down the runway" said Capt. McWhirter, and he pulled the aircraft back up. On the second approach to the field a flare went off to the right of the plane. As the plane dipped toward the runway tracer bullets went screaming past the wing. Once again the plane came in too far down the runway and the Captain pulled it up. "Without good lights on the ground and with our lights off, this is a hairy landing to make," said Capt. McWhirter.

More tracers went flying past the aircraft as it came into position the third time. Over the intercom Capt. McWhirter, told his co-pilot, Lt. Bob Murdock, 24, from Thornton, PA, "we're going to have to come in lower; let's make this one good." The approach was perfect, the landing was rough but successful. The word was passed to the plane's Loadmaster, Sgt. Elbert Thompson, 21, from Washington, D.C., to open the ramp as they taxied to the end of the runway. The MOB team wasted bearly a second. They drove their fork-lift off the plane and hurried to a bunker nearby.

The Loadmaster told the pilot the load was clear and the ramp was coming up. The plane moved quickly down the runway on takeoff roll; "80, 90, 100," the co-pilot was reading off the plane's speed. In what seemed like hours, but in reality only a moment or two, the Hercules had landed, been unloaded and was once again airborne.

Two members of an Air Force maintenance team had checked aboard the aircraft at Kontum for a lift to Pleiku. "We were really happy you guys made it," they said, "after your plane touched down, we saw a couple of artillery hits at the end of the runway."

Three other aircraft loaded with ammunition and supplies shuttled in and out that night and Kontum was once again open.

The Flying Jenny

by Charles E. Malcom

War stories? I've seen it at close hand. World War II, European Campaign, Korean police action, you name it You might say the action took place right under my hooves. Hooves? Oh, I forgot to tell you, I'm a Jenny, a flying, red Jenny; mascot of the 815 Tactical Airlift Squadron. Until 1967 they were known as Troop Carrier Squadron and had existed since 1952 as such, but the unit has a distinguished record of combat service as a heavy bombardment squadron in World War II.

In the spring of 1944, the squadron was sent to Italy. There, flying B-17s, it participated in many historic operations, including the famous raids on the Ploesti oil fields. Before its deactivation in September 1945, the squadron received two Distinguished Unit Citations.

The War Department brought the unit out of mothballs in the winter of 1952, to help in the Korean Conflict. The unit was now designated a troop carrier and was flying C-110 Flying Boxcars. It was at this time that I was officially adopted by the squadron. By the time the truce was negotiated, we had taken part in many large troop movements around the Korean peninsula.

After this conflict had ended, it became my duty to see that my boys were kept in a state of constant combat readiness, so we could counter any communist move in the Far East.

My image was seen participating in operational readiness exercises throughout the area. NATO maneuvers in Korea, SEATO deployments in Thailand, Australian forces in Sydney, all knew the sign of the Red Jenny. My portrait was painted on other aircraft, tanks, howitzers and buildings. Even an occasional tipsy bar patron has had to explain the Red Jenny stenciled on the seat of his pants.

Then, another change in 1958, we switched from C-119s to the present C-130s. After the conversion period to the modern tactical transport was over, my charges once again demonstrated they were always ready by continually receiving superior ratings in their annual inspection.

Life then went pretty smoothly until 1964. It was then I received my first "Purple Heart" decoration. An associated squadron of C-130s, caught my men napping and horsenapped me. I was spirited to Okinawa where there was a big sacrificial ceremony in my honor, I was deeply grateful until they told me I was the sacrificial "ass." The ceremony over, I was ransomed back to my boys, a little the worse for wear, I had a hole in my head and so was awarded the "Purple Heart."

I could have gone awhile quietly in life, but such was not my fate, for in early 1965, I was decorated with a cluster to my Purple Heart when a determined wind blew me from my perch high atop the squadron headquarters and I bent my tail.

In July 1968, a group of reserves from the States horsenapped me again. However, since they're TDY here at Tachi they only got as far as their hangar with me. I spent a comfy night there and upon popular demand was back at my old home the next day with proper attendance of all concerned and even allowed their mascot "a little green man" to ride on may back about an hour.

Yes, I've had a very full life, and I'm just getting in my prime. I'm still here, at my regular post on Tachikawa Air Base, Japan, watching over my children. Come over and stop in for a chat sometime. Oh, the tales I can tell!!!

An Loc, RVN

by Edward N. Brya, Brigadier General, USAF (Ret.)

In April 1972, as the war in Southeast Asia was winding down the North Vietnamese Army launched their spring offensive. This caused a major build up and reinforcement of Air Force, Navy and Marine Air from both the CONUS and Pacific Theater.

In late 1971 and 72 we had been withdrawing forces from South East Asia. The C-130 wings at Clark, Mactan, and Naha had been deactivated and the airplanes returned to the States. The 374 TAW at CCK in Taiwan, with it's four squadrons, was the only PACAF asset for C-130 airlift in sea. At the start of the spring offensive the wings had 27 aircraft, 43 crews and 260 maintenance personnel deployed to Vietnam, or in-country as we called it.

During April we surged to 44 aircraft, 60 crews and 370 maintenance personnel.

In response to the NVA spring offensive our short field operations increased from the occasional landing at an out of the way field to major operations on a 24 hour basis. Early in April we brought ARVN troops from the 3,000 foot strips in the Delta to Bien Hoa for defense of the Saigon area. Other aircraft went to long forgotten places to pull out men and equipment. We operated in the DMZ and Quang Tri until the enemy forced us out. The major areas of operation were at Kontum, in the highlands, and in the south surrounding Bien Hoa and Saigon. During the month of April we made 358 short field landings including 51 at night.

Let me go back a few months to January when I arrived PCS from Pope to CCK for my third tour in SEA. As I was signing-in at wing headquarters, the vice commander, Col. John Dunn, saw me and said, "Major are you an IP?" After I said "Yes Sir" he stated, "Your going to Stan-Eval," and I said, "No sir I don't want to do that I just want to fly my 15 month tour and go home."

People were rotating back and forth so many times that the experienced C-130 crew members had all found a way not to go back again. Arriving in mid-January I was the first IP to come into the wing since October - this was an isolated tour and 80% of the crew members were on a 15 month rotation cycle.

Well, I went to STAN EVAL and on my upgrade ride to be an evaluator, I busted the two pilots and navigator for airdrop procedures. They had gone through the pipeline, CCTS at Little Rock and RTU at Pope, over to CCK and on to the routine air land mission in-county. Once in a while an experienced crew would be sent into short fields but by and large it was milk run flying as the war wound down.

Back home at CCK they would do some basic training, get the laundry done and if their birthday cycle was wrong, and their check ride from CCTS-RTU had expired, they would have to get a TAC or instrument check.

Unbeknownst to me, the tactical training was more orientation than standardization. As I debriefed that poor crew's squadron commander and later the wing DO, who by the way was a SAC pilot with no airlift background, I said if they ever had to make an airdrop in Vietnam they would be lost.

Around April 10, I went in-country with our wing commander, Colonel Andy Iosue. After we flew a couple of short field missions he told the mission commander to ensure only the best qualified pilots were going to the tough LZ's, especially at night.

The solution was to go back to a system of pilot qualification used earlier in the war. Category I, II or III pilots. Category I were IPs who could go to any a C-130 strip. CAT II were good steady aircraft commanders who could go to the mid range strips around 5,000 feet and CAT III was for the newer pilots who did the rest. When Col. Dunn told me I was going to STAN EVAL what I really told him was that I was a Category III Pilot, only qualified for Saigon, Bangkok, U-Tapao, Clark and Yokota.

In the south the centerpiece battle of the spring offensive was fought in Bien Long Province. The communists boasted that An Loc would become the seat of government for the liberated provinces.

Col. Ray Bowers in his excellent *Air Force History of Tactical Airlift* describes the battle for An Loc as "the most trying time of the war for the C-130 crews."

An Loc was a small provincial capital, 60 miles Northwest of Saigon, which lies on a plateau surrounded by plantations with tall rubber trees. In early April the area was cut off and surrounded by the NVA. The northern half of the town was captured...the ARVN and their American advisors were forced into a small area in the southeast corner of town. The command element was inside a bunker which was located by a soccer field 200 meters square.

The VNAF had tried without much success to resupply the beleaguered troops; during this time, a C-119 was lost to ground fire.

On Saturday, April 15, two C-130s were sent in. They used the then approved method of a descending slow down into the DZ. They were briefed by 7th Air Force to approach up the road from the south, as all the VNAF drops had been flown, and drop on the soccer field. The first aircraft made a successful run in taking only a couple of hits from ground fire. The second aircraft came in approximately 15 minutes later and was under constant fire. One 51 cal. round came through the right hand circuit breaker panel, killed the engineer and went on to shatter the windows on the left side of the cockpit. Other shells ripped the cargo compartment and ignited part of the ammo load. The loadmasters jettisoned the load, which landed on the DZ.

Number one and two engines were shut down. The navigator and co-pilot were both wounded and incapacitated. With the loadmasters fighting the fire and manually cranking down the landing gear, the pilot, Bill Caldwell, managed to get the aircraft back to Tan Son Nhut.

That day Col. Iosue and I had been out flying a leaflet drop. We returned to Saigon shortly after Capt. Caldwell landed. He directed me to get with our Chief Nav., Maj. Bob Highly and plan a better way - and that the three of us would fly it the next day.

That night we got together with the Airborne FACS and devised some ways which would hopefully get us through with minimum damage. The FAC would serve as our combat controller.

To avoid the appearance of the C-130 from the same heading, we drew a circle around An Loc and laid out six different inbound tracks to the DZ. The FAC would assess the situation and choose the track which would be the safest for us to enter. He also gave us a recommended outbound track to escape on.

We proceed to an orbit point approximately 10 minutes from the drop zone. In this orbit, at a safe altitude of 5 to 10,0000 feet, we completed the 20, 10 and 6 minute checklists.

After depressurizing, the bleed valves were closed, this item was not mentioned anywhere in our manuals except for assault landings and takeoffs.

We were having a problem with the AFT cargo door. The up lock did not always work and thus the door would not stay up. While still in the orbit, we opened the door from the back of the airplane and left the aux pump on.

After the loadmaster completed his checklist. The load was hot, all restraint had been removed, it was being held in only by the CDS gate. But if the aircraft were hit and the load needed to be jettisoned, the ramp could be opened and the gate cut.

Another new checklist item we instituted, was for the pilot and co-pilot to lock their shoulder harness to prevent one of them from falling on the yoke if he was hit.

When the FAC cleared the crew into the DZ. They would start a low level dash at approximately 100 feet and 250 KTS. Inside two minutes, the slowdown was started. While maintaining the low altitude, the power was reduced to idle. Flaps were lowered on air speed to the proper CDS setting. At approximately 170 KTS the ramp was started down by the engineer.

Thirty seconds out the pilot attained his 600 foot altitude and airspeed.

In order to identify the drop area, the navigator drew three circles around the DZ. A slowdown, a one minute warning, and a release line.

The pilot and co-pilot were too busy flying the aircraft to be looking for the DZ. The engineer was watching his panel so the navigator was the only one who could make the drop. Of course during all of this time from two minutes in, the aircraft was under attack from ground fire.

As soon as the load was clear the pilot increased airspeed, descended, and turned to his escape heading.

During the escape maneuver, the FAC, who was flying up behind the 130 during the run in, gave directions to help the escape. He would tell you to break left or right as required; then clear you to pop up when out of the threat area.

Once out of the area, the crew reviewed the checklist and turned off the green light which was invariably left on.

Unfortunately, we were required to return again and again to the same DZ, at roughly the same altitude and airspeed. Even though we made the drops that day, they had us in their sights and the aircraft were hit.

Two days later, the first plane in was hit in the right wing by 23 mm while about one minute out. They continued on, made the drop and then turned on their escape. They tried to climb for bailout but couldn't get high enough. Proceeding on toward Lai Khe, the pilot, Doc Jensen, realized he couldn't make it. The intercom system was out. Number three engine was out, number one and two engines were in flight idle, number four engine was at max power, but still the aircraft was in a right turn. The reason was a fire on the right wing which had already burnt off the aileron and flap. The pilot, still unable to hear anything decided to put it into a swamp. Fortunately two Air Cavalry choppers saw them impact and immediately rescued the crew.

The only serious injuries were the loadmasters and they were hurt because of the flying debris in the cargo compartment. Here we learned to take the unnecessary equipment out; all the seats, stanchions, chains, devices, tool boxes, etc.

While we were flying around the country our resourceful crews were acquiring, some defensive equipment. Later on the supply system finally authorized us ballistic helmets, flak vests and some armor plated vests, however in the early stages of the battle, we were scrounging for equipment anyway we could.

Let me tell you about the way the loadmaster went to war. He would put that armor vest on, take a flak vest apart, tape it around his legs, lay chains on the floor, put the garbage can on the chains, get in and from that position activate the static line retriever as back up for the drop.

Next we tried the MLRADS or the Mid Level Radar Air Drop System which used Charlie Brown Reef Cutters on the parachutes.

We felt we could increase the accuracy of the deliveries by employing the SAC MSQ radar bomb procedures, which the B52s used for the ARC light mission, and the C-130 for the commando vault mission, our BLU 82 10,000 pound bomb. The navigator would give a ground speed and drift angle to the MSQ site and they in turn could place you within 50 feet of a requested point by means of a GCA. However we still had to compute an accurate carp for use by the MSQ site. The problem as always was the wind, so we took readings each thousand feet on the way up.

We made the drops from 8,000 feet using the 40 second cutters. A few of the bundles hit the DZ but a majority fell long. We found out later the chutes were opening early due to improper rigging of the cutters.After two days MACV stopped the high drops and directed us to go back low.

Down we went. But his timc wc went at night and we had gunships to help us. The gunship was much better for our purposes than a fighter. He could loiter in the area and deliver continuous fire on a target while we made our run-in.

A side note on how we got the gunships. I was over at 7th AF one morning and saw a friend from Pope - Howard Rowland (who by the way is here at the convention,) he was pulling a 30 day TDY from Ubon as the gunship liaison. The rest is history.

Because the crews couldn't see the obstacles, at night we flew 500 feet above the terrain. We hoped that they wouldn't see us until we were at the target. But of course with a bright moonlit night, they saw you anyway.

We had pretty good success putting the bundles on the DZ from low altitude. But once we got in two or three they wanted more. A requirement for 10 sorties per night was laid on.

During this time Kampon Treach in Cambodia came under attack and was surrounded. We ran a drop to the DZ in the daytime and the aircraft sustained 86 hits. The next night we took in three successful sorties. But this same night we lost our second plane at An Loc. He was hit and downed approximately two kliks southwest of town.

The following night the FACS would not clear any more 130s into the An Loc area. They considered the fire to be murderous. Twenty-three and 37 mm were active; 51 cal. was everywhere.

Meanwhile at Kamphon Treach the first aircraft in was hit by 51 cal. from a four-sided box pattern. The next two airplanes were sent back by the FAC. We had lost any element of surprise.

New ground rules were laid down by 7AF, we would have either gunships or fighters for all night drops and fighters for all day drops.

Then 7th AF directed planning for a 10 ship daylight, standard altitude, one minute in trail airdrop to be supported by fire power from A-37s. Col. Iosue considered this plan suicide and when he was unable to get MACV to shut it off, somehow got word to the CINC at PACAF and Gen. Clay turned it off.

Again the drops were renewed using night low level procedures. However, some day drops were scheduled to make up for those we could not complete at night.

In a few more days we lost the third airplane, this time just east of the town. By now the higher echelons of command were convinced of the danger inherent in our mission and we were allowed to go back to high altitude.

But due to defenses of the enemy such as 37 mm and the new SA - 7 Strella Missile, we were being forced to go to higher altitudes.

The SA-7 downed a C-119 gunship over An Loc and hit a C-130 gunship at 7,500 feet.

By this time help was on its way from the States. TAC deployed two 130 squadrons, the 61st from Little Rock and the 36th from Langley. The 61st was an AWADS squadron which along with a newly developed high velocity airdrop, using thousand pound bundles, fitted to fifteen foot slotted parachutes, provided the needed airdrop accuracy.

The successful airdrops turned the tide and by mid May the NVA had ceased ground attacks on An Loc. However highway 13 from Saigon remained closed until late June.

From mid-April to mid-May C-130 crew made 57 low level and 90 mid or high level drops at An Loc. During that time at least 56 airplanes were hit plus the five we lost. Seventeen crew members were killed or MIA and another 10 wounded.

At night planes who knew they were hit would come back to Saigon, shut down, and discover fuel streaming out of all four main tanks.

While the battle of An Loc was a major event in the history of tactical airlift, perhaps the most important outcome was a series of conferences that led to development of an airlift tactics manual. Representatives from PACAF, USAFE and TAC drafted MCM 3-4 which recorded and improved upon the tactics developed at An Loc.

It was events such as this and those experienced by the fighter community that led TAC to develop the Red Flag activities that we are so familiar with today.

One final story from the manual writing phase. As we researched into the available documents and talked with older pilots who had been there, we heard the same experiences from Dien Bien Phu, Korea and WWII. The crime was they were only hanger talk and had never been put down on paper as authorized tactics. With the sophisticated weaponry of the 70s and 80s, the tactics of Nadzab, Corrigidor, Sicily and Normandy - which we continued to fly at demonstrations and exercise for our military and government leaders, were not only outdated but suicidal.

The preceding account of tactical airlift activities, which took place at An Loc, RVN, during April 1972, was delivered to the "Old-timers Seminar" at the Airlift-Tanker Convention and Symposium 1994 in Seattle Washington, on Friday, October 28, 1994.

The speech is taken from first hand experiences during that campaign. These remarks were initially presented to an airdrop working group hosted by the Tactical Airlift Center at Eglin AFB, Florida, in June 1972.

This account was also provided to Col. Ray Bowers who used it as a source document for his book Tactical Airlift, published by the Office of Air Force History 1983.

Researches of this and other Southeast Asia battles may have uncovered additional facts which could shed new light on the events. But from my perspective, and the notes I kept, this is what took place.

Rescue!

by Sam McGowan

It was the wee hours of Sunday morning two weekends before Thanksgiving, 1964. I was sleeping in my bed on the third floor of the aircrew barracks at Pope. My radio had been tuned to the Grand Old Opry and I had just dropped off to sleep when the sirens went off. About the same time, the telephone down the hall started ringing. There was no doubt about it, in a TAC troop carrier wing in 1964 the combination meant something was up. The wing was expecting an ORI, but TAC headquarters had sense enough to kick those off early on a Monday, or just after the clubs had closed after a Saturday night on an Air Force base!

Within minutes those of us living in the barracks were on out way to our respective squadrons, which in my case meant the 779th Troop Carrier Squadron building. Those of us living on base were the first to arrive, and as soon as we walked in, the duty officer put us to work on the call-out sheet. My flight engineer thought I was joking when I reached him at home. I had to call him twice to convince him that we really were on alert. We were assigned to a non-operational crew since we were both still in student status, along with most of the men in the 779th. Our parent unit the 464th Troop Carrier Wing, had just competed transition into the brand new E-model version of the C-130 and the 779th had been included as an afterthought. My new squadron had been activated to provide the nucleus for a brand new concept, what today is known as Combat Talon, but all of that was still in the future and none of us were even aware of it. What it meant to me in 1964 was that the creation of a fourth C-130 squadron at Pope afforded me an opportunity to exchange my tool box for a flight suit and cross-train out of OMS into the loadmaster career field.

It wasn't long before the rest of the squadron began trickling in, and not long after that before everyone began realizing this was not a TAC practice alert. Something really big was brewing, and most of us had an idea what it involved It was said in TAC in the 1960s that all a C-130 crewmember had to do to know where he would be going next was to read the paper, and the headlines for several weeks had been devoted to events in the former Belgian Congo. The Congo held special interest for those of us at Pope because the 464th had been maintaining a detachment of C-130s there since the preceding June as part of Operation LEO, a Strike Command Mission that had been sent to Africa at the request of the Congolese government after a group of tribesman calling themselves "Simba" revolted. In August the Simbas took a number of whites living in the areas under their control as hostages and seized the U.S. Consulate in Stanleyville. Since that time the whole world had watched in fear as the Simbas made one threat after another against the United States and the hostages if the LEO force was not removed from Africa.

As things turned out, it was not those of us at Pope at the time who would be involved in Operation DRAGON ROUGE, but rather our buddies who were TDY to Evreux, France with the Pope Rotational squadron to the 322nd Air Division. Shortly after the Simbas seized the hostages, the Pentagon and several other military organizations began working on a rescue plan. Though several ideas were advanced, the one ultimately chosen was that worked up by the United States Air Forces, Europe and the Belgian Army. The USAF/Belgian plan called for an airborne assault on the Stanleyville Airport by Belgian paratroopers, with the American participation provided in the form of a squadron of C-130s to airlift the rescue team to Africa and make the drop. Because the Pope crews were flying the long-range version of the Hercules, the E-model squadron at Evreux was selected.

At the moment that the alert sirens went off at Pope, our comrades at Evruex, all from the 777th and 788th squadron, were crew resting after being called back to France from airlift missions throughout the USAF region. Late that afternoon the crews reported to the flight line and began launching just before nightfall, with instructions to climb to 2,000 feet then open a sealed manila envelope that had been handed to each navigator and proceed as directed. Their next stop was Klinebrogel, Belgium where each crew loaded Belgian paratroopers and their equipment then proceeded to a refueling stop at Moron Air Base, Spain, then on to Ascension Island where the force would spend the next several days waiting for "the word." At Ascension the DRAGON ROUGE rescue force was joined by a TAC "Talking Bird" C-130 with a special communications package to maintain contact with Washington and Brussels.

On Saturday before Thanksgiving the Talking Bird teletypes received a message – DRAGON ROUGE was to proceed to African soil. Early the next morning the flight of Tactical Air Command C-130s landed at Kamina Air Field in the Congo as the early morning mist lifted from the veldt. Again, the crews went into the "crew nest," if such is a proper term. No quarters existed on the base and the only place to sleep was in the airplane, which was not unusual for TAC's combat crewmembers. Unlike MATS airlifters, who could not function without clean sheets and a shower, TAC C-130 crews were prepared to accept anything. Early Monday evening another message came through; the rescue mission was on. That evening the crewmembers tried their best to sleep, knowing that the next morning they would be going into action.

Carrier Wing, Pope AFB, North Carolina, airlifted a unit of Belgian paratroopers to Stanleyville, Republic of the Congo, to free several hundred civilian hostages, both Congolese and citizens of at least 18 foreign countries, who were being held under the threat of death by Congolese rebels. Nov. 24, 1964. Here wounded hostages from Stanleyville are being carried from a USAF C-130 at Leopoldville, Republic of Congo. (Courtesy of Sam McGowan.)

Just as the sun peeked out of the veldt, a Cuban pilot employed by the Central Intelligence Agency made a strafing pass down the Stanleyville runway in a B-26 Invader. Right behind the Invader came the first Hercules. As it crossed the airfield, 61 Belgian troopers leaped into the morning sky, following their commander, Colonel Charles Laurent. Five C-130s dropped 320 troopers as each airplane made two passes, first to drop troops and the second to drop bundles and allow the jumpmasters to jump. The assault was not without opposition. The C-130 crews had been briefed to expect only sporadic rifle and automatic weapons fire, but the Simbas turned out to have .50 caliber machine-guns. Each of the five C-30s took hits, though fortunately none were serious. After dropping their troops, four of the five airplanes headed for Leopoldville to refuel, while Chalk One, flown by Captain Huey Long and his crew, and carrying Lt. Col. Burgess Gradwell of the 322nd Air Division who had been sent along as mission commander, remained overhead. While orbiting at 2,000 feet, against the advice of the navigator, Lt. Charles Coble, Chalk One suddenly began taking hits from a .50 caliber that knocked out some hydraulics. Long took his damaged airplane to Leopoldville as well.

Exactly 45 minutes after jumping from the lead C-130 Colonel Laurent radioed that the airfield was secure and suitable for landings. Chalks Seven through Eleven landed and discharged their troops, then took off again for Leopoldville. Chalk Six, flown by Captain Mack Secord, came in later; the crew had lost a life raft after taking off from Kamina and went back for the spare airplane. Secord was told to remain on the ground at Stanleyville along with Chalk Twelve, flown by Captain B. J. Nunnally's crew, which was configured for air evac and included an Air Force doctor as part of the medical crew.

After the first C-130s landed, the Belgians dispatched a rescue team for the Victoria Hotel where the hostages were being held. The Simbas realized a rescue effort was underway and began marching their captives toward the airport. Evidently the Simba leaders realized the jig was up, and planned to turn the whites over to the Belgians unharmed. However, the Simba rank and file were hopped-up and hung-over after a night of drinking and smoking hemp. When they realized their officers intended to release the hostages unharmed, they decided to take matters into their own hands. The Simbas shot their own officers, then began a systematic slaughter of the hostages, picking women and children as their first targets. Fortunately, after the first volley of shots, the Belgian paratroopers rounded a street corner and the Simbas fled. nevertheless, some 30 white lay dead and dying and scores were wounded.

With the hostages now safe, though many were injured, an airlift was mounted to transport them out of Stanleyville. Mack Secord's airplane was the first to take-off. As he was taxiing out, a pair of Simbas ran out of a field of tall elephant grass and fired a burst of automatic weapons fire into the wing of the Hercules. Secord ignored the warnings from Nunally that his airplane had been damaged and took-off for Leopoldville where he landed with no flaps. The airlift continued for the rest of the day and the next as more than 2,000 whites and non-African people of color were brought out of the city.

Two days after DRAGON ROUGE, a second mission was flown a DRAGON NOIR to the airfield at Paulis on Thanksgiving Day, where the paratroopers jumped blind into fog. The C-130 pilots landed in a cloud of fog and red dust that obscured the runway and made the landings extremely difficult. As had happened at Stanleyville, the Simbas at Paulis killed or wounded some of their captives, but the rescue was overall successful.

Back at Pope, pandemonium ranged on the base when we heard what our buddies had done. The heart of every airman on the base was filled with pride. But there was a down-side. Though the C-130s and their crews belonged to Tactical Air Command, and were under Strike Command Control while in Africa, they were TDY to 322nd Air Division, which in 1964 had recently come under control of the Military Air Transport Service. At the time, MATS was lobbying fiercely for the creation of an airlift command, and the MATS PIO put out a news release giving credit for the rescue to MATS, though the only MATS personnel involved was Colonel Gradwell, the second loadmasters on each crew who came from the Fifth Aerial Port Squadron at Evruex and a pair of combat controllers, also from 5th APS. Eventually, the credit was properly given to TAC, but not before a lot of hard feelings erupted among TAC C-130 crews.

Those who took part in the mission were awarded Air Medals, and were selected for that year's MacKay Trophy for the most meritorious flight of the year. The trophy was presented during ceremonies at Pope the following May, including a parade that the men of the airlift squadrons were proud to participate in.

Commando Vault

by Sam McGowan

It was a hot summer afternoon in North Carolina. I was lying on a section of troop seats in the back of a C-130E during a training flight over the Carolina sand hills. With nothing to do except make my periodic checks of hydraulic quantities and scan the engines, I was thinking. Recently I had read in the *Air Force Times* of the Air Force's new concept of arming transports, specifically the antiquated C-47s that were spread all over the world in base flight detachments. I was wondering just how the C-130 could be armed, how it could be made to drop bombs. After all, the airplane reminded me of a B-24!

That was in the summer of 1965. Four years later I reported into the 29th Tactical Airlift Squadron at Clark for a 15 month tour as a loadmaster. When I met my new squadron commander, I found my assignment was welcome. The 463rd wing at Clark was getting a huge influx of new loadmasters, all coming from MAC C-141 squadrons, and most with only experience as "pallet pushers." I'd only been in "Mother MAC" for a year when I got my orders to Clark,. Before that I'd been at Naha flying A-models, after getting my start in E-models at Pope. I may have been a "Mac weenie" all right, but I was one with a couple of thousand hours of TAC C-130 experience, and more than two years of Vietnam experience on top of that. I quickly realized that the squadron saw me as an asset.

My first trip out of Clark was with a crew ordinarily commanded by Major Howard Seaboldt, though the aircraft commander was in-country at the time serving a tour as airlift mission commander with 834th Air Division. The AC for this trip was the squadron ops officer, but the navigator, flight engineer and loadmaster were from the original crew. During our 16 days in-country, I learned that the crew had pioneered the delivery of 10,000 pound bombs from C-130s, and that the Air Force had approved their routine use in Vietnam to create helicopter landing pads. Now, that was right up my alley!

My first crew turned out to be a bummer. The aircraft commander refused to take his crew's advice, and after taking an airplane with a low tire into a forward airfield and blocking the runway, he was removed from operational flying and sent to a ground job, leaving me without a crew. That was when I was notified that I had been chosen to be one of the bomb-qualified loadmasters. I went in-country, to Cam Rhan, where the 463rd had moved its operating location from Tan Son Nhut, with a bomb-qualified crew from the 774th. The loadmaster and I went way back; we were in the same batch of mechanics who cross-trained out of the OMS squadron back at Pope to become loadmasters back in 1964.

My checkout involved making four drops under the supervision of an M-121 qualified instructor loadmaster. We made our first drop near Katum. The FAC came back with a BDA (bomb damage assessment) of at least 100 KIA. That was when I realized that while we were making instant LZ's, we were doing it in places where there were large concentrations of NVA. On our second day in-country, there were no bomb drops scheduled so the crew, commanded by a Major Devers, was scheduled for a day of tactical airlift flying. Since I was in-country specifically to check out on bomb drops and needed no instruction in cargo hauling, I decided to sit this one out. I'd brought my guitar in-country with me and the night before I'd had a lot of free drinks pushed in my direction. I've had mixed emotions ever since about my decision, because the crew was landing at Katum that day when they were caught in the sights of an NVA quad-fifty that put a dozen rounds of .50 caliber through the airplane. No one was hurt, but I still wonder if I would have been if I'd been aboard. One round came up right by the left paratroop door, which is where I had a habit of standing during take-offs and landings. (C-130 loadmasters preferred to be in the back, strapped into the door, because most often if anyone survived a crash, it would be the loadmaster who was near the tail.) The airplane was pretty badly shot up.

Loading a BLU-82 15,000 pound bomb aboard a C-130. (Courtesy of Sam McGowan.)

After checking out as an M-121 qualified loadmaster, I went to my own bomb crew. We would spend about half of our time in-country flying bomb missions. For certain reasons, the AC on this crew shall remain nameless. On one mission we had Brigadier General John Herring, the new commander of 834th Air Division aboard. We had made our first drop and I was in the back re-rigging for the second when I felt the airplane swerve. We had lost an engine. A C-130 will fly just as well on three as on four, so we kept right on with what we were doing and made our second drop. Afterwards, my AC decided to take the general down for a close look at the results of our work. his aide, a C-123 pilot who probably came from MAC, wasn't too impressed that my AC had taken his boss down to low altitude in a C-130 with one engine caged! But that wasn't what got him. The blow fell when my crew went in-country with another loadmaster aboard and no one bothered to inform them that their first in-country stop would be at Da Nang instead of Chu Lai, where every 463rd shuttle input mission had been stopping for at least two years.

As a result of that episode, my entire crew was busted down to student status, but since I wasn't with them I was excepted. I was a bomb-qualified loadmaster without a crew. One afternoon I was standing at the counter in operations when Chick Anderson, the engineer on the crew I had gone in-country the first time with, came over with his AC, a hawk-nosed major with the name "Seaboldt" on his name tag. I'd seen him once before, right after he came back from the States from a mod-bird trip, but I'd never met him.

"Sam, Howie and I've been talking. Mike (the crew's loadmaster) just left for the States. We want you to join our crew."

Now, this was a bolt out of the blue. "Seaboldt's Flying Circus" was a crew with a reputation. Everyone knew Howie, and even though there were some officers who thought he was a disgrace, no one could accuse him of being a bad pilot. Every pilot in the wing admired Howie, though perhaps grudgingly in some cases. If there was a difficult job to be done and the Circus was around to do it, they got the job. Howie had been copilot on the crew with Major Bob Archer when they encountered a pair of officers from Systems Command who picked them to help with the M-121 tests. Archer had gone to Stan/Eval and Howie had been given the crew. Both Archer and Seaboldt had SAC background in B-47s and B-52s at Barksdale; now they were superb C-130 pilots after more than three years each at Clark. I was about to join a crew with a reputation, and consequently I would achieve a reputation of my own.

I'll not go into the adventures I had with Howie

and our crew, like the incident with the mouse at Sydney, Australia, but confine this solely to the M-121 mission. When we went in-country, if there was a bomb mission to be flown, we flew it. Typically, it would take about three hours of flying to drop two bombs, depending on where the target was. After the second drop, HILDA, the 834th command center, would put us on whatever mission was most urgent. This meant that we would spend the rest of the day flying tactical airlift missions into the "hot spots," places like Katum, Tonle Cham, Bu Dopa and Quan Loi. On more than one occasion, we would drop a couple of bombs to launch an Army operation, then spend the rest of the day carrying ammo and fuel into a nearby forward field to support the troops who had just assaulted into the LZ we had created for them.

Now, one might think that using a C-130 to drop bombs was a rather futile idea. Actually, the Hercules made an excellent bomber. The technique of using the same ground radar used to control B-52 strikes, combined with a delivery method allowing pinpoint accuracy, gave the C-130 bombers the lowest average circular error of any airplane type dropping bombs in Vietnam! I have seen bombs that I had just released explode squarely on the top of mountain peaks no bigger that a house. I've got a picture of it. We could hit targets that B-52s couldn't see, and that fighter pilots were lucky to come within a mile of. The Army was so pleased with the results of Commando Vault that they and the Air Force developed a more powerful 15,000-pound bomb the designated as the BLU-82.

During Desert Storm, the Air Force used MC-130s from the Special Operations Command to drop BLU-82s. In Vietnam the C-130 bombers were "plain vanilla" airlifters, and the crews flew airlift missions when they were not bombing. No "Special Operation" supermen here! Just us trash-haulers.

Commando Vault continued throughout the American involvement in Vietnam as the C-130 bombers became an integral part of the war. Bombs were dropped to kick-off the Cambodian incursion in 1970, and the South Vietnamese invasion of Laos during Lam Son 719. An outgrowth of the bombing mission was the development of the GRADS ground-directed cargo release system which ultimately saved the South Vietnamese force at An Loc in 1972.

Incidentally, anyone who has watched the Discover Channel "Wings" program on the C-130 has seen Howie Seaboldt in action. He is the pilot featured in the segment on the C-130 bombers. Some people who have seen it and haven't seen me in years think the loadmaster is me. It is actually Dick Klemm, who took my place in Stan/Eval when I came back to the States to fly C-5s with "Mother MAC." Klemm and I resemble, but I'm better looking.

Some of those who worked on and flew "Ol' Shakey" are (standing L to R) MSgt. Ralph Wright, TSgt. Kenneth Vaughn, Lt. Col. Bob Symmes, Col. Clark Olander, Lt. Col. Bob Schorr, SMSgt. Alvin Adams, MSgt. Gordon Bloodworth, (Kneeling) MSgt. Arthur Koon, MSgt. Jimmy Stanton, SMSgt. Louis Boos and CMSgt. Frank Beachman. (Courtesy of Bob Schorr.)

Old Shakey's Final Mission

by Lt. Col. Stephen A. Ashley
From The Mobility Forum July 1993

9 October 1986 was another beautiful day in the Great Northwest. There was an early fall breeze, and the sky was clear. The fresh snow on the Cascade Range glistened in the afternoon sun. Majestic Mount Ranier towered over McChord Air Force Base, Washington, and the Puget Sound countryside.

I was working in the 62nd Military Airlift Wing safety office when word began to spread throughout the headquarters. The Douglas C-124 *Globemaster II* destined for McChord's Air Museum (MAM) would be arriving within the hour. I had heard much during my airlift career about the old workhorse. To those who flew this airlifter, she was affectionately called "Old Shakey."

The C-124 was operational at McChord from 1951 until 1969, when it was replaced by the C-141. The 62nd's *Globemasters* had flown in support of the French at Dien Bien Phu in 1954, the United Nation's peacekeeping troops in the Congo in 1960, and many other worldwide contingencies.

Today's mission would be the last for the C-124. Efforts to obtain a *Globemaster* for MAM began in December 1981. After several false alarms, an airplane in restorable condition was located at the Detroit Institute of Aeronautics in Willow Run, Michigan, with the help of the U.S. Air Force Museum (USAFM). The school had been using the airplane for instruction in aircraft systems, but would be willing to swap it for a more modern aircraft. A trade was finally approved by the USAFM—a T-39 Sabreliner for the old *Globemaster*.

Acquiring the aircraft was relatively easy compared to moving the giant airplane to McChord. It took over $40,000 in contributions to fund the project. A volunteer maintenance team from MAM, led by Master Sergeant Don Egan (Ret), worked for many months to refurbish the aircraft and get it airworthy. Lieutenant Colonel Gary Caldwell of McChord's 4th Military Airlift Squadron was selected to command the ferry flight. Lt. Col. Caldwell selected the rest of his crew, and began mission planning for the flight.

The flight was delayed several times due to weather and engine problems—not engine maintenance, but availability. The only airworthy set of Pratt and Whitney R-4360 engines were on a C-124 displayed at Travis AFB, CA. These engines had already been "borrowed" to ferry a *Globemaster* to Charleston AFB, and the mission was several months behind schedule. The flight from Willow Run, planned for mid-May 86, slipped several times until finally everything was ready for flight.

Lt. Col. Caldwell and crew departed Willow Run for a short test flight (gear down) to nearby Selfridge ANG Base, Michigan. The plane was jacked and a gear retraction test accomplished. A few other write-ups were also cleared. With all systems go, then-CINCMAC General Duane Cassidy gave approval for the one-time, non-stop flight to McChord.

As the huge cargo plane labored into the air from Selfridge's runway, everyone on board knew they were now part of aviation history. This would be the last flight of a C-124.

Mount Ranier and the Cascade Range of Western Washington provided a breathtaking background for Old Shakey's arrival. The old warrior was met and escorted across the mountains by a C-141B and a C-130E from the 62 MAW. As the vintage aircraft approached the field, white oil-smoke began spewing from the number one engine. Lt. Col. Caldwell and crew shut the engine down on final approach. The perfect landing set off a round of thunderous cheers from the large crowd gathered to watch the end of this historic flight.

After swapping out the Travis engines, the aircraft was repainted in the markings of the Military Air Transport Service and the 62d Troop Carrier Wing. The *Globemaster II* is now on display as a member of McChord's Air Museum, where future generations of airlifters will be able to see this magnificent old bird and relive part of our proud mobility heritage.

C-130 Gives Second Boost to Gemini 3 After Orbit

A Charleston-based C-130 Hercules provided another "boost" for the Gemini-3 spacecraft after it completed its historic orbits around the earth March 23.

The Hercules - assigned to the 76th Air Transport Squadron - flew to Jacksonville where it picked up the spacecraft and returned it to Cape Kennedy. Heading the nine-man crew was Capt. Harold G. Wiechman. The Gemini-3 had been taken from its pick-up point in the Atlantic to Jacksonville on the Aircraft Carrier *Intrepid.*

Returning spacecraft to Cape Kennedy has become a routine mission to aircrews of the 1608th Air Transport Wing based in Charleston.

In May, 1963, the wing airlifted astronaut Gordon Cooper's capsule *Faith 7* from Hawaii. Earlier, they brought home John Glenn's *Friendship 7* and Scott Carpenter's *Aurora 7.*

A Key to Success

Teamwork keeps operations in Haiti successful

by MSgt. Russ Maheras

As Operation Restore Democracy went into its 11th day, Oct. 3, Air Mobility Command aircraft had flown more than 350 missions to Port-au-Prince International Airport, airlifting more than 8,379 tons of cargo and 13,342 passengers.

A key to the operation's success had been the more than 60 aircraft maintenance and logistics personnel assigned to support the 436th Tanker Airlift Control Element, which has been operating Haiti's largest airport since the operation began Sept. 19, said Capt. Teresa Stith, senior logistics officer here.

Maintainers from Dyess AFB, Texas, and Travis AFB, Calif,; supply specialists from McChord AFB, Wash., and Dover AFB, Del.; along with contracting personnel from Dover, round out the logistics team at Port-au-Prince.

Although the bulk of traffic has been "the big three (C-5s, C-141s, and C-130s)," said Stith, her troops have marshaled a variety of other fixed-and rotary-winged aircraft- everything from Boeing 747s to Army CH-60 Blackhawk helicopters. Maintainers have handled more than 400 aircraft.

Because of the great influx of troops and equipment into the airport, a big challenge for Stith and her people has been to fit all incoming aircraft into ever-shrinking ramp space.

"When we first got here, we did a parking plan. But almost every day our available space gets smaller and smaller," said Stith, who is normally the maintenance supervisor for the 60th Component Repair Squadron at Travis. "The Army needed convoy parking space - we needed an area for equipment marshaling - it all adds up."

Space is so tight, "sometimes we have to close the (ramp) road if a wingtip will go over it."

"This delays traffic, but we have to do it," Stith said.

During the heaviest days, Stith estimates her maintainers marshaled 40 to 50 cargo aircraft.

Another problem for the maintainers is the close proximity between fixed-and rotary-winged aircraft operations. "It's a big safety issue, and we have to work with the TALCE to keep on top of it," Stith said.

But the biggest challenge, Stith said, has nothing to do with aircraft operations. At the beginning of the operation, crew chiefs worked in high-heat, high-humidity conditions while wearing full combat gear- including flak vests. "The guys were just cooking out there, but we've since gotten relief on the flightline," Stith said.

Living conditions are austere as well. Rodent and insect infestation plague some areas. Cramped conditions are a problem everywhere, and for most, there is little relief from the heat.

Like other deployments to austere locations, the troops learn to appreciate some of the things they normally take for granted. SrA. Allan Tender, a C-141 maintainer, is a good example.

"All I'm looking forward to is getting that kitchen set up (in tent city) so I can get my first hot meal."

Yet despite the obstacles, the Port-au-Prince maintainers are working as a team and getting the job done.

"Considering the conditions, we're doing great," said Stith. "Everyone gets along real well. We're getting everyone involved on the flightline, even the folks who normally work in the back shop."

Lyons agreed. "There was some tension at first, because you don't know what the other guy can do. But now we're a team."

(Maheras is assigned to 436th Airlift Wing Public Affairs Office, Dover AFB, Del.)

Esoteric Airlift/ Tankers ops

by Tom "Fig" Newton, Col USAF (Ret)

Upon graduation in December 1943 with Pilot Class 43K at Ellington Field, TX, I was ordered to Gowen Field, Idaho for RTU training in B-24 high altitude bombing. After completing this training we delivered a B-24 H from Topeka, Kansas to Stone, England in May 1944. We were in a crew replacement pool awaiting assignment to an operational unit.

Early one morning, around 3:00 a.m., three of the crews were awakened (ours included) and told to be ready to move out in one hour. We were trucked to a train station and boarded without any of our questions being answered as to what and where. After a few hours we were escorted off the train onto a small train platform, in the rain, and then loaded into 6-bys (trucks) again. I was lucky to get to sit in front with the driver who was the first to shed light on our destination.

We were going to Harrington Air Drome from where Black B-24s were flown on night missions in support of the "underground resistance organizations" over continental Europe.

We were assigned to the 492nd Bomb Group (H) and our crew to the 858th Bomb Squadron (H). We went into flight training, almost immediately flying simulated night missions over England. Our operational drop altitude for material or personnel was 400 feet above ground, measured with a radio altimeter which usually was accurate.

When we had filled all the squares we were sent individually with experienced crews to get a taste of the "real thing." We then, as a crew, were declared "ready to go."

The air crew briefings were held in the afternoon around 2 o'clock. Targets, drop zones, were identified by a set of coordinates only and the time the reception party would have the target open. Usually 20-30 minutes except in hot areas when it would be only 10-15 minutes. A description of the "reception to be expected" i.e. four fires inline, or three fires in triangle, etc. And usually there was supposed to be a person at the drop zone with a light that was supposed to reply to our flashing a code with his code. I do not recall ever seeing one. After the briefing was over, our entire crew met and discussed the location of our target. Usually there were two ways to reach there. The short way straight across the English Channel, or go South and enter into Southern France which was less heavily defended. If any one of the crew felt uneasy that night with the short route, we would go south. Our route planning begin with the open time of the target and work back to take off time. En route our bombardier sat on cushions in the nose of the aircraft doing pilotage with road maps and calling out what might be obstacles in our path. Our navigators had a radio triangulation device called a "G" box, and then went to dead reckoning when out of radio range, which was most of the time. When we estimated we were at the coordinates of the target at 400 foot altitude, we would make a very large 360 degree left circle, and if no response we would make one to the right. Nearly always there would be a visible reception signal before completing the second circle. The physical setup of the ground signal designated the desired direction of final approach, and we would set up our drop pattern in accordance with that, terrain allowing. I should mention here that weather was never a factor considered in whether to go or not. We went. If the reception party was going to hang it all out to be there, we would too. In hilly or mountainous terrain we frequently went on our navigators word to spiral down in clouds to reach our drop altitude, sometimes in valleys just below the cloud bases.

Our cargo loads were packed in long canisters fitted with a small parachute mostly for stabilization. These hung in the bomb bays on the bomb racks. Everything imaginable was carried. The load was usually salvoed on one pass to reduce the exposure time for the reception on the ground, and us. A B-24 humming around at 400 feet in the middle of the night would signal out the "bad guys." Our planes had no nose, waist or ball guns. The hole for the ball turret was covered with a folding plywood door. When we carried personnel to drop, they, on arrival at the target would sit with their back to the front of the aircraft with their legs and feet hanging out of the ball turret. We called this the "Joehole." On the greenlight the "dispatcher," as he was called, would shove the person out. The jumpers wore special parachutes which opened okay from the 400 foot drop altitude. Only one jumper went on each pass. So we rarely carried more than two jumpers. These people, some of them female, were super organizers and planners. Some of them made several trips into the continent.

One element of our organization flew C-47 aircraft at night and made landings in all sorts of places to recover these organizers so that they could go again. Well, my hat's off to them! And to the people with the radios hidden away who made this whole thing work.

Late in 1944, our group began flying fuel into fighter strips in Belgium and France. Mogas in Fighter drop tanks on the bomb racks and AV gas from our wing tanks. These were daylight trips and mostly hohum.

In December our crew volunteered to go to Italy for another go at it. In January 1945, we were stationed at Brendisi. We were called the 2641st (Prov) Bomb Group and 839th (Prov) Bomb Squadron. We were sharing the base with an English group which had been there quite some time doing the night thing. A volunteer American squadron had come up from North Africa also. We operated into the Po Valley of Italy and into Yugoslavia/Austria. By late April we were flying some daylight missions into northwestern Italy near the Swiss border. We lost the only aircraft to enemy fighters, that I know of, on one of these missions. The crew was jumped by some over zealous Italian pilots who didn't know they were supposed to surrender.

I completed my tour on April 30 and left for Cannes, France for R&R on May 1. On May 7 I was listening to AF radio in the passenger compartment of a C-47 en route back to Italy when the end of the war was announced.

I flew back to the United States for drumming out of the Corps in June after declining an invitation to volunteer to join a group of excitement seeking aviators going to the Pacific to pursue the drop business.

I did not realize at the time that I had been in the airlift and tanker business since neither term had been used. Airlift was Troop Carrier or Air Transport Command. I was in Active Reserve Troop Carrier until recall for duty in the Korean thing in May 1951.

It may be of interest to any of today's crop of airmen that I arrived at Stone, England with a total of 336 hours and 40 minutes of flying time. I returned to the United States with a total of 714 hours and 20 minutes flying time.

Operation Deep Freeze. 1964. Top row (L to R): Duane Todd, Doc Pennington, George Thompson, Cecil Cheevers, Dave Pelletier and Mike Walcanski. Bottom row (L to R): Don Hampton, Barney Elansbeek, Hap Allen, R.K. Smith, Perry Gentry, and Willie Glover. Named "Coffee's Grinders" after mission commander Lt. Col. Robert Coffee. (Courtesy of David Pelletier.)

BIOGRAPHIES

ALBERT EUGENE (GENE) ADCOCK, born March 11, 1937, in Christopher, IL. Joined the service in January 1955, classified as a combat controller. Military locations or stations at which he served were: Sewart, Tachikawa, Tan Son Nhut, Dyess, England, Udorn, Forbes, Little Rock, Scott Air Force Bases. Battles in which he participated included first CCT in Vietnam, 1965; and three tours Project 404, Palace Dog.

Memorable experiences: Project NCO "Operation Close Look." Coordinated first and only three ship formation, Ground Proximity Extraction System "drop," Ft. Campbell, 1962. Modified the first modular platform for ground extraction. Suggested the system be modified to replace the extraction hook with an extraction parachute to simplify ground operations. A year later the first LAPES drop was made.

Established and operated the Forward Air Guide School at Udorn RTAFB, Thailand in 1969. By the end of SEA hostilities the school had trained hundreds of Lao, Thai and Cambodian students. Awards: Two Bronze Star Medals and six Air Medals.

Retired Jan. 31, 1977, as CMSgt, from MAC IG.

He is married and has two sons. Today he is Director, U.S. Government Programs, Night Vision Equipment Company, Emmaus, PA.

ANDREW B. ALDERSON, born Sept. 29, 1960, in St. Charles, MO. In 1978, he graduated with honors from Winfield High School in Winfield, MO. He holds a bachelor of science in fisheries and wildlife from the University of Missouri and a master's of science in operations management from the University of Arkansas. As a C-130 pilot at Howard AFB, Panama, Captain Alderson was part of the initial cadre for a one-of-a-kind airlift mission. He was also a C-130 formal training instructor pilot and the chief pilot of squadron standardization and evaluation at Little Rock AFB, AR. He is currently assigned to HQ AMC, Command Presentations. During his career, Capt. Alderson has been selected as the company grade officer of the quarter many times. He was also chosen as one of MAC's Outstanding Young Leaders of 1985. Capt. Alderson is a senior pilot with over 3,300 flying hours. His decorations include the Meritorious Service Medal, the Air Medal with seven oak leaf clusters, the Air Force Commendation Medal, and the Humanitarian Service Medal. Capt. Alderson is married to the former Giselle Roquebert of Panama City, Panama. They have one child: Yvonne.

JAMES W. ALEXANDER, Colonel, USAF, (Ret.)., graduated from East Carolina University in 1955, and began his airlift career in 1956, with the 21st TCS in Tachikawa, Japan. From 1958 to 1967, he was a C-130 Instructor Pilot and Flight Examiner with the 61st TCS, 17th TAS, 346 TAS and 4442 CCTW. His Vietnam tour was with the 776 TAS in 1967-1968. Following staff tours at Langley, Eglin and a joint command tour in Thailand, he was assigned as squadron commander of the 38 TAS in 1974. He was the 463 TAW Deputy Commander/Ops from 1977-1978, and the 374 TAW Wing Commander from 1980-1981.

He completed his military career as Chief of Staff, 23 Air Force, Scott AFB in 1985. Since 1986, he has been vice president of operations for the Metropolitan Nashville Airport Authority. He is married to the former Glenda Merriman and has three children and four grandchildren.

LIONEL D. ALFORD JR., born Jan. 23, 1958, in Melbourne, FL. He was awarded a bachelor of science in chemistry from Pacific Lutheran University in 1980, and a master's of science in mechanical engineering from Boston University in 1984. He completed Squadron Officer's School in 1987.

Alford was a distinguished graduate of AFROTC Det. 900 in 1980. He attended UPT at Laughlin AFB, TX, in class 82-03, and was assigned as a FAC to the 20 TASS in the OV-10A. He moved with the squadron and helped open the 27 TASS at George AFB, CA. In 1985, he was assigned to the 1401 MAS Det. 1 at Offutt AFB, NE. There he served as a C-21A instructor and CINCSAC pilot. Alford was transferred to the 39 TAS at Pope AFB, North Carolina in 1987 and flew the C-130E as an instructor pilot. While at Pope AFB, he was the Assistant Chief of Wing Training for the 317 TAW and the acting chief of Training during Desert Storm. Alford attended Air Force Test Pilot School class 91A and was assigned as an experimental test pilot to the 4953rd Test Squadron at Wright-Patterson AFB, OH. He is currently an instructor pilot in the T-39 and actively flies a host of different aircraft and simulators supporting Air Force developmental testing.

Among his many awards and decorations are: the Air Force Meritorious Service Medal, Air Force Commendation Medal, Air Force Achievement Medal and the Humanitarian Service Medal.

Alford and his wife, Tammy have four children: Tanna, Christin, Erin and Devon.

JULIAN D. ALLEN, attended the National War College, Ft. McNair, Washington, DC, Class of 1993. Lt. Col. Allen hails from Jackson, MS, where he was born Jan. 11, 1954. After graduating from Council Manhattan High School in 1971, he attended Louisiana Tech University where he earned a BS degree in professional aviation, May 1975, and an MA degree in human relations and supervision, August 1976.

He was commissioned a second lieutenant in May 1976, as a Distinguished Graduate of Tech's AFROTC program. In March 1977 Lt. Col. Allen entered active duty at Columbus AFB, MS, as a student in Undergraduate Pilot Training Class 78-04. After designation as an Outstanding Graduate in March 1978, he remained at Columbus as a T-38 Instructor Pilot and Class Commander until May 1980. From Columbus he was assigned to the 6th Military Airlift Squadron, McGuire AFB, NJ, flying the C-141. In February 1981, he became Aide-to-the-Commander, HQ 21st Air Force, McGuire AFB. Upon selection to the Air Staff Training Program (ASTRA), Lt. Col. Allen moved to the Pentagon as the executive assistant to the Assistant Secretary of the Air Force (Manpower, Reserve Affairs, and Installations) from September 1982 through October 1983. He then returned to the C-141 at Charleston AFB, SC, where he flew with the 20th Military Airlift Squadron and served in the Wing Command Post and Current Operations Division until July 1985, when entered Air Command and Staff College at Maxwell AFB, AL. After designation as a Distinguished Graduate in June 1986, Lt. Col. Allen returned to the Pentagon where he served under the Air Force Deputy Chief of Staff, Personnel, as both a personnel programmer and as a Congressional Liaison to the House and Senate Appropriations Committees. In December 1989, Lt. Col. Allen once again returned to C-141 operations with his assignment to McChord AFB, WA, where in April 1990, he assumed command of the 8th Airlift Squadron.

In June 1992, he relinquished command and reported to the National War College. Lt. Col. Allen has over 3,400 flying hours and his awards and decorations include the Meritorious Service Medal, three oak leaf clusters, the Air Force Commendation Medal, the Air Force Achievement Medal, the Combat Readiness Medal, the Southwest Asia Service Medal, two bronze stars, and the Kuwaiti Liberation Medal. Lt. Col. Allen is married to the former Eva Carpenter of Columbus, MS, and they have a son, Randy, and a daughter, Jamie.

JAMES I. BAGINSKI, "BAGGER," born Feb. 18, 1932, in Baltimore, MD. He joined the USAF April 14, 1954, serving as a pilot in C-119 troop carriers. He was discharged Dec. 1, 1984, with the rank of major general.

His memorable experiences include Typhoon Hunter, Laos, Ankhe, Cambodia and Saigon evacuation.

Awards/Medals: Order of the Sword, Distinguished Service Medal, Bronze Star, Superior Service Medal, Meritorious Service Medal with oak leaf cluster, Air Medal with oak leaf cluster, Army Commendation Medal, Air Force Commendation Medal.

His background includes 31 years of increasingly responsible experience in the USAF in areas of operations, training, personnel, aerial port activities, administration, facilities management, logistics services and long range planning. Accelerated advancement to top level management. Director of Deployments. Joint Deployment Agency (now part of U.S. Transcom).

Education: MS degree from George Washington University (public administration); BS degree from University of Maryland (animal husbandry); Air Command and Staff College, Air University National War College, National Defense University.

Since retirement from the USAF, he has consulted for AMTEC Systems Corp, Lockheed Tactical Airlift, Motorola Data transfer device, Bendix Air Transport & Avionics, CONTEL, CSC, Burdeshaw Assoc. and GE USTRANSCOM Baseline Study.

WILLIAM M. BAILEY JR., born Aug. 16, 1928, in Knoxville, TN. Enlisted Sept. 1, 1948, USAF Troop Carrier TAC AFISC, U.S. Navy 7th Fleet MAC. Military locations at which he served include Germany, Rhein-Main, Sewart AFB, Maxwell, Alameda NAS Narton AFB, Saigon 834th A.D. MAC Norton Travis. Participated in Vietnam, KHE-SANH TET.

Memorable experiences include the wonderful people he has worked with. C-130 and C-141 assignments, tour with USN 7th Fleet. Gen. Bill Moore, JR. SQDN, Comdr. 14th MAS. Founder member Airlift Association.

Awards: B/S-V MSM 30LC AM-26LC OUA 30LC, ARM 10LC, GCM, AO, GER, NDSM, AFEM, 25510LC, WHSM Rep VN with Palm, RUN Camp, AOM, NDSM w/ 1BSS, VSM w/8BSS, CRM, RVCM, PUC, MHA, CRM w/ 10LC, AFOUA w/30LC AND V, AFEM, w/5BSS, AFM 900-3/AM w/20LC SOG-106 HQ7AF, 68/BSM SOG-66 HQPAF, 68/AFCM w/20LC SOG-143 HQ22AF, 69/REP of Vietnam Gallantry Cross w/PALM SOG-505, DOFAF, 72/MSM-30LC.

Discharged Jan. 1, 1975, with the rank of colonel regular U.S. Air Force.

Married Georgia McArthur in 1955. They have a daughter, Virginia and two sons, William III and Herbert C. Now retired, residing in Moreno Valley, CA, planning to move back south.

DARIUS V. BAKUNAS, born March 20, 1937, in Klaipeda, Lithuania. He emigrated to the United States in 1949, and entered the Air Force 1960. He received a bachelor's degree in business administration from the Municipal University of Omaha in 1964, and a master's of aerospace operations management from the University of Southern California in 1970. He is a graduate of the Squadron Officer School, Air Command and Staff College and the Air War College.

At the time of his retirement in 1989, he was chief of Deployment Operations for the United States Transportation Command at MacDill AFB, FL. He presided over the first transportation command managed and coordinated, real world deployment of joint forces to overseas contingency locations. His varied career included 16 years with the Military Airlift Command consisting of line flying assignments and five years at the Command Headquarters.

His many decorations include the Defense Superior Service Medal, Legion of Merit and the Vietnam Service Medal with three Bronze Service Stars. He currently resides in Tampa, FL, with his wife Linda. They have two children: Laura and Kenneth.

MELVIN R. BARRETT, is the Lockheed marketing manager for airlift programs at HQ AMC. This includes the sale of new aircraft and modifications. He provides Lockheed daily interface with the HQ AMC Staff in operating over 600 Lockheed-built aircraft. Mel was the HQ

MAC marketing manager for CAE-Link Corporation's C-130 Aircrew Training System and simulators. Mel spent 26 years in the Air Force with 12 years of operational flying assignments as a navigator and pilot in eight different aircraft (Airlift, Tankers, Forward Air Control). He worked 14 years as a staff officer at HQ MAC, Air Staff, and the JCS. He is a command pilot with 5,000 hours and 600 hours of combat time.

Mel has a master's in management, and a bachelor's in business. He has completed Armed Forces Staff College and AWC. He is a life member of the Airlift Tanker Association, Air Force Association, and the Order of Daedalians. Mel resides in Belleville, IL, with his wife, Cheryl.

ROGER D. BASKETT, Colonel, is the Deputy Chief, Joint Operations Division, United States Transportation Command, Scott Air Force Base, IL.

Col. Baskett was born Feb. 7, 1945, and graduated from Chillicothe High School, Chillicothe, MO, in 1963.

Following commissioning in April 1970, Col. Baskett attended pilot training at Craig Air Force Base, AL. He served a tour with the 41st Tactical Airlift Squadron as an aircraft commander, instructor pilot, and flight examiner.

Col. Baskett was assigned to the 21st Tactical Airlift Squadron, Clark Air Base, Republic of the Philippines in October 1986. During this assignment with the 374th Tactical Airlift Wing, he served as an assistant squadron operations officer, chief of Wing Standardization and Evaluation, the 21st Tactical Airlift Squadron Commander and as the wing assistant deputy commander for Operations.

Col. Baskett returned to the United States in July 1991, to attend the Army War College, Carlisle Barracks, PA.

He is a command pilot with over 3,900 hours in the C-130A/B/E/H.

Col. Baskett was promoted to colonel April 1, 1992. He is married to the former Mary Evelyn Hamrick of Chillicothe, MO. They have two children, Melanie and Jeffrey.

GORDON D. BATES, born Sept. 18, 1921, in Cameron, IL. Enlisted June 13, 1940, in Chanute Fld, IL, serving USAAC/USAAF/USAF. Military locations at which he served include American, EAME, Asiatic Pacific Campaigns, Troop Carrier 42-44, ATC/MATS 45-49. Flew with supplemental air carrier 1961-1972, flew commuter Guam 1979-1982. Retired June 30, 1960, with 20 years of active service and the rank of major (graduated S/Sgt pilot 1942).

Memorable experiences include first Glider Tow Mission of WWII; flew Col. Gen. Jurgen Von Arnem from the Tunisian Front, with British escorts to Oran Algeria after he surrendered the German Forces; and flew parts of Atomic bomb to Pacific with couriers, dropped on Hiroshima or Nagasaki.

Awards: Air Medal with oak leaf cluster (Bronze), American Defense Service Medal, American Campaign, European African-Middle Eastern Campaign with three Bronze Stars, Asiatic-Pacific Campaign with one Bronze Star, Air Force Longevity Service Award with four Bronze Oak Leaf Clusters, Medal for Humane Action.

He married Jan and has a daughter Marcia, and a grandson, Curtis. Retired from flying in 1982.

DON A. BENNETT, graduated from Theodore Roosevelt High School in San Antonio, TX, in 1968. He entered the Air Force that same year and completed Jet-Over-Two Aircraft Maintenance Technical Training in 1969.

He was a distinguished graduate and Commandant Award winner at the 21st Air Force Leadership School, Dover AFB, DE, in 1978. While assigned to Headquarters, Military Airlift Command, Scott AFB, IL, he initiated and spearheaded the command action that led to the approval of the Aircraft/Munitions Maintenance Badge. He earned his enlisted aircrew wing as an initial cadre aircrew member of the C-23A *Sherpa,* which supported the European Distribution System out of Zweibrucken AB, FRG. During his stay in Germany he was selected SNCO of the Quarter twice and SNCO of the Year for the 10th Military Airlift Squadron in 1985. He was also selected by 21st Air Force as the Outstanding Aircrew Training of the Quarter for the 301st Air Refueling Wing and SNCO of the Quarter for the 40th Air Division. He served during the Persian Gulf War at Monte de Marsan, France and Riyahd, Saudi Arabia.

He earned an associate degree in applied science from the Community College of the Air Force in 1983. He attended the Tactical Air Command NCO Academy, Bergstrom AFB, TX, in 1981, and the SNCO Academy, Gunter AFS, AL, in 1987.

His decorations include the Meritorious Service Medal with one oak leaf cluster, Air Force Commendation Medal with one oak leaf cluster, Air Force Achievement Medal, National Defense Service Medal with one star, Humanitarian Service Medal, Southwest Asia Service Medal with one star, Vietnam Gallantry Cross with Palm. He is currently serving on active duty as a chief master sergeant and is assigned to the 91st Air Refueling Squadron, Malmstrom AFB, MT. He married to the former Carole Lynn Stetson of Lindsay, MT. They have one daughter, BriAnna.

HERMAN G. BENTON, enlisted in the Army Air Corps Jan. 24, 1941, at Ft. Riley, KS. After a year of training and aircraft maintenance, his remaining career was as a flight engineer. He spent 10 years with the Ferrying Command, Air Transport Command and Military Air Transport Service on B24, B17, C54, C121A, C124 and C135 type aircraft. He spent six months as line chief for the 11th Air Transport Squadron at Rhein-Mien AAB Germany during the Berlin Airlift.

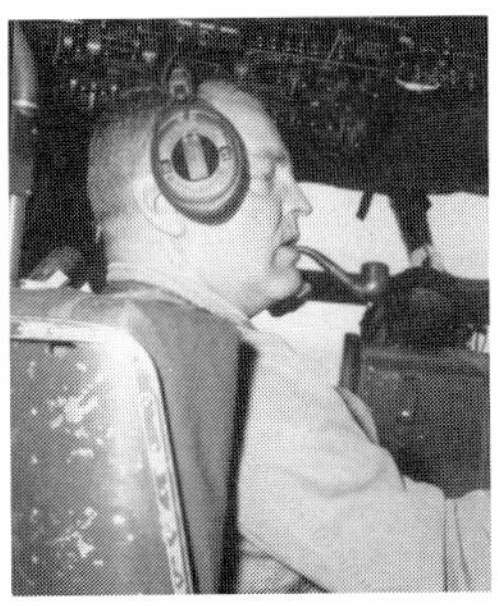

The remaining years were spent with the Strategic Air Command on B29 and KC97 type aircraft. After a tour of combat on a B29 during the Korean War, he returned to March AFB California spending 10 years as a KC97 engineer with the 320th Air Refueling Squadron.

He married Eileen McKernan at Dearborn, MI, Sept. 4, 1943. They have two children and four grand-children.

He retired as chief master sergeant Nov. 1, 1963.

ANDREW W. BIANCUR, born in the small town of Leicester in upstate New York. He graduated from Baldwin High School, Baldwin, Long Island, NY, in 1956, and attended the Air Force Academy. He graduated in 1960, with a bachelor's degree in engineering, received a masters degree in comparative government from Webster University in 1976, and a master's degree in international relations/international law from Yale in 1980. He completed Squadron Officers School, Armed Forces Staff College and Industrial College of the Armed Forces.

Andi retired from the United States Air Force in Oct. 1, 1986, with the rank of colonel (0-6). During his career, his association with airlift included: flying the C-118 and C-130 (initial cadre) with the 29th Airlift Squadron and the C-141 with the 30th Airlift Squadron, both at McGuire AFB. While there he completed a tour in Special Operations Scheduling at HQ 21st Air Force and was project officer for Operation Deep Freeze in 1968, the first year jet aircraft "went to the ice." He acted in the capacity as Air Force Advance Agent for Presidential Support. After a tour at HQ MAC 1975-1977, he commanded the 11th Aeromedical Evacuation Squadron flying C-9s. Following tours with the Joint Chiefs of Staff and as vice commander of the 92nd Bombardment Wing (Heavy), where he fly KC-135s and B-52s, he returned to HQ MAC as the director of Training from 1983-1986. While there, the Administrative Airlift Mission matured with the acquisition of C-12s and C-21s, and C-5, C-130, and C-17 training was successfully transferred to civilian contract.

Among his awards and decorations are: Legion of Merit with one oak leaf cluster, Defense Meritorious Service Medal, Meritorious Service Medal with two oak leaf clusters, and Bronze Star and four oak leaf clusters.

Memorable experiences include setting flight record Hickam to McGuire in C-130E non-stop 13 hours. Project officer Operation Deep Freeze 1968, first air jet aircraft into Antarctica.

He and his wife, Joan, reside in Salt Lake City, UT. They have two grown children, Beverly and Christian. Today he is program manager, Evans & Sutherland Computer Company-Simulation Division.

CLAIR D. BIXEL, born Feb. 14, 1919, in Rauchtown, PA. Joined the U.S. Air Force Jan. 15, 1942, serving with 333rd Bomb Gb, 314th TC Wg, 26a Assault Sq., 50th TC Sq., 16th TC Sq., 55th STRAT WG SAC. Served Topeka Army Airfield; Sewart AFB, TN; and Forbes AFB, KS.

Memorable experiences include flying as engineer instructor with lowtime crews in B17 and B24Aeft. Troop drops at minimum altitude over Korea. SAC RECON. missions world wide.

Discharged Nov. 1, 1969, with the rank of CM Sgt.

Married Kathryn Dawson, Oct. 13, 1943, Carbondale, KS. They have eight children and 11 grandchildren. He retired Oct. 30, 1969, with 27 years of active duty.

CARLTON L. BJERKAAS, Colonel, is the director of Program Management and Integration, Headquarters, Air Weather Service, Scott Air Force Bass IL. This directorate directs engineering, acquisition, modification and installation of specialized equipment used at more than 260 locations worldwide to provide environmental support to the Air Force, Army, and other designated agencies. It manages life cycle logistics for peacetime, contingency, and wartime operations.

Col. Bjerkaas was born April 17, 1948, in Fergus Falls, MN. He graduated from Roosevelt Park High School in 1966, and received his bachelor of science degree in mathematics from the University of North Dakota in 1970, a master's of science degree in meteorology from the Massachusetts Institute of Technology in 1977, and a master's in public administration from Auburn University at Montgomery in 1983. Col. Bjerkaas is a graduate of Squadron Officer School and Air Command and Staff College.

Col. Bjerkaas was commissioned through the Reserve Officer Training Corps as a distinguished graduate in 1970 and assigned to San Jose State College for basic meteorological training. Upon completion in 1971, he reported to Homestead Air Force Base, FL, where he served as a forecaster and wing weather officer to the 31st Tactical Fighter Wing. In 1973, he was assigned to the 54th Weather Reconnaissance Squadron at Andersen Air Force Base, Guam, where he served as an Aerial Reconnaissance Weather Officer, an instructor ARWO, a flight examiner ARWO, assistant standardization ARWO and altimetry officer.

In 1975, Col. Bjerkaas was selected by the Air Force Institute of Technology for graduate study. After graduating from Massachusetts Institute of Technology with a specialty in radar meteorology in 1977, he was assigned to the Weather Radar Branch, Meteorology Division. Geophysics Laboratory, Hanscom Air Force Base, MA, where he served as a research meteorologist. In July 1982, he went to Air Command and Staff College.

Moving to Scott Air Force Base, IL, in July 1983, he served as weather systems program manager and Chief, Support Branch, Systems Division, Directorate of Operational Requirements and Test, DCS for Plans and Programs, Headquarters, Military Airlift Command. He was assigned to Andrews Air Force Base, MD, in July 1985, as chief of Aerospace Environmental Requirements Branch, 2nd Weather Squadron with concurrent duty as Chief, Aerospace Environmental Requirements Division, Office of Staff Meteorologist, Headquarters, Air Force Systems Command. In July 1987, Col. Bjerkaas assumed command of Detachment 19, 15th Weather Squadron, Lajes Air Base, Azores, with concurrent duty as staff weather officer to the Commander, U.S. Forces Azores (USFORAZ) and to the Commander, 1605th Military Airlift Support Wing.

In 1989, Col. Bjerkaas was assigned to Headquarters, Air Weather Service at Scott. He has served as the Assistant Chief of Staff for Automated Weather Distribution System (AWDS) and Next Generation Weather Radar (NEXRAD) Integration, assistant DCS, Program Management, Assistant Chief of Staff, executive officer to the commander, and director of Resource Management. He assumed his current position on July 20, 1992.

Col. Bjerkaas has been awarded the Master Meteorologist Badge and basic Officer Aircrew Member Wings. His military awards and decorations include the Meritorious Service Medal with three oak leaf clusters, the Air Medal, and the Air Force Commendation Medal with one oak leaf cluster.

He was promoted to colonel May 1, 1992, with the same date of rank.

Col. Bjerkaas is married to the former Ester

Calingasan of Astoria, Long Island, NY. They have three sons: Kris, Eric and Todd.

LOREN F. BJORNSON, Lieutenant Colonel, is the chief of Training Support for the 1st Special Operations Wing, Tactical Air Command, Hurlburt Field, FL.

Col. Bjornson was born in Grand Forks, ND, on June 26, 1934. He was graduated from Minot High School, Minot, ND, in 1952, and obtained his bachelor of science degree from North Dakota State University in 1956. Col. Bjornson was commissioned a second lieutenant through the AFROTC program in 1956. In February 1957, he entered active duty at Lackland Air Force Base, TX, and was assigned to the 3700th Pre-Flight Training Group. In April 1957, he was assigned to the 3607th Navigator Training Squadron at Ellington Air Force Base, TX, where he received his wings in March 1958.

He was assigned to the 815th Troop Carrier Squadron at Ashiya Air Base, Japan in 1958, and moved with the squadron to Tachikawa Air Base, Japan, in 1960, where he served as a flight instructor.

He returned to the United States in 1962, and was assigned to the 774th Troop Carrier Squadron at Sewart Air Force Base, TN, and in 1963, to the 772nd Troop Carrier Squadron at Langley Air Force Base, VA, serving as a flight instructor at both locations.

In January 1965, he was transferred to the 21st Troop Carrier Squadron at Naha Air Base, Okinawa, where he served as a flight commander and flight instructor. In April 1966, he became a standardization/evaluation flight examiner for the 315th Air Division at Tachikawa Air Base, Japan.

In February 1969, he was assigned to the 7th Airborne Command and Control Squadron at Udorn Royal Thai Air Force Base, Thailand, where he again served as a standardization/evaluation flight examiner.

He returned to the United States in February 1970, and was assigned as air operations officer, 4442nd Combat Crew Training Wing, Little Rock Air Force Base, AR. He assumed duties in October 1970 as executive officer for the 4442nd Wing Commander. In February 1972, he became executive officer for the 834th Air Division, Little Rock Air Force Base, AR.

Col. Bjornson returned to the Far East in August 1973, and was assigned as Commander, 37th Air Base Squadron, Fuchu Air Station, Japan and served as base commander.

In November 1974, he was assigned to the 475th Air Base Wing, Yokota Air Base, as chief of Social Actions.

In July 1975, he assumed command of the 475th Air Base Squadron, Tachikawa Air Base, Japan and was Base Commander. In July 1977, he returned to the United States and was assigned to the Armament Development and Test Center, Eglin Air Force Base, FL, as Vice Chief of Staff.

In August 1979, he was assigned to Hurlburt Field, FL, and presently serves as chief of Training Support for the 1st Special Operations Wing.

Col. Bjornson is a Master Navigator with over 7100 flying hours, nearly all of it in C-130 aircraft. He has flown 343 combat sorties and has logged 1192 combat hours. His military decorations and awards include the Meritorious Service Medal with one oak leaf cluster the Air Medal with eight oak leaf clusters; Air Force Commendation Medal with one oak leaf cluster; Air Force Outstanding Unit Award with Combat V Device and six oak leaf clusters; Combat Readiness Medal with three oak leaf clusters; Armed Forces Expeditionary Medal with four service stars; National Defense Service Medal; Vietnam Service Medal with 11 service stars; Republic of Vietnam Cross of Gallantry with Palm; and the Republic of Vietnam Campaign Medal.

He was promoted to the grade of lieutenant colonel on Sept. 1, 1973, with a date of rank April 15, 1973.

Col. Bjornson is married to the former Eloise Ramsey of Memphis, TN, and they have five children: Skip, Laura, David, Katie and Chris. Retired March 1, 1985.

JESSE L. BLANKENSHIP, born Dec. 19, 1934, in Ragland, AL. Received a BS degree in economics from the University of Tampa.

Joined the U.S. Air Force April 30, 1956, serving with the 7405 Ops Sq., 7406th Ops Sq., 332 TCS, 773 TCS, (HQ TAC, Det. 1), 4409th Ops Sq. Military locations or stations at which he served include: Lackland AFB, Harlingen AFB, Sewart AFB, MacDill AFB, Homestead, Rhein-Main AB Germany, and Hurlburt Field, FL. Retired May 1, 1983, with the rank of lieutenant colonel. He and his wife Janice, have two sons and three grandchildren.

RICHARD W. BLATTER, USAF, Ret., Colonel, commissioned in 1959, through AFROTC, at Kent State University. After graduating from pilot training class 61-F at Laredo, he served as pilot and flight examiner in various aircraft, but primarily in the C-130. Other assignments included Chief Standardization and Evaluation 351 Strategic Missile Wing; editor, *Combat Crew Magazine;* Commander, 16th Tactical Airlift Training Squadron; and MAC liaison to United Nations Command in Seoul.

He retired in 1987 as director of Safety at HQ MAC. He is a command pilot with 5800 hours, 600 of which were flown in combat, and has been awarded the Legion of Merit and the Distinguished Flying Cross. He is employed at McDonnell Douglas Aerospace on the C-17 program. Col. Blatter is married to the former Peggy Lou Nell of Lodi, OH. They have two daughters, Cristina and Tamela.

JAMES R. BOMBERGER, graduated from Ephrata High School, Ephrata, PA, in 1951. He entered the Air Force the same year and had a distinguished career of 23 years, and retired from the Air Force on June 30, 1974. After completing basic training James was a security policeman for a short period of time and thereafter was a loadmaster for his entire military career. Upon retirement he was employed by AAR Brooks & Perkins where he has also had a distinguished career, and continues with this firm as director, military marketing for AAR Manufacturing Group.

Some additional information about his education and specific accomplishments follow: He is a graduate of the Non-Commissioned Officers Academy, Aircraft Loadmaster Course (C-124, C-130, C-141 and C-5 Aircraft), Advanced Air Transportation Course, Technical Transportation of Regulated Items Course, and Advanced Marketing Course, University of Michigan.

The last six years of his military career was served as staff loadmaster, HQ Military Airlift Command (MAC) Scott AFB, IL. Duties involved development of procedures for handling, processing, loading, and aerial delivery and storage of cargo on a worldwide basis. In other words, this position was the highest of all loadmaster positions in MAC's worldwide system.

AAR Brooks & Perkins is a major manufacturer of unit load services (ULDS), like pallets and containers and in-plane air cargo systems for both military and civil aircraft as well as helicopters.

When James first joined AAR his initial job was senior sales engineer, Advanced Structures Division, where he promoted new cargo systems for the YC-14 and C-130.

He was elevated to manager, Military Marketing and Sales, with responsibility over products made by several plants like cargo pallets, container, fiber products, shelter/tent flooring, cabinets, and improved pallet designs.

Finally, he was awarded the Air Force Meritorious Service Medal and MAC Award for Outstanding Achievement in Nuclear Safety for improvements on the C-141 Nuclear Airlift Program.

JONATHAN J. BORTNER, Colonel, USAF, Vice Commander, 39th Special Operations Wing, Royal Air Force Alconbury, United Kingdom.

Col. Bortner was born Oct. 3, 1944, in York, PA. He graduated from William Penn Senior High School and obtained his bachelor of science degree in political science from Gettysburg College, Gettysburg, PA, in 1967. In 1977, he received his master's of arts degree in human relations from Webster University at Little Rock, AR. His military education includes Squadron Officer School, Air Command and Staff College and Air War College.

Col. Bortner was commissioned through the Air Force Reserve Officer Training Corps in 1967. In 1967, he attended navigator training at Mather Air Force Base, CA. Upon graduation, in 1968, he completed F-4 Weapons System Operator Training at Davis Monthan Air Force Base, AZ. He served a tour at Da Nang Air Base, Republic of Vietnam, from 1969-70. In 1970, he was assigned to Yokota Air Base, Japan, and after a unit move to Kadena Air Base, Okinawa, he continued training officer and instructor duties in F-4s. He graduated from pilot training in December 1972.

His first operational pilot assignment was with the 50th Tactical Airlift Squadron, Little Rock Air Force Base, AR, where he served as co-pilot, instructor pilot and assistant operations officer. In 1979, he was reassigned as the C-130 Program Manager and Airlift Advisor to the American Embassy, Office of Military Cooperation, in Cairo, Egypt. In 1982, he was the Tactical Airlift Director, 21st Air Force Operations, McGuire Air Force Base, NJ.

Upon completion of Combat Talon School at Hurlburt Field, FL, in 1983, he became director of Operations for the 7th Special Operations Squadron, Rhein-Main Air Base, Germany. He took over as commander of the 7th SOS in June 1984. From September 1986 to June 1988, he was the assistant deputy commander of Operations for the 1st Special Operations Wing, Hurlburt Field, FL. In July 1988, he served as Chief, Operational Requirements Division and Chief, Requirements and Training Division, United States Special Operations Command, MacDill Air Force Base, FL. He assumed the duty of deputy commander for Operations, 39 SOW, Rhein Main, Germany in July 1991, and moved to RAF Alconbury with the wing in January 1992. He has been the 39 SOW vice commander since August 1992.

Col. Bortner is a senior pilot with more than 5,000 military and civilian flying hours, including 212 combat hours. His military decorations include the Defense Superior Service Medal, Distinguished Flying Cross, Defense Meritorious Service Medal, Meritorious Service Medal with three oak leaf clusters, Air Medal with seven oak leaf clusters, and the Joint Service Commendation Medal with one oak leaf cluster.

He was promoted to his present rank on March 1, 1989.

He is married the former Sharon Hoffman. They have three daughters, Kim, Stacy and Kristin.

PHILIP A. BOSSERT JR., Captain, USAF, assistant professor, Military Art and Science Division, United States Air Force Academy. Capt. Bossert was born in New York City, NY, on July 6, 1959. He attended Massapequa High School, in Massapequa, NY, graduating in 1977. Capt. Bossert graduated with Military Distinction from the United States Air Force Academy in 1982, earned his pilot wings at Reese AFB in 1983, and attained a master's degree in economics, Magna Cum Laude, from Old Dominion University in 1986. His first assignment was at Langley AFB, VA, where he was a CT-39B aircraft commander and C-21A instructor pilot. Following Langley, he served at McGuire AFB, NJ, as a C-141B instructor pilot and wing combat plans officer.

Capt. Bossert is currently course director of MAS-440, Joint and Combined Operations, an academic course all senior cadets must take to graduate. He also flies the UV-18B Twin Otter which supports the Academy's Wings of Blue parachute demonstration team.

Capt. Phil Bossert has over 3,700 hours of flying time, including combat time from Operations Just Cause, Desert Shield and Desert Storm. He was selected as the 1990 McGuire AFB Company Grade Officer of the Year, the 1990 MAC Operations Security Officer of the Year, was a recipient of the 1991 Airlift Association Young Leadership Award, and has won one first place, one third pace, and six honorable mentions in the *MAC Flyer* writing contests between 1988 and 1993.

He is married to Capt. Anita M. Shorosky, a hospital administrator. They have one son, Andrew.

CHARLES BOTULA III, "PACKY" Colonel, USAF, (Ret.), was Commander, Middle East Liaison Region, Civil Air Patrol-USAF with headquarters at Andrews AFB, MD, until his retirement in January 1994.

Prior to assuming that position in September 1992 he was commander and chief pilot, Supreme Allied Commander Europe (SACEUR) Flight Section, Chievres Air Base, Belgium from August 1989 until August 1992.

Col. Botula is from Riverhead, NY. He graduated from the State University of New York at Buffalo in February 1968, with a bachelor's degree in history. He has a master's degree in management and supervision from Central Michigan University. He is a graduate of Squadron Officer's School, Air Command and Staff College, National Security Management and the Air War College.

Col. Botula received his commission through the Reserve Officer Training Corps in 1968. Following pilot training at Webb AFB, TX, in 1969, he was assigned to the 17th Special Operations Squadron flying the AC-119 Gunship. His operational assignments include C-130Es with the 776 Tactical Airlift Squadron. Ching Chuan Kang (CCK) Air Base, Taiwan, C-130B(II)'s at Rhein-Main Air Base, Germany and C-9s at Rhein-Main Air Base, Germany. In 1979, following 10 years of consecutive overseas duty he was assigned to the 89th Military Airlift Wing, Andrews AFB, MD, as a VC-9C special mission select pilot. He has been a flight examiner in both the C-130 and the C9 and has two complete tours of duty in Vietnam with over 1,100 hours of combat time. Following an assignment to the Air Command and Staff College in 1983, Col. Botula moved to Headquarters Military Airlift Command where he served as Chief, Contract Airlift Capability Survey Division in the Directorate of Air Transportation. He was Commander, 55th Aeromedical Airlift Squadron, Rhein-Main Air Base, Germany from 1986 to 1989.

Col. Botula is a command pilot with more than 8,000 flying hours including over 5,000 hours in the C-9. His military decorations include the Defense Superior Service Medal, the Distinguished Flying Cross with one oak leaf cluster, the Meritorious Service Medal with three oak leaf clusters, the Air Medal with 13 oak leaf clusters and the Humanitarian Service Medal.

Col. Botula is married to the former Susan Kaye Johnson from Mahnomen, MN. They reside in Upper Marlboro, MD.

GORDON F. BRADBURN, born Nov. 15, 1919, in Thawville, IL. He received his BA degree from De Paul University and his MA degree from the George Washington University.

Joined the USAF Dec. 27, 1941, serving with the SAC, TAC, MATS, Alaskan Air Command, Air Rescue, Air Commandos Joint Staff, JCS 14th Air Commando Wing, SEA. Military locations and stations at which he served include Sewart AFB, Maxwell, LADD, Taipei, Pentagon, Bolling.

Memorable experiences include: Short field landings, Swift Strike III, 500th man rescued by Jolly Greens in Sea, November 1966; Air Attaché Taiwan, 1960-1962; Vice COMDR 314 TCW, COMDR 1st Air Commando WG 14th ACW, Nha Trang.

He was discharged with the rank of colonel.

He and his wife, Patricia Anne, have three children and five grandchildren. As a civilian he was employed as chairman of the board, American Stripping Company, retiring Dec. 1, 1970.

RAYMOND A. BRANDT, born March 15, 1930, in Chicago, IL. He received his education at the University of Illinois.

Joined the USAF in November 1951, serving with the 75th and 61st Troop Carrier Squadrons. Military locations and stations at which he served include: Columbus, MS; Enid, OK; and Smyrna, TN.

Memorable experiences include: being Distinguished Graduate of USAF Pilot Training; Operation Snow-Bird; Elmendora AFB, Alaska, 1955.

He was discharged in March 1957, with the rank of first lieutenant.

He and his wife, Bedelia "Dede" Anne Coode, have six children and six grandchildren. As a civilian he held the position of vice president, Bell Federal Savings. He retired from the Reserves in March 1969.

ALTON P.H. BREWER JR., born Feb. 5, 1933, in Mt. Holly, NC. The family moved to Oak Ridge, TN, in 1943; Al graduated from Oak Ridge High School in 1950, and enlisted in the Air Force in 1951. He entered Aviation Cadets in 1953, and was commissioned in March 1954. His first operational assignment was in SAC fighters in Great Falls Montana. Subsequently Al earned his degrees from Chapman College in California and a master's in business management from Webster University in St. Louis, MO.

Al was an Aircraft Commander and Standardization Pilot in C-47, C-130 and C-141 aircraft in a flying career accumulating over 10,000 hours, spanning 26 years of active flying. He was the senior Aircrew Standardization Officer for two squadrons and two airlift wings. During his tenure as commander the 15th MAS was credited as the best Airlift Squadron in MAC, recording seven laudatory findings on the annual IG inspection.

Assigned to Headquarters, MAC in 1976, and as director of Current Operations he emphasized lift efficiency and international force projection through airlift.

Retiring in 1983, with 32 years service, he holds the Legion of Merit, Distinguished Flying Cross, and other decorations. He lives in Fairview Heights, IL, with his wife Carolyn. They have four children and three grandchildren.

CALVIN REED BROWN, M.D., Colonel, USAF, (Ret.), born July 11, 1926, in Koosharem, UT. Joined the service Oct. 12, 1943. Served with the Army, Navy, Coast Guard, Marines, MS Public Beach Service, National Guard and Air Force. Military locations and stations at which he served were Ft. Benning, GA; Ft. Douglas, UT; Camp Roberts, CA; Camp Grohn, Germany; Thailand, Yokosuka, Japan; Attu Alaska.

Memorable experiences include serving in every Branch of the service, Army, Navy, Marines, Coast Guard, U.S. Public Health Service, National Guard. Retired as colonel Air Force after 43 years. Discharged July 11, 1986.

Awards: Bronze Star, Purple Heart, ETO, ATO, Good Conduct Medal, Vietnam Service Medal, National Guard Ribbon, Combat Infantry Badge, Army of Occupation Ribbon, Korean Service Ribbon, Coast Guard Medal, Pacific Theater Ribbon.

He married Barbara Jenkins, Dec. 24, 1948. They have children: Cindy, Randy, Lisa, Tammy and nine grandchildren. Today he is a general surgeon, and involved with politics and church work.

DONALD D. BROWN, Major General, USAF, (Ret.), born in Montreal, Canada. He graduated from Columbia University in New York in 1954, where he received a bachelor of arts degree and his commission from ROTC. In 1965, he was granted a master's of business administration degree in production management from Syracuse University. He was a distinguished graduate of Squadron Officers' School and completed Air Command and Staff College and the Industrial College of the Armed Forces.

Brown served as command[illegible] 30th MAS, McGuire AFB, NJ, the 62nd MAW at McChord AFB, WA and as assistant DCS Logistics, assistant DCS Operations, DCS Plans and DCS Operations in two tours at MAC HQ at Scott AFB. He retired as Commander, 22nd Air Force at Travis AFB, CA.

Among his many awards and decorations are the Distinguished Service Medal with oak leaf cluster; the Legion of Merit with oak leaf cluster; the Distinguished Flying Cross with oak leaf cluster; the Bronze Star and the Air Medal with four oak leaf clusters. He has over 12,000 hours flying time, including 147 combat missions in Vietnam.

Brown and his wife, Joan, reside in Steilacoom, WA.

DONALD P. BROWN, Lieutenant Colonel, USAF, (Ret.), was born June 5, 1944, in Coalville, UT. He graduated from Brigham Young University in 1969, with a degree in Chinese and was commissioned through ROTC.

Brown graduated from pilot training at Vance AFB, then served nine years (1970-79) as a C-141 and C-5 pilot at Travis AFB. In 1975, he was chosen MAC and USAF nominee for Ten Outstanding Young Men in America Award. He received a master's degree from Naval Postgraduate School in 1981. He was an Air Force Research Associate at Hoover Institution, Stanford University (1985-86) and graduated from SOS, Air Command and Staff College and Air War College.

During eight years duty at HQ USAF, Brown escorted several senior USAF and Chinese leaders, including Chief of Staff and Secretary of Air Force and Chinese Air Force Commander on exchange visits. At retirement in March 1990, he was Deputy Chief, Pacific/East Asia Division, under Assistant Chief of Staff, Plans and Operations.

Brown and his wife Laurie reside in Salt Lake City, UT, where he is director of Asia/Pacific Marketing for Simulation Division, Evans and Sutherland Computer Corporation. They have five daughters: Michele, Lisa, Janice, Renae and Pamela.

IVAN DOYLE BROWN, Colonel, USAF, (Ret.), born in Ramsdall, TX, Aug. 9, 1931. He received a civil engineering degree and commission as a distinguished AFROTC graduate from Washington State College in 1954. He completed an aeronautical engineering graduate degree, was an Air Command and Staff College Distinguished Graduate and received special thesis recognition at the National War College. He earned pilot wings in 1955, and accrued 4,400 hours in transport aircraft including 500 Vietnam combat hours. Col. Brown held many operational, staff, engineering, management and command assignments. Notable assignments include Commander, 316th and 463rd Tactical Airlift Wings, Deputy Inspector General (MAC) and Chief of Staff, 21st Air Force. Under his command, the 463rd TAW received the highest ORI grade ever awarded to such a unit. He received numerous medals including Air Medals, Distinguished Flying Cross and Legion of Merit.

Col. Brown retired Aug. 1, 1980. He is married to the former Jean Scott; they have five married children and 10 grandchildren. He is currently an engineering projects manager for Martin Marietta Corporation in Washington, DC.

WILLIAM T. BROWN, born Nov. 12, 1919, in Carbondale, WV. He attended college for two years. Joined the U.S. Air Force Jan. 17, 1941, served with the Airlift 15th TC Frankfort, Germany. Military locations and stations at which he served were USA, Germany and

Japan. He retired in 1966, with the rank of SM/Sgt.

As a civilian he was employed by Mooney Acft Company, Aircraft Insp.

CHRIS BRUCE, born Oct. 24, 1948, in San Francisco, CA. He joined the U.S. Air Force in November 1971, served C-130 NAV, C141 Pilot. Military locations and stations at which he served were Pope, 1972-75; Williams, 1976; McChord, 1977-80; RAF Mildenhall, 1980-84; Scott, 1984-88; Travis, 1988-91. Battles he participated in included South East Asia, Nickle Grass; Just Cause, Panama; Desert Shield/Storm.

Awards: Meritorious Service Medal and Air Medal.

He retired in September 1991, with the rank of lieutenant colonel.

He and his wife Lynn, have two daughters, Tracy and Kelly. He is now a pilot with United Air Lines, based in Denver.

EDWARD N. BRYA, flew various models of C-130 in tactical Airlift, rescue and special operations. SEA tours flying C-130s at Naha and CCK in addition to an Arc Light tour in B-52s.

During the 1972 NVA Spring Offensive he designed and flew the low and high altitude airdrops that relieved the siege of An Loc. Career highlight came during the mission to Hanoi that returned the first American POWs.

After staff tours at TAC, MAC and Air Staff he returned to Dyess AFB as 463 TAW/DO. He commanded the 39 ARRW at Eglin AFB and the 317 TAW at Pope AFB.

Promoted to brigadier general he moved to Ft. Bragg as Deputy Commanding General, Joint Special Operations Command and then to MacDill AFB as director of Operations, U.S. Special Operations Command.

In 1991, he and Beth retired to Beaufort, SC, where he serves as executive director of the Arnold Air Society.

FREDRIC N. BUCKINGHAM, Brigadier General, USAF, born Dec. 24, 1941, in Kendallville, IN, and graduated from Leesburg (IN) High School in 1960. He earned a bachelor of science degree in business finance from Indiana University in 1964, and a master's degree in business administration from Golden Gate University in 1976. He completed Squadron Officer School in 1971, Air Command and Staff College in 1978, Air War College in 1983, and the University of Michigan's executive program in 1988.

His numerous assignments generally centered around C-130 Tactical Airlift and included a tour at Edwards Air Force as Senior YC-14 Operational Test Pilot, 1976-1978. He commanded the 345th Tactical Airlift Squadron, Yokota Air Base, Japan, 1980-1982; and both the 463rd Tactical Airlift Wing at Dyess Air Force Base, TX, 1986-1988; and the 317th Tactical Airlift Wing at Pope AFB, NC, 1988-1989. As vice commander of 21st Air Force, McGuire Air Force Base, NJ, 1989-1991; he served as Theater Commander of Airlift Forces (COMALF) for both Operation Just Cause in Panama and the initial phase of Operation Desert Shield in Saudi Arabia. He is currently vice commander of the Warner Robins Air Logistics Center at Robins Air Force Base, GA. Buckingham is a command pilot with over 6,000 flying hours.

He and his wife Diane have a son, Todd.

GEORGE H. BURGER, Lieutenant Colonel, born Jan. 19, 1947, in Grand Rapids, MI. He is a 1970 graduate of Michigan Tech University and received his commission in 1971 from Officer Training School (OTS). He holds a master's degree and is a graduate of Pilot Training, Squadron Officer School, Air Command and Staff College, Air War College, Airlift Operations School, Combat Aircrew Training School, and the Air Force Battle Staff School.

During Desert Shield/Storm, he deployed to Saudi Arabia as the chief of Command and Control for the MAC Airlift Control Center (ALCC). With limited personnel and communications assets, he organized and initiated a command and control system for the ALCC in less than 12 hours. After two months in the desert, he was reassigned to U.S. Forces Command (FORSCOM) as the TRANSCOM/MAC liaison officer.

At FORSCOM, he kept CINCFOR updated on all Desert Shield/Storm transportation to include airlift missions as well as ship movements in support of the U.S. Army.

His decorations include the Meritorious Service Medal with two oak leaf clusters, the Air Force Commendation Medal, and the Army Commendation Medal. He is a command pilot with more than 2,500 flying hours.

He is married to the former Betty Roysden of Tifton, GA. They have three children, Heather, Christopher and Matthew.

ANTHONY J. BURSHNICK, Lieutenant General, Vice Commander in Chief, Military Airlift Command, with headquarters at Scott Air Force Base, IL.

Gen. Burshnick was born Jan. 12, 1937, in Forest City, PA, where he graduated from high school in 1954. He earned a bachelor of science degree in public administration from Auburn University in 1976. He completed Squadron Officer School in 1964, Armed Forces Staff College in 1971 and Air War College as a distinguished graduate in 1976. The general also completed the executive development program at Cornell University in 1980 and Harvard University's International Security Forum in 1986.

He enlisted in the Air Force in 1954, and served as an electronics instructor at Chanute Air Force Base, IL, until entering the academy in 1956. Upon graduating in June 1960, he was commissioned as a second lieutenant and then completed pilot training at Spence Air Base, GA, and Vance Air Force Base, OK. After receiving his pilot wings in September 1961, he was assigned to the 40th Military Air Transport Squadron at McGuire Air Force Base, NJ, as a C-135 pilot. In January 1965, he transferred to the 19th Air Commando Squadron in the Republic of Vietnam as a C-123 pilot.

Upon his return to the United States in January 1966, he served as a C-135 instructor pilot with the 40th Military Airlift Squadron at McGuire. In January 1967, he returned to the Air Force Academy as an air officer commanding and chief of the Cadet Scheduling Division. The general attended the Armed Forces Staff College from January 1971 to July 1971. He then was assigned as war plans officer at MAC headquarters. This was followed by an assignment in January 1973, to Washington, DC, as an Air Force liaison officer to the U.S. House of Representatives. In June 1975, he was selected to attend the Air War College.

Gen. Burshnick served as assistant deputy commander and, then, deputy commander for Operations, 443rd Military Airlift Wing, at Altus Air Force Base, OK, from July 1976 to April 1978. He next was assigned as vice commander of the 63rd Military Airlift Wing at Norton Air Force Base, CA. In May 1979 the general took command of the 475th Air Base Wing, Yokota Air Base, Japan.

Returning to the United States in July 1980, he was assigned to Travis Air Force Base, CA, where he served initially as commander of the 60th Military Airlift Wing. In March 1982, he became vice commander of 22nd Air Force. In June 1982, he returned to the Air Force Academy as commandant of cadets and in July 1984, he was assigned as Deputy Chief of Staff for Plans at MAC headquarters. In October 1986, the general was assigned as director of personnel programs, Office of the Deputy Chief of Staff, Personnel, Headquarters U.S. Air Force, Washington, DC. He became Assistant Deputy Chief of Staff for Plans and Operations at Air Force headquarters in June 1987. He assumed his present duties in September 1988.

The general is a command pilot with more than 5,000 flying hours in aircraft including T-29s, T-39s, C-47s, C-118s, C-123s, C-131s, C-135s, C-141s and C-5s. His military decorations and awards include the Distinguished Service Medal, Legion of Merit, Distinguished Flying Cross, Meritorious Service Medal with three oak leaf clusters, Air Medal with 16 oak leaf clusters, Air Force Commendation Medal with one oak leaf cluster, Air Force Outstanding Unit Award with V device and four oak leaf clusters, Combat Readiness Medal, Good Conduct Medal, National Defense Service Medal with one service star, Armed Forces Expeditionary Medal with three service stars, Vietnam Service Medal with three service stars, Air Force Longevity Service Award Ribbon with seven oak leaf clusters, and Republic of Vietnam Campaign Medal.

He was promoted to lieutenant general Oct. 1, 1988, with same date of rank.

Gen. Burshnick is married to the former Karen Maynard of Toms River, NJ. They have two children, Kristine and Joseph.

KATARINA S. BENTLER, born July 6, 1961, in Beverly Hills, CA, to Thea and Peter Bentler of Los Angles, CA. After graduating from the United States Air Force Academy, she attended UPT at Columbus AFB, MS. She was assigned to the 46th AREFS, K.I. Sawyer AFB MI, flying KC-135As, including missions in support of the U.S. Forces during the Iran/Iraq war. Her next unit of assignment was the 350 AREFS, where she became the first female KC-135/Q instructor pilot at Beale AFB CA. Later, as the chief of Tanker Tactics for the 9th SRW she flew missions in support of Desert Shield. Her awards include the Aerial Achievement Medal, AF Commendation Medal, National Defense Service Medal, Southwest Asia Service Medal with three Bronze Stars, Armed Forces Expeditionary Medal, Air Force Outstanding Unit with Valor and two oak leaf clusters, Combat Readiness Medal with two oak leaf clusters, and others. Capt. Bentler left active duty in October 1992 and is currently serving as a reserve officer for the Civil Air Patrol in New Mexico.

JAMES E. CALLAHAN, Colonel, (Ret.), enlisted in the USAF in 1956. In 1958, he entered the Aviation Cadet Program and received his pilot rating and commission as a second lieutenant in March 1960. His first operational flying assignment was to the 429 ARS, Langley AFB, VA, flying the KB-50J tanker.

He was later assigned to the 420 ARS at RAF Sculthorpe, United Kingdom. In 1964, Jim was reassigned to the 778 TCS at Pope AFB, NC, to fly the C-130E aircraft. While at Pope AFB he qualified as an instructor pilot and flight examiner. In 1971, he became the USAF airlift advisor to the Royal Saudi Air Force.

In 1972, he was assigned as a staff officer to Headquarters USAFE. While in Germany, he held positions as: airlift control officer, operations briefing officer and branch chief, exercise planning officer and Chief Airlift/Support Standardization and Evaluation Division.

Jim returned to the U.S. to fly the C-130H at Dyess AFB, TX, in 1977. He was assigned to the 463 TAW, 772 and 774 TAS and was commander of the 773 TAS. In 1980, he was reassigned to the 317 TAW at Pope AFB as assistant deputy commander for Operations and later became deputy commander for Operations. He left Pope AFB in 1983, to become the commander 616 MAG and DCS Airlift, Alaskan Air Command, Elmendorf AFB, AK.

Jim retired from the Air Force in November 1986, with over 8,000 flying hours and over 1,000 hours of combat and combat support time.

His decorations include the Legion of Merit, Meritorious Service Medal with two oak leaf clusters, Air Medal with four oak leaf clusters, Joint Service Commendation Medal and Air Force Commendation Medal.

Jim and his wife Bonnie reside in Buffalo, NY, where he's employed by Sierra Technologies. Their daughters, Jamie and Jodi, are Air Force officers and call home where the Air Force sends them.

DICK CALTA, Colonel, graduated from Iowa City High School in 1963, and from the University of Iowa in 1967, an AFROTC distinguished military graduate. He earned a master's of science degree from the Air Force Institute of Technology. He completed Squadron Officer School, Air Command and Staff College, Armed Forces Staff College and Air War College.

He served in logistics assignments in Hawaii, Turkey, Greece, Scott AFB, IL, and McChord AFB, WA. He was a distinguished graduate from the aircraft maintenance

officer course and the 375th Aeromedical Airlift Wing Junior Officer of the Year in 1977. While commanding the 62nd Organizational Maintenance Squadron at McChord AFB, the win won the CINCMAC Daedalian Trophy for aircraft maintenance. At HQ MAC, he developed the Global Decision Support System (GDSS) for the Logistics Readiness Center. Twice he was a wing deputy commander for maintenance. At McChord AFB he formed and command the 62nd Logistics Group.

Col. Calta's military decorations include the Defense Meritorious Service Medal, the Air Force Meritorious Service Medal with three oak leaf clusters, and the Air Force Commendation Medal with one oak leaf cluster. He and his wife, Linda, have two children.

FARRELL D. CAMMACK, born June 21, 1942, in Blackfoot, ID. Attended Idaho State University. Joined the U.S. Air Force Jan. 29, 1961, serving with the 314 TCW-50th TCS and 18th TCS, C-130A. Stationed at Sewart AFB, TN.

Discharged Jan. 29, 1965, with the rank of E-4, worked in aircraft maintenance and as a loadmaster. As a civilian he is the general manager of Wade Farms Potatoes.

JOSEPH E. CAREY, born July 22, 1936, in Butler Township, PA. He attended Glrakdville High School and Ford School of Business. He joined the U.S. Air Force June 21, 1954, serving with the 1401st AB Wing, 6167th Support Sq., 314th Troop Carrier Wg.

His memorable experience is traveling to and from Korea on a troop ship. He was discharged Dec. 31, 1957, with the rank of A/2C.

He was married to Mary Helene Tunevage, who died on June 13, 1993. As a civilian he was employed with Pennsylvania Department of the Auditor General, as a auditor. He retired Dec. 31, 1985.

ALBERT N. CAROTENUTO, born March 25, 1927, in Jersey City, NJ. Enlisted for Air Combat Crew Training in October 1944, and was called to active duty in the summer of 1945, and did his basic training and B-24 School at Keesler Field in Biloxi, MS. After re-enlisting in December 1945, he was assigned to the United States Air Forces in Europe (USAFE) as an airplane and engine mechanic.

He was shipped to the 15th Troop Carrier Squadron at Rhine/Main where he gained his crew chief and aerial engineers status. His assignment was servicing the European Air Transport Service (EATS) mission until April of 1948, when the "little" Airlift began and served in C-54 aircraft as flight engineer in Operation Vittles until rotation in December of 1948, attaining the rank of staff sergeant before discharge.

He received his BS degree from Montclair State Teacher's College majoring in the physical sciences in 1953, and his MS degree from the University of Pennsylvania in 1959. He taught chemistry and physics for 36 years and is currently enjoying retirement with his wife Susan.

SHIRLEY M. CARPENTER, "SAM", Major General, USAF, is mobilization assistant to the Commander in Chief, Air Mobility Command with headquarters at Scott Air Force Base, IL.

General Carpenter was born July 22, 1937, in Orange County, VA. He attended public schools in Williamsburg, VA, graduating from James Blair High School in 1956. He received a bachelor of arts degree from the University of Virginia in 1960, and a master's of science degree in counseling and guidance from Troy State University in 1976. The general completed Squadron Officer School in 1967, Air Command and Staff College as a distinguished graduate in 1976, and the National War College in 1982.

Upon graduation as a distinguished graduate of the University of Virginia's Air Force Reserve Officer Training Corps program in June 1960, he was commissioned as a second lieutenant. The general earned pilot wings in March 1962, as a member of the first all-jet undergraduate pilot training class at Moody Air Force Base, GA. His first assignment out of pilot training was with the 44th Air Transport Squadron, Travis Air Force Base, CA, flying C-135B aircraft for the Military Air Transport Service. After upgrading to aircraft commander in the C-135B, he transitioned to the C-141A in February 1965. Shortly thereafter, he upgraded to aircraft commander, then to flight examiner, in the Starlifter.

In November 1968, the general separated from active duty and was employed by Trans World Airlines as a pilot/flight engineer. In June 1969, he joined the newly formed Reserve Associate unit at McGuire Air Force Base, NJ. The unit was transitioning to the C-141A aircraft and later became the 514th Military Airlift Wing. When furloughed by Trans World Airlines in 1970, he became an air reserve technician at McGuire Air Force Base, working as a squadron standardization officer and, later, as acting wing standardization officer. In 1975, while attending Air Command and Staff College, General Carpenter wrote a research paper on low-level windshear that influenced the Military Airlift Command's aircraft low-level windshear procedures.

The general's next assignment was to the 702nd Military Airlift Squadron, McGuire Air Force Base, as an operations and training officer. He was recalled by Trans World Airlines in June 1977, but retained his Reserve status. In February 1979, he assumed command of the 702nd Military Airlift Squadron and served in that capacity until April 1981, when he was assigned as vice commander of the 514th Military Airlift Wing, McGuire Air Force Base.

While attending National War College, Gen. Carpenter was again recognized for his research efforts by directing a research team to examine personnel aspects of the Civil Reserve Air Fleet. The results of this study had a significant impact on the future management and perception of the Civil Reserve Air Fleet program. The paper was selected for the Air Force Association's Award for Excellence in Research and Writing.

In July 1982, Gen. Carpenter was recalled to active duty to serve as the Reserve adviser to the Commander in Chief, Military airlift Command, Scott Air Force Base. In this position, he was actively involved in many vital total force issues, to include active-reserve force-mix studies and strategic airlift conversions into the air reserve forces.

In May 1985, he was selected as deputy to the Chief of Air Force Reserve, Headquarters U.S. Air Force, Washington, DC. For the next four and one-half years he assisted the chief in planning, programming and budgeting for the Air Force Reserve. He separated from active duty to assume his present position in October 1989.

The general is a command pilot with more than 9,000 flying hours, including 83 combat support missions to the Republic of Vietnam. Awards and decorations: Distinguished Service Medal, Legion of Merit, Meritorious Service Medal, Air Medal, Air Force Commendation Medal, Air Force Outstanding Unit Award with three oak leaf clusters, Combat Readiness Medal with four oak leaf clusters, National Defense Service Medal, Armed Forces Expeditionary Medal, Vietnam Service Medal with three service stars. Armed Forces Reserve Medal, Small Arms Expert Marksmanship Ribbon, Republic of Vietnam Gallantry Cross with Palm, and Republic of Vietnam Campaign Medal.

In his civilian capacity, Gen. Carpenter is 727B Check Airman for Trans World Airlines.

His professional affiliations include membership in the Air Force Association, Airlifters Association, Reserve Officers Association, Order of the Daedalians and National War College Alumni Association.

He was promoted to major general April 22, 1990, with same date of rank.

Gen. Carpenter is married to the former Ruth Shoemaker of Bridgeton, NJ. They have four children: David, Dale, Scott and Kristina.

GEORGE M. CARROLL, is Lockheed Aeronautical Systems Company's Director for Airlift Requirements, responsible for marketing airlift programs to U.S. Government agencies. This includes the sale of new aircraft, modifications and/or maintenance of existing airlift aircraft.

George began his Lockheed career in 1956, in the Engineering Flight Test Division where he remained for 10 years.

From 1966 to 1971, George was a sales representative in Marketing, working C-141 and C-130 programs. From 1971 to 1985, George ventured out of the aerospace industry and became National Sales Manager for Bausch and Lomb.

George rejoined Lockheed in 1985, and became involved in developing marketing strategies associated with the introduction of the C-130 J to the U.S. Air Force.

George graduated from Georgia State University. He also completed graduate studies at Syracuse University in sales management and marketing.

KEN R. CARSON, Major, USAF, was born Jan. 13, 1958, in Oklahoma City. He graduated from Midwest City High School in 1976, and graduated from the University of Oklahoma with a commission in the USAF.

He holds an MBA from Embry-Riddle University. He completed Army Command and General Staff College in 1993.

From 1982-87, Carson piloted weather reconnaissance missions in the WC-130 at Andersen AFB, Guam and Keesler AFB, MS. In 1987, he cross trained into the C-5 and served at Dover AFB, DE. He has piloted the WC-130 through 43 penetrations of typhoons/hurricanes and flown strategic airlift missions in Operations Just Cause and Desert Shield/Storm. He currently serves on the staff at United States Transportation Command, Scott AFB, IL. He was a recipient of the Airlift Association Young Leader Award in 1987.

Decorations include the Meritorious Service Medal with two oak leaf clusters, the Air Medal with one oak leaf cluster. He and his wife, Jan from Norman, OK, have two sons, Travis and Brandon.

JAMES MARVIN CASEY, Colonel, USAF, (Ret.), born Everton, MO, Nov. 19, 1927; BS in business administration, University of Missouri, 1951; BS in industrial engineering, University of Oklahoma, 1964; MPA, Golden Gate University, 1976. He enlisted in the Navy in January 1946, for two years. Following 22 months in the Navy he returned to the University of Missouri.

James took the advanced Air ROTC courses at the University of Missouri. He was commissioned in the Air Force upon graduation in June 1951. In July he reported to Tinker AFB, OK, to work until his pilot class started in September. Flight training started at Malden AB, MO. In November 1951 he married Shirley Rusk of Oklahoma City.

After completing primary flight training James and Shirley moved to Vance AFB at Enid, OK, in June 1952. Their first child, Jama Lynn, was born Aug. 23, 1952. James completed basic flight training in the B-25 in December 1952, and left for Japan in January 1953. He flew C-46s in the 43rd Troop Carrier Squadron between Japan and Korea. In February 1954, he was assigned to the 53rd Fighter Group, Sioux City AB, Iowa in the Air Defense Command. Base Ops had a C-45, C-47, C-82 and a C-119 to support deployment of the fighter squadrons.

James and Shirley's second child, Larry Denton, was born in Sioux City March 2, 1955. James had been assigned the job of budget and accounting officer at the Air Base. He attended the Air Force school at Sheppard AFB, TX, in 1957. While at this school Marcina was born June 18, 1957, in Oklahoma City where Shirley and the children were visiting her folks. Lacinda was born Dec. 3, 1958, in Sioux City. In April 1959, James received orders to Ankara, Turkey. His family accompanied him to Turkey. His assignment in the Joint United States Military Mission for Aid to Turkey (JUSMMAT) was Budget and Accounting officer.

During the summer of 1961, James and his family returned to the States. The first assignment was to Montgomery, AL, for Command and Staff College for one year. In 1962, James, now a major, received an assignment to the University of Oklahoma for an engineering degree. They lived in Midwest City, OK, close to Tinker AFB for this assignment. Following graduation in 1964, the next assignment was to Sewart AFB, TN, for flying transition in the C-130E aircraft. Following training James was assigned to the 50th Tactical Airlift Squadron at Sewart AFB and went overseas in December 1965, to fly tactical airlift in Vietnam.

He returned from the airlift base in Taiwan in March 1967, and was assigned to the 313th Tactical Airlift Wing, Forbes AFB at Topeka, KS. Lt. Col. Casey worked in tactical operations and then as squadron commander of the 38th Tactical Airlift Squadron.

In September 1971, James was promoted to colonel and transferred to Washington, DC. James was assigned to the Joint Staff in the Pentagon as Chief, Policy Division of the Logistics Directorate, J-4. They bought a house in Oxon Hill, MD, where they lived until June 1974. Jama remained in Topeka at Washburn University to finish her degree.

In June 1974, Col. Casey and his family moved to Shaw AFB, SC, where his assignment was director of airlift for 9th AF. In 1975, all tactical airlift functions were transferred to Military Airlift command. Col. Casey chose retirement rather than another move. Lacinda finished high school, Shirley finished her degree in finance, and Col. Casey finished his masters degree in pubic administration during 1976. Also during this year Jama was in law school, Larry and Marcina were in college.

DUANE H. CASSIDY, General, USAF, Commander in Chief of U.S. Transportation Command and Military Airlift Command, with headquarters at Scott Air Force Base, IL. As commander in chief of Transportation Command, he is responsible for global land, air and sea transportation for all U.S. fighting forces. As commander in chief of Military Airlift Command, he is responsible for airlift in support of unified and specified commands during war, periods of crisis, and contingencies. The general also commands special operations, rescue, weather, aeromedical evacuation, audiovisual, and operational support airlift forces for Department of Defense agencies throughout the world. He is executive director of the Single Manager Operating Agency for the Department of Defense Airlift Service.

Gen. Cassidy was commissioned a second lieutenant in 1954, following aviation cadet training. He then attended navigator training at bases in Texas. His initial operational assignments in the Air Force were to the Military Air Transport Service: first to the Air Weather Service's 6th Weather Group flying B-25s and then to Air Rescue Service's 49th Air Rescue Squadron. During these assignments he participated in numerous Rescue and Weather Reconnaissance missions, including the Hydrogen Weapons Test in 1956, at Eniwetok Atoll in the Marshall Islands. His navigator assignments ended after two years as a MATS C-121 line crew member at Charleston Air Force Base, SC. He entered pilot training in 1958.

Gen. Cassidy was assigned to Strategic Air Command following his graduation from pilot training and flew B-47s at McCoy, Little Rock and Lincoln Air Force Bases. In 1965 he transferred to Minot Air Force Base, ND, and served with the 810th Strategic Aerospace Division, whose mission included B-52 bomber and Minuteman missile operations. In 1968, he was reassigned to the Republic of Vietnam, serving first with 7th Air Force's Tactical Air Control Center and then with the Military Assistance Command Vietnam Directorate of Public Affairs as an air briefer to the Saigon Press Corps.

Reassigned from Vietnam in 1969, Gen. Cassidy returned to the Air Force airlift mission and was assigned to Headquarters MAC as executive to the Deputy Chief of Staff, Operations, and later as executive aide and pilot for the MAC Commander. In 1972, he assumed command of the 8th Military Airlift Squadron, McChord Air Force Base, WA. He entered Air War College in 1974. Upon graduation, Gen. Cassidy again served at HQ MAC as Assistant Chief of Staff. In 1976, he was reassigned to the 63rd Military Airlift Wing, Norton Air Force Base, CA, serving as vice wing and wing commander. He returned to Military Airlift Command headquarters in July 1980, and served initially as Assistant Deputy Chief of Staff for Operations. In August 1981, he became the command's Deputy Chief of Staff for Operations. In 1983, he took command of MAC's 21st Air Force at McGuire Air Force Base, NJ, for one year. During this period MAC was heavily involved in support of United States operations in Lebanon and Grenada. In August 1984, Gen. Cassidy became the Air Force's Deputy Chief of Staff, manpower and personnel at the Pentagon. The general assumed command of Military Airlift Command in September 1985, and of U.S. Transportation Command upon its activation in Oct. 1, 1987.

Gen. Cassidy is a command pilot and senior navigator and holds the parachutist badge. He has over 8,000 flying hours. In addition to Air War College, he has completed the Industrial College of the Armed Forces and Squadron Officer School. He holds a bachelor's degree from the University of Nebraska at Omaha and a master's of science degree from Troy State University; he has attended the Kellogg Graduate School of Management, Northwestern University, and Harvard University's Program in national and international security affairs. His decorations include the Distinguished Service Medal, Legion of Merit, Bronze Star Medal, Defense Meritorious Service Medal, Meritorious Service Medal, and the Air Medal.

Gen. Cassidy and his wife, Rosalie, have four children, Michael, Diane, Susan and Patrick and three grandchildren.

DEAN CASTLEMAN, born in Cape Girardeau, MO. Joined the service in November 1950, Pekin, IL, serving with Air Force, as command pilot. Military locations in which he served include James Connally AFB, TX; Keesler AFB, MI Charleston AFB, SC; Korea, Libya, Vietnam, Scott AFB, IL.

Memorable experiences include MAC Squadron Commander at Wheelus AB, Libya when Al-Qaddafi assumed power in 1969.

He retired as Director, Studies and Analysis, Military Airlift Command. He retired Aug. 1, 1978, with the rank of colonel.

Awards: Legion of Merit, Distinguished Flying Cross, Bronze Star, Air Medal with five oak leaf clusters.

He and his wife Lita, have two daughters Lori and Linda. Today he is manager, Boeing office in Fairview Heights, IL.

KAREN T. CHRISLEY, Chaplain, Lt. Colonel, from Roseville, OH, received her doctor of ministry from Graduate Theological Union at Berkeley, CA, in 1987. She has completed Squadron Officers School, Air Command and Staff and is currently enrolled in Air War College. She is presently assigned chief of Pastoral Care Ministries at Langley AFB, VA. In 1977, Chaplain Chrisley was the first woman chaplain in the Air Force assigned to an Air Lift Squadron, the 36th TAS at McChord AFB, WA, and was designated as an honorary Airlifter. In 1982-85, Chaplain Chrisley was the chaplain for the 7th MAS at Travis AFB, CA. She was a charter member of two Airlift Associations, at Travis AFB, and at Lajes Field, Azores. Among her awards and decorations are the Meritorious Service Medal with three oak leaf clusters, the Air Force Commendation Medal, and the Air Force Outstanding Unit Award with two oak leaf clusters. Chaplain Chrisley and her husband, the Reverend David Yingling, now reside in Newport News, VA.

JOHN CHRISTMAS, Master Sergeant, retired at Dyess AFB, TX, on Dec. 1, 1990, where he was assigned in 1976. Previous duty locations beginning in 1970, included Beale AFB, Thailand, and Loring AFB before he retrained into the loadmaster career field. At Dyess, he was assigned to 1 MAPS and served in several key positions, then was transferred to the 463rd Wing and served as chief of Loadmaster Training and later chief of Combat Tactics and Techniques, Loadmaster Section, and than served in the 773rd TAS, as assistant resource manager, Loadmaster Section.

His decorations include the Meritorious Service Medal with two oak leaf clusters, Air Force Commendation Medal, Air Force Outstanding Unit Award with one bronze oak leaf cluster, Combat Readiness Medal with four Bronze Oak Leaf Clusters, Air Force Good Conduct Medal with one Silver Oak Leaf Cluster and National Defense Service Medal.

Today, John and his wife Sharon and son Joey own and operate a family business in Clarkesville, a small town in the mountains of northeast Georgia.

Memorable Experience: In 1980, two pilots, one navigator myself and another loadmaster, all from Dyess AFB, comprised a six-month Mobile Training Team to Cairo, Egypt. I remained an additional two months to wrap-up loadmaster training. Our goal was to instruct, qualify and upgrade Egyptians crewmembers in C-130 tactical airdrop operations in their aircraft.

Among the many adventures shared with our Egyptian friends the following experience typifies our stay.

During an initial training mission events didn't go quite as planned. My student had just called "One Minute Checks Complete" when I observed through our open ramp and door what appeared to be numerous mushroom clouds rising off the desert floor directly under our formation. Everyone viewing this simultaneously arrived at the same conclusion, we were receiving fire! A formation "No Drop" and an "Escape" maneuver were initiated immediately. As the formation was safely blocking in the Egyptian Squadron Commander, a veteran of many air campaigns, was, as usual on the ramp, awaiting our return. Our (IP) instructor pilot, anxious to have a face-to-face confrontation with the Egyptian Commander, exited the crew door in a flash. We hurriedly tried to catch up him to prevent an "international incident" from occurring. As we approached the Egyptian Commander and the IP, we overheard the Commander calmly reasoning to the IP that if no one was injured and all the aircraft returned undamaged, then all was well. He ended with the common Moslem phrase "God wills." No matter how animated the IP became, the same calm reply was given by the Commander, "God wills."

Finally, realizing he could never convince the Commander to share his concern about the incident, the IP relented and we joined the veteran Commander inside for debriefing, along with the traditional hot Egyptian tea.

The "coordination ball" had been dropped, and dropped with a bang! The Egyptian Army began a live artillery fire, at the same moment we were completing our "One Minute Checks." Their "Impact Zone" had also been designated as our "Drop Zone!"

The eight months I spent with our Egyptian Airlift friends was the highlight of my career, inspite of their inspirational approach to everything. To be a part of upgrading a country's airdrop capabilities plus, have the challenge of achieving this with local resources and directives to include, salvaging and integrating leftover Russian airdrop equipment/vehicles into our airdrop methods was definitely a Loadmasters dream come true.

HARVEY M. CLARK, Master Sergeant, USAAF, enlisted in June 1947. Received basic training at Lackland AFB, TX, (known then as SAC field) initially trained as a radio operator at Scott Field, IL, and Rhein/Mien, German; later trained into AVIONIC maintenance.

Served in several major commands maintaining cargo, fighter, bomber tanker and RECON aircraft. Although he has many memorable experiences the ones that stand out are related to airlift and tanker aircraft; namely serving at Tempelhof Air Base Berlin during the airlift where he also participated in the Airlift Memorial dedication, and leading the AVIONICS maintenance team that participated in the 1961 SAC combat competition and won the best KC-97 award.

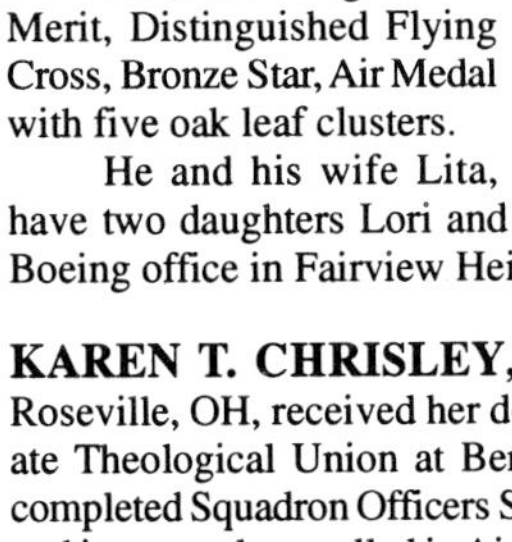

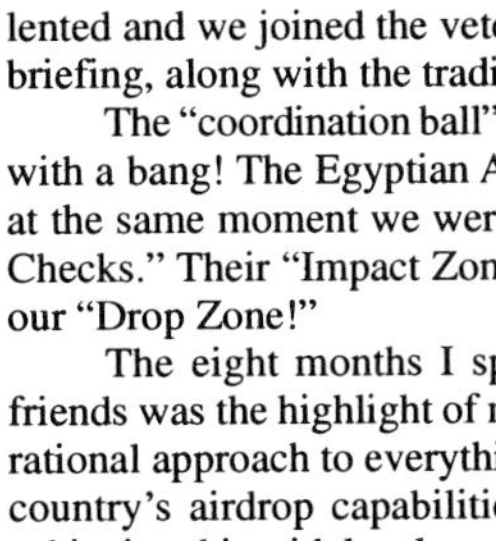

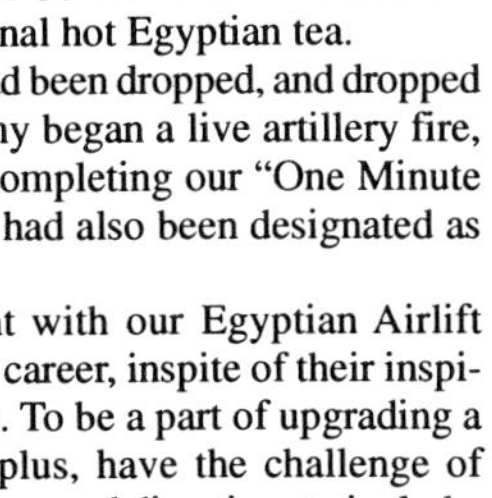

Retired in 1967, having served at 13 duty locations in six states and overseas locations in Germany, France and England.

He is married to the former Julia Ollington of London, England. They have one son and two grandchildren.

HOWARD M. CLARK JR., was born July 2, 1944, in Ft. Jackson, SC. He was commissioned on June 5, 1966, in the ROTC program at Union College. He was inducted into the USAF. He had pilot training at Craig AFB from December 1966-67, with class 68-D.

Clark's memorable experiences and locations and stations include: Ranchhand Operations, 1968-69 in RVN; FAC Mission, June-December 1969, Refueling missions January - April 1971 "Young Tiger," Airlift operations from January 1972 - 1974 and airlift for Operation Desert Storm January - April 1991. He was stationed at Bien Hoa, Vietnam, Loring AFB, Willow Grove, ARS, PA.

Awards/Medals: Distinguished Flying Cross with oak leaf cluster; Bronze Star Medal; Air Medal with 19 oak leaf clusters; Air Force Commendation Medal; Distinguished Presidential Unit Citation; Air Force Outstanding Unit Award with Valor Device and three oak leaf clusters; Combat Readiness Medal with 7 oak leaf clusters; National Defense Service Medal with one star; Vietnam Service Medal with six stars; Southwest Asia Service Medal with 2 stars; Air Force Longevity Service Award Ribbon with 5 oak leaf clusters; Armed Forces Reserve Medal with 1 oak leaf cluster; Small Arms expert Marksmanship Ribbon; Air Force Training Ribbon; Philippine Presidential Unit Citation; Republic of Vietnam Gallantry Cross with device; Republic of Vietnam Campaign Medal.

He achieved the rank of lieutenant colonel. He is married and has one son. Clark is a Jet Aviation/Executive Air Fleet, 1976 - president, check captain, AFRES - 913 Airlift GP, vice commander.

ROLLAND F. CLARKSON JR., graduated from Atchison Kansas High School in 1956. He earned a BBA and was a distinguished AFROTC graduate from Washburn University. He also earned a MA in guidance and counseling from Troy State University and completed SOS, ACSC, ICAF and Army War College.

He received navigator wings in 1961, graduated from ECM school in 1962, and from pilot training in 1966. Sent to SEA in 1967, he served with the 776TAS, 815TAS, 35TAS and 314TAW and flew 1300 combat hours. He had staff assignments at Headquarters MAC and the Pentagon. He served as operations officer 62TAS and 50TAS, commander 40TAS, vice and commander 317TAW and as vice 23AF. Colonel Clarkson retired in 1990, Assistant Deputy Chief of Staff for Operations Plans, Deputy Chief of Staff Plans and programs, Headquarters MAC.

He received Legion of Merit, Distinguished Flying Cross, Meritorious Service Medal and Air Medals. Ron and his wife, Marilyn, reside in O'Fallon, IL. They have two children Randall and Candice.

THOMAS R. CLAYTON, Captain, is flight examiner, 18th Airlift Squadron, McGuire AFB, NJ. Captain Clayton was born in Youngstown, OH, on Aug. 10, 1959. He attended Mercer High School in Pennsylvania, and excelled in many

school activities. As a starting first baseman, he helped his American legion baseball team in a third place finish in the state tournament. After graduating from high school in 1977, Capt. Clayton attended Slippery Rock University of Pennsylvania and worked part-time during college to pay for flying lessons. Upon completion of Officer Training School, Lackland AFB, TX, and undergraduate pilot training, Columbus AFB, MS, Capt. Clayton began initial C-141 qualification training at Altus AFB, OK, where he was a distinguished graduate. He is married to the former Barbara Dioletta and has one son. He is active in the base and local community serving as a basketball coach and Little League baseball coach. He was selected as Company Grade Officer of the Year for the 18th Airlift Squadron.

THOMAS J. CLELAND, born Jan. 27, 1920, in Los Angeles, CA. He graduated from Herbert Hoover High School, Glendale, CA, in 1938, and Glendale College in 1940. Cleland is a graduate of the Columbia University and Georgia State University Executive Management Programs.

Cleland learned to fly with the Civilian Pilot Training (CPT) Program in 1939. He joined Lockheed in 1940 and spent three years with the 8th Air Force in the European Theater (ETO) with the P-38. Cleland managed Lockheed's Product Support Programs on the C-130, C-141 and C-5 Aircraft for over 30 years.

He is a life member of the Airlift Association, OX5 Aviation Pioneers, P-38 National Association, the Society of Logistics Engineers, the Air Force Association and the Navy League of the United States. He is a U.S. Navy Veteran.

Cleland is widowed and resides in Marietta, GA. He has one daughter, Kimberly.

NEWTON V. COLE, Colonel, (Ret.), was born Sept. 7, 1928. He graduated from Refugio Texas High School in 1944 at age 15 and Texas A&M in 1947 at the age of 18. He attended Southwestern Baptist Theological Seminary, Fort Worth, TX, where he received the bachelor of divinity, master's of religious education and master's of divinity. He is a graduate of George Washington University with a MS in public administration, and received the doctor of ministry from San Francisco Theological Seminary, San Anselmo, CA. He attended Air Command and Staff College in 1965.

Cole entered the Air Force Chaplaincy in 1951, during the Korean War, serving as the youngest chaplain in the armed forces, at age 22. He was promoted to the rank of colonel in 1974. His overseas assignments were in Newfoundland, England, NKP Thailand, and Yokota Japan, where he served as Staff Chaplain U.S. Forces Japan and Fifth Air Force, 1979-1982. He was Command Chaplain, Military Airlift Command from 1982 until his retirement in 1987, after 36 years of service.

His awards and decorations include the Legion of Merit with two oak leaf clusters, the Airman's Medal, the Meritorious Service Medal and the Air Force Commendation Medal. Cole and his wife Linda are the parents of six children and seven grandchildren.

He resides in Houston, TX, in the winter months where he directs military chaplaincy matters for the Baptist Convention of Texas. He pastors the community Chapel in Tin Cup, CO, during the summer. He and his wife are avid trout fishermen.

GREGORY P. COOK, Captain, graduated from high school in Coeur d'Alene, ID, in 1979, then earned a political science degree from the University of Idaho in 1983. After pilot training at Columbus AFB, MS, he flew C-21s at Scott AFB, IL. From 1987 to 1993, he flew C-5s at Travis AFB, CA, where he participated in many contingency operations, humanitarian relief efforts, and other special missions. These included Operations Just Cause and Desert Shield/Storm. He is presently a plans and issues officer at Air Mobility Command Headquarters.

He earned a master's degree in International Relations from Webster University in 1987, and attended Squadron Officer School in 1990. He has received honors as a 1992 Airlift Association Young Leadership Award Winner, an Outstanding Young Man of America, and squadron Company Grade Officer of the Year three times.

His decorations include the Aerial Achievement Medal, Air Force Commendation Medal, Air Force Achievement Medal, Combat Readiness Medal with two oak leaf clusters, Armed Forces Expeditionary Medal, and the Southwest Asia Service Medal with three Bronze Stars. He and his wife Tammy have two daughters, Tiffany and Victoria.

There is an experience as an airlifter in the United States Air Force that I will never forget. It occurred on Jan. 19, 1991. We had been at war with Iraq for just two days, during which time Saddam Hussein had continuously lobbed SCUD missiles at Saudi Arabia and Israel. I had narrowly missed one such attach at Dhahran, Saudi Arabia, on January 17, and now I was headed to Tel Aviv, Israel, with a C-5 full of Patriot Missile launchers to protect that nation from Iraqi tyranny. Keeping Israel out of the war and maintaining a strong allied coalition were our country's top priorities, and I was proud to be a part of that effort. Yet it took a regular in the Israeli armed forces to give the significance of the mission a truly human perspective. For two nights, he told me, he and his family had huddled in their small apartment as air raid sirens and explosions shattered the darkness around them. With each alarm, they donned their chemical masks and waited, not knowing whether the next sound they heard would be an Iraqi messenger of death or the "all clear" signal that danger had passed, if only temporarily. To them, each night had seemed like a lifetime, with apprehension, fear and feelings of hopelessness growing ever larger. Suddenly for him, however, there was hope, for America had arrived to save the day. "Thank God for America ... Thank God the A-team is here!" I heard him exclaim again and again, which made me realize that only America had the capability and the will to respond as we did to protect that nation. I was proud to be a part of America's team.

DANIEL J. COONAN III, born on Dec. 10, 1944, in Worcester, MA. He graduated from Cathedral High School in Rapid City, SD, in 1962, and from the U.S. Military Academy at West Point, NY, in 1966. Coonan graduated with distinction from SOS in 1971, and completed his master's from Central Michigan University in 1977. He completed ACSC in 1977 and AWC in 1982.

A career combat controller since 1966, Dan served as a team leader at McGuire, Tan Son Nhut, Naha and Forbes, and as a CCT Staff officer at McGuire, Ramstein and Scott. He was a services squadron commander in 1983, and then, Director, Combat Control and Pararescue, HQ MAC. He was base commander at Scott and retired in September 1992.

He holds the Legion of Merit, the DMSM, and six awards of the Meritorious Service Medal. Dan and his wife, Fran, reside in Belleville, IL. They have two children, Rebecca and Danny.

NICHOLAS P. COSTA, Colonel, (Ret.), born Dec. 18, 1939, in Salem, OH. He received his bachelor's degree in

education from Ohio State University and master's degree in guidance and counseling from Louisiana Tech University. He completed Air Command and Staff College and Air War College.

A master navigator, he flew over 4,300 hours in the C-130E, including combat missions to Southeast Asia. His other assignments include air attaché to Bucharest, Romania, 314th Field Maintenance Squadron Commander at Little Rock AFB, and professor of aerospace studies at the University of Wyoming.

Col. Costa retired in August 1991, after serving as deputy commander for resource management for the 63rd MAW at Norton AFB, CA. His decorations include: Legion of Merit, Distinguished Flying Cross, Meritorious Service Medal with three oak leaf clusters, and Air Medal with two oak leaf clusters.

Costa and his wife, Amber, have two children, Nicole and Susan.

ROBERT F. COVERDALE, Lieutenant General, USAF, (Ret.), is director of Transportation, St. Clair County, Belleville, IL.

Gen. Coverdale was born Sept. 24, 1930, in Amherst, OH, and graduated from Amherst High School in 1948. He attended Ohio Wesleyan University in Delaware, OH, and received a bachelor of arts degree in 1952. He graduated from the Armed Forces Staff College at Norfolk, VA, in 1967.

He was commissioned a second lieutenant through the Air Force Reserve Officers' Training Corps program and received his pilot wings after completing training at Vance Air Force Base, OK, in November 1953.

His first assignment as a C-119 pilot at Sewart Air Force Base, TN, with the 75th and 62nd Troop Carrier Squadrons, was the beginning of his long career in airlift. While at Sewart he also served as a C-130 pilot, chief of the command post, aircraft scheduling officer and as the 314th Troop Carrier Wing's Chief of Training.

Gen. Coverdale was assigned to Detachment 1, 315th Air Division, Naha Air Base, Okinawa, in September 1962. His positions there included chief of standardization with the 6315th Operations Group and operations officer of the 817th Troop Carrier Squadron.

He departed Okinawa in July 1965, and was assigned at Langley Air Force Base, VA, serving first with the 4450th Standardization Evaluation Group and later with Headquarters Tactical Air Command as a staff officer in airlift concept and doctrine.

In August 1967, after graduation from the Armed Forces Staff College, he was assigned to Detachment 4, 314th Air Division of Don Muang Airport, Thailand, where he served as operations officer. He returned to the United States in August 1968, and was assigned to Headquarters U.S. Air Force, Washington, DC, where he became chief of the Employment Test Branch for the deputy director for operational test and evaluation.

Gen. Coverdale became vice commander of the 516th Tactical Airlift Wing at Dyess Air Force Base, TX, in March 1971. He assumed command of the wing, redesignated as the 463rd Tactical Airlift Wing, in June 1972. Gen. Coverdale was named commander of the 314th Tactical Airlift Wing at Little Rock Air Force Base, AR, in November 1973, and took command of the 834th Air Division there in July 1974.

In June 1975, he became commander of the 317th Tactical Airlift Wing as well as commander of the newly established U.S. Air Force Airlift Center at Pope Air Force Base, NC. He was named Deputy Chief of Staff for Plans, Headquarters Military Airlift Command at Scott Air Force Base in August 1977, and became the Command's Chief of Staff, in December 1979. From May 1980 to July 1981, Gen. Coverdale commanded the 22nd Air Force at Travis Air Force Base, CA. He assumed duties as vice commander in chief in August 1981, and retired from the United States Air Force on Sept. 1, 1985.

In November 1985, Gen. Coverdale assumed the position of director of Aeronautics for the State of Illinois. In this position he was responsible for state wide Airport Planning and Engineering Services; Financial Management; Airport Capital Development and Improvement; Land Acquisition; Flight Department Operations and Maintenance; Airport Safety Inspections and Aerospace Education.

On Sept. 4, 1990, Gen. Coverdale became director of Transportation for St. Clair County, IL.

Gen. Coverdale is a command pilot with approximately 8,000 flying hours in various types of transport aircraft. He flew 46 combat missions in Southeast Asia and accumulated 162 combat flying hours. Gen. Coverdale also has a parachutist rating, which he received in 1975, after completing Airborne Jump School with the 82nd Airborne Division at Ft. Bragg, NC. His military decorations and awards include the Defense Distinguished Service Medal, Distinguished Service Medal, Legion of Merit with one oak leaf cluster, Meritorious Service Medal with one oak leaf cluster, Air Force Commendation Medal with oak leaf cluster, Republic of Vietnam Gallantry Cross with palm and Republic of Korea Order of National Security Merit.

Gen. Coverdale is married to the former Norma Tate of Huntingdon, TN. They have three children: Lisa, Amy and Tate.

EMERY J. CRANE, born Oct. 2, 1928, in Henry County, IA. He entered Aviation Cadet Training in 1949, and was commissioned as a pilot in 1950. During his career, he has flown over 11,000 hours in a number of airlift aircraft including the C-5A. He retired from the Air Force in February 1978, as a colonel, where he was serving as the director of safety, Military Airlift Command. He was stationed at Brookley AFB, AL 50-52; Haneda AB, Japan 52-54; Charleston AFB, SC 54-60; Command and Staff College 60-61; Directorate of Plans, HQ USAF, 61-64; Hickam AFB, HI 64-67; Andrews AFB, MD 67-70; Phan Rang AB, Vietnam, 70-71; Dover AFB, DE 71-74; and Scott AFB, IL 74-78. He is married to the former Janet Smith of Chicago, IL. They have three sons David, Alan and Steve, and a daughter Kathy. He is active in community and Kiwanis supervising a summer camp program for disadvantaged school children and tutoring at the local elementary school.

RUSSELL D. CRANE, was born Dec. 28, 1917, in Henry County, IA. He received a bachelor's degree in chemistry from Iowa Wesleyan College in 1939.

Entering Aviation Cadet Training in September 1942, he retired in February 1971 with the rank of colonel.

Some of his more significant assignments were: Commander 50th Air Transport Squadron, Commandant MAC NCO Academy, Deputy Commander for Operations 62nd Troop Carrier Wing, Commander 374th Tactical Airlift Wing, and Deputy Chief of Staff Personnel Military Airlift Command. While under his command, the 374th TAW received a Presidential Unit Citations an Air Force Outstanding Unit Award, and a USAF Flying Safety Award for its actions in South East Asia.

Among his awards and decorations are: Legion of Merit with one oak leaf cluster, Distinguished Flying Cross, Air Medal with two oak leaf clusters, Air Force Commendation Medal with one oak leaf cluster, Presidential Unit Citation, and the Air Force Outstanding Unit Award with three oak leaf clusters.

Crane and his wife, Doris, reside in Tucson, AZ. They have two adult sons, Barry and Jeffrey.

FREDERICK W. CRESSMAN III, "RICK" commissioned in June 1974, through ROTC at Ohio University Athens, OH. I attended pilot training at Webb AFB, TX, from January 1975, until December 1975 in class 76-05.

After receiving my wings and attending survival training I reported to Little Rock AFB, AR, to fly the C-130. I was assigned to the 61st TAS. I worked as a Command Post controller from 1980-1981. I was then assigned as an RTU instructor in the 62nd TAS at the "Rock" from 1981 until January 1983.

From January 1983 to June 1986, I served as the Air National Guard Advisor to the Wyoming Air National Guard in Cheyenne, WY.

From that assignment I reported to Yokota AB, Japan to fly C-130's in the Pacific. I was assigned to the 345th TAS from 1986 until 1989. I served in many capacities while there; chief of Resource Management, chief of Standardization, Assistant Operations Officer, and chief of DOO.

June 1989 found me back in the states working for the 89th Military Airlift Wing. I served as Flight Commander flying the C-12C, checking out military attaché's in the plane. After a year and a half of this I moved over to the 99th ALS to fly the C-20. I'm currently serving as an instructor, in the C-20, and I am the chief of Training for the Squadron. My wife and I and our two children, son Kyle and daughter Erin, reside in Waldorf, MD.

I have over 6,000 hours flying time and I am a command pilot. My wife is the former Diane Barrett Knudsen of Sheffield Lake, OH.

EDWARD J. CROCKETT, born February 18, in Martin, TN. He attended Tennessee State University in Nashville, 1947-1948. Joined the U.S. Air Force in July 1948, serving with the 314th TCW, FEAF, Strike Command, 7370 FLT SER France. He served in Japan, England, France, and Greenland.

Memorable experiences include calling a softball game in Greenland, the field was solid ice. Was stationed at Sewart AFB, TN, 10 of my 20 years.

Discharged in August 1968, with the rank of SST.

He and his wife Catherine D., have two children and five grandchildren. As a civilian he was employed with State of Tennessee City of Martin. As of August 1994, still working for the city.

After retirement in 1968, I worked 10 years at the University of Tennessee in Martin, and five years at Van Ayer Manor Nursing Home in Martin, TN. Since 1985, I've been employed by the City of Martin with Martin Parks and Recreation. In 1991, joined Headquarters 104th Battalion, Tennessee Defense Force and 1993, was promoted to the rank of first sergeant. I am currently a member of the Jacksonville Lodge #50 of South Fulton, TN; and member of American Legion University Post #69 of Martin, TN; member of McCabe United Methodist Church; was chairman of trustee board, board member of Pastoral Advisory Committee, administrative church board, usher board president of United Methodist Men and a member of the building committee during the years of 1993-1994 for the building of a new church.

WILLIAM E. CULPEPPER, "ED", born Sept. 7, 1931, in Connellsville, AL. Attended the University of Alabama, Birmingham, for three years. Joined the Air Force Oct. 20, 1950, serving with the 55th TCSq., 37th TCSq., 817th TCSq. (Blue Tail Flies). Locations at which he served include Donaldson AFB, SC; Sewart AFB, TN; and Ashiya, Japan.

His memorable experiences include training French air crews to fly C-119s at Clark AFB, could not speak French and they could not understand English. These French crews flew missions from Hai Phong to Dien Ben Phu.

He discharged from active duty Oct. 19, 1954, and the Guard June 30, 1961, with the rank of T/Sgt.

He and his wife, Joyce Kendrick Culpepper, have three children and four grandchildren. As a civilian he is employed as a traffic manager, sales representative, and international sales manager.

RONALD E. CUNEO, is the president and chief executive officer of HFSI, a worldwide systems integrator and supplier of information systems, hardware, software, technical and maintenance service to the federal government, NATO and other prime contractors.

Since 1969, Mr. Cuneo has served in marketing and

executive management positions, initially with the information systems division of General Electric Company and later with Honeywell Inc., where his experience spanned federal and commercial sectors. In 1976, Mr. Cuneo directed one of Honeywell's major defense information systems programs, the Worldwide Military Command and Control System. He subsequently served as national market director for Honeywell's Xerox program; and in 1981, he became vice president of corporate accounts. In 1982, Mr. Cuneo was appointed vice president and general manager of Honeywell Federal Systems, where in eight years he brought in more than $1.5 billion in federal contract awards. Under his leadership, the company evolved from its traditional role as a provider of proprietary systems to a full-service provider of integrated information systems solutions.

Following the sale of Honeywell Federal Systems in 1990, Mr. Cuneo became president and chief executive officer of the company, renamed HFS In. (HFSI). As CEO, he has positioned HFSI as a major contender in the federal systems integration and services marketplace, achieving prominence in diverse, high-growth markets, such as security, EDI, and multi-vendor maintenance. Under his guidance, HFSI has built a $40 million technical services business.

Mr. Cuneo holds a bachelor's degree in accounting and taxes from Pace College and a master's degree in management technology from American University. He is affiliated with numerous professional associations and serves on the board of directors of AFCEA-International and The Riggs National Bank of Washington, DC, and Riggs National Corporation.

JOHN J. D'ANIERI, was born May 13, 1929, in Long Island, NY. He graduated from Queens Vocational High School, New York, in 1946. In March 1947, he entered the U.S. Army Air Corps, he served as a member on C-54 Aircraft, Air Transport Command (ATC) during the Berlin Airlift. He completed his Air Force active service in 1952. In 1956, he joined the U.S. Air Force (Active) Reserve and spent the next 17 years as an aircraft loadmaster on C-119 and C-124 aircraft.

Military courses completed include: Army Intelligence Analyst School, (Special Forces); Depot Operations Management Course, Army NCO/SGM Academy; Aviation Supply/Maintenance Material Management Course and Career Supply Management Course, Naval Officer Supply Corps School.

In 1973, he transferred to the Army Aviation Technician Program and was assigned to various commands including: Army Security Agency (AVN) Army Special Forces Logistics and Army Chemical Warfare Group. Prior to retiring he was attached to the Air Force as the first and only SGM (E-9) Army Ground Liaison Officer (GLO). He was liaison between the Commander 93rd TFS, Homestead AFB and the 18th Airborne Corps, Ft. Bragg, where he served as Air Land Battle Coordinator, retiring on May 13, 1989.

Among his many awards and decorations are: Legion of Merit, Joint Forces Achievement Medal (DOD), Army Commendation Medal, Army Achievement Medal, Korean Service Ribbon, Medal for Humane Action (Berlin), Presidential Certificate of Appreciation for over (30 years) of service.

D'Anieri has been involved with the Civil Air Patrol since 1944, and is presently inspector general for the Southeast Region, USA. Recently promoted to logistic element manager at Naval Air Warfare Center. D'Anieri and his wife Barbara now reside in Mount Dora, FL.

CATHERINE M. DANZY, Technical Sergeant, was born in Shirley, MA, on June 5, 1958. She graduated from Pemberton Township High School in 1977 with a 3.5 grade point average. She entered the Air Force on Dec. 7, 1977. Her assignments include the 1978th Communications Group, Howard AFB, Panama, 1978-80; Travis AFB, CA, 1980-81; McGuire AFB, NJ, 1981-82; Hellenikon AB, Greece, 1982-86; Andrews AFB Presidential Radio Station as a shift supervisor, NCOIC, and now the superintendent, Presidential Radio Operations, 1990-present. Sergeant Danzy has 61 credit hours towards her CCAF degree in Information Systems Technology. She completed all military education for which she is eligible. She is the recipient of the Air Force Commendation Medal with one oak leaf cluster; Headquarters Military Airlift Command Radio Operator of the Year, 1991; 89th Communication Group NCO of the Quarter and NCO of the Year, 1991; John L. Levitow Award, AFCC NCO Academy; NCO of the Year, 1st Combat Communications Group, distinguished graduate honors and the Commandant's Award, NCO Leadership School; and senior airman below the zone. She is married to Technical Sergeant Oran L. Danzy, 89th Mission Support Squadron, Andrews, AFB, MD, and they have two children, Nicholle and Larry.

JOHN E. DAVIS, born in West Virginia on July 4, 1929. After one year in college, he entered the U.S. Air Force in 1948. He was commissioned in 1951, through the Aviation Cadet Pilot Program.

The majority of his career was in tactical airlift, flying C-82s, C-119s, and C-130s. His assignments include one year in support of the Korean conflict and five years in support of Vietnam. He has flown more than 1,200 combat hours. During 1972-73, he was commander of the AC-130 gunships in Thailand.

In 1975, he became Commander, 314th TAW, at Little Rock AFB.

His decorations include the Legion of Merit, three DFCs, 10 Air Medals, the Bronze Star, and two individually awarded Vietnam Crosses of Gallantry.

He retired in 1977. In 1979, he became the chief pilot of a Corporation until again retiring in 1991. He has more than 15,500 pilot flying hours.

During his career he continued his education, earning a bachelor's degree and a master's degree in personnel management.

He has four daughters and two grand daughters.

PRESTON H. DAVIS, born Dec. 18, 1940, in Milwaukee, WI. He joined the U.S. Air Force June 5, 1963, serving as a pilot. Military locations at which he served include Travis AFB (four tours); Phan Rang, Nua Trang (each six months); Scott AFB; the Pentagon; Altus AFB; Andrews AFB; and Kelly AFB. He was discharged Sept. 30, 1990, with the rank of colonel.

His memorable experiences include flying the C-133 was and remains a great thrill in his life, second only to flying the C-5. He enjoyed working at MAC HQ trying to make the C-17 program go and enjoyed the Air Staff work to approve the C-5A wing modification program and getting it to the production approval. Finally, being the C-5 system manager was a great job. Awards/Medals: Presidential Unit Citation with three oak leaf clusters, and Vietnamese Cross of Gallantry with palm.

He and his wife, Carole have been married 25 years. They have three children: Jennie, a lawyer; Sonny going to the airlines; and John, a kid forever. Today he is principal analyst at Arinc Research Corp.

JAMES R. DE ANGELO JR., Lieutenant Colonel, was born April 4, 1950, in Utica, NY. He earned his bachelor's degree in industrial engineering from Oklahoma State University in 1973, and his master's degree in business administration in 1983, from Webster University.

In 1974, DeAngelo received his navigator wings and was assigned to Clark AB, Philippines, where he flew over 2,500 hours in the C-130E, to include the Vietnam refugee evacuation. In 1983, while assigned to 62nd TAS, he was a member of the best C-130 airdrop crew and won the Bullseye Award - best airdrop in MAC's worldwide airdrop competition. In 1984, he worked as a war plans officer at HQ MAC.

After graduating from Air Command and Staff College in 1987, he remained on the college's staff as Chief, Defense Planning and Operations Branch. In 1989, he became the Chief, Operations Curriculum Division at the Airlift Operations School and is presently the Air Mobility School's director of Curriculum.

DeAngelo is a master navigator with more than 4,300 flying hours. His military awards include the Meritorious Service Medal with three oak leaf clusters, Air Force Commendation Medal with one oak leaf cluster, Humanitarian Service Medal, Combat Readiness Medal with two oak leaf clusters, and National Defense Service Medal with Bronze Star.

WILLIAM HOWARD DELACEY, Colonel, USAF, (deceased), Commander of 839th Air Division, Sewart AFB, died of a heart attack while hunting. He was a native of Utica, KY. He began his military career in November 1939, when he entered flying training. After advanced training he was assigned as flight training instructor at Kelly AFB, TX, then transferred to Barksdale AFB, LA. He then moved to Hendrix AFB, FL, to instruct B-17 students. He was then assigned to the first B-29 Superfortress wing at Great Bend AFB, KS. From there he went overseas in April 1944.

In the China-India-Burma theater, he served as squadron commander of the 686th Bomb Sq. In 1946, he returned to the U.S. and was sent back to the Pacific. After a brief tour, he went to Germany. During the Berlin Airlift, he was stationed at Templehof AB, Berlin, where he served as chief of Operations, wing executive officer, and base commander.

In 1949, he returned to the U.S. to attend the Command and Staff School at Maxwell AFB, AL. He first came to Sewart AFB, in January 1950, as commander of the 50th Troop Carrier Sq. But with the outbreak of the Korean conflict he departed Sewart with his squadron for overseas duty with the 314th Troop Carrier Group in Japan and Korea.

In addition to serving as commander of the 50th during his tour in Korea, he also served as liaison officer to the Fifth Air Force and Eighth Army for the 316th Air Division, detachment commander at Kimpo Air Field, at Seoul, and commander of the 314th TCG before his return to Sewart in 1951.

In 1956, he again left the base to attend the National War College at Ft. McNair, VA. From there he was assigned to the Pentagon. In 1960, he returned for the third time to Sewart as commander of the 314th Troop Carrier Wing. In 1962, he attended the language school at Ft. Myer, VA, prior to his assignment as chief of the Air Force Section, Military Assistance Advisory Group, Portugal. He returned to Sewart from Portugal.

As the division commander, Col. DeLacey had overall responsibility for not only the units at Sewart, but also those stationed at Pope AFB, NC.

He had received many decorations including the Legion of Merit, the Distinguished Flying Cross with two oak leaf clusters, the Air Medal with four oak leaf clusters, the Bronze Star Medal with one oak leaf cluster, the American Defense Service Medal, the Army Occupation Medal with Berlin Airlift Device, the Asiatic-Pacific Campaign Medal with five Bronze Service Stars, the Distinguished Unit Citation with one oak leaf cluster, the United Nations Service Medal, the Korean Service Medal with seven Bronze Service Stars, the Republic of Korea Presidential Unit Citation, the National Defense Service Medal with one Bronze Service Star, and the Air Force Length of Service Award with five oak leaf cluster.

He is survived by his father, Charles DeLacey, Utica, KY; his widow, Mrs. Mary Ina DeLacey, Smyrna; two sons, First Lieutenant William H. DeLacey Jr., Norton AFB, CA; and Charles; and four daughters, Mrs. William Hornovich, Detroit, MI; twins Mary A. and Betty L., and Linda S., all of Smyrna.

HARRY S. DENNIS, born May 4, 1921, in Morristown, NJ. He served with the U.S. Air Force, as command pilot, Comdr. Sq. level and wing level, HQ USAF Staff. Military locations at which he served were Europe, Pacific and CONUS. He participated in ETO D-Day, etc., 25 months in theater.

Memorable experiences include 47 bombing missions (B-26) ETO as bombardier, pilot training later.

Awards: Legion of Merit, Distinguished Flying Cross Air Medal with seven oak leaf clusters, Air Force Commend with two oak leaf clusters, plus assorted campaign medals.

He was discharged Dec. 31, 1970, with the rank of colonel.

He has three sons and one daughter, the oldest son is a former Air Force pilot C-141 and C-130.

He retired from the Hilton Head Island, golf, golf and more golf.

SEVERINO DI COCCO, came to this country from his native Italy in July 1933. He graduated from Pennington Central High School in 1945, and was drafted into the U.S. Army in January 1946 for 18 months. He was immediately switched to the Army Air Forces and was in the first Air Force flight to begin and complete Air Force basic training at Lackland AFB. In December 1948, he re-enlisted in the USAF for Operation Vittles (Berlin Airlift) and was assigned to the 53rd Troop Carrier Sq., which had C-54s, as a personnel specialist. During his off-duty time and on weekends, he made numerous flights with his unit into Berlin assisting in the off-loading of food and coal cargo at Templehof AB. In July 1950, his group was sent on 30 days TDY to McChord AFB to fly supplies and troops to Korea. In December 1950 his Group was sent on further TDY to Ashiya AB, Japan. It was during this time that his group became known as Norwood's Nomads after their group commander, Col. Frank Norwood. He recalls the unit aircraft he arrived in was quickly modified by removing the additional fuel tanks in the forward cargo area and made its first mission to Korea the same day to assist in the evacuation of U.S. troops from Hungnam who were retreating from the Chosin Reservoir area. Again during the next 16 months he made numerous trips on his own time with air crews into nearly every airstrip in Korea assisting in unloading cargo such as food rations, 50 gallon drums of fuel, 500 pound bombs, troops being returned into combat from R&R in Japan, etc. There were several months in which he made more trips into Korea than some of the air crews.

In October 1951, his Group was made PCS from Rhein Main AB in Germany to Ashiya AB, Japan. So after 21 months of TDY he was returned to the U.S. in April 1952. From January 1953, to June 1955, he was assigned to the 31st Air Transport Squadron C-97s with MATS at Westover AFB and then moved to McGuire AFB with HQ Atlantic Division MATS. From March 1956 to May 1960, he was assigned to the 60th Troop Carrier Wing at Dreux AB, France which had both C-119s and C-123s. During this time his unit supported the 1958 crisis in Lebanon. After assignments as a military training instructor at Lackland AFB and personnel supervisor at Dover AFB (C-124s and C-133s) and a tour at Lajes Field, Azores, he was assigned to HQ COMD USAF at Bolling AFB where he retired as a member of the Command IG Team on Feb. 1, 1975.

He holds the Meritorious Service Medal with oak leaf cluster, Air Force Commendation Medal with two oak leaf clusters, U.S. and ROK PUCS, United Nations Service Medal, Korean Service Medal with five Bronze Stars, Medal for Humane Action and numerous other service awards. He is also a Distinguished Honor Graduate of the MAC NCO Academy.

He is married to Anna Maria Bopp and have five children. Both of his sons are currently Air Force officers.

GEORGE E. DOCKERY, was born Dec. 30, 1933, in Evansville, IN. He graduated from Reitz High School in Evansville and received a bachelor's degree in history and political science from the University of Evansville in 1955. He completed Armed Forces Staff College in '970, and received a master's in management and business administration from Webster College in 1978.

Dockery retired from the United States Air Force in December 1985, having achieved the rank of colonel. During his last assignment he served as Deputy Chief of Staff, Operations for 22nd Air Force, Military Airlift Command at Travis Air Force Base, CA. The vast majority of his career was spent in airlift operations including troop carrier and tactical airlift assignments in C-119s and C-130s at Sewart AFB, TN; Ashiya AB, Japan; Naha AB, Okinawa and CCKAB, Taiwan. He served in airlift command and control positions with HQ PACAF, TAC and MAC.

His awards and decorations include service in Vietnam, Air Medals and Legion of Merit with one oak leaf cluster. Dockery and his wife Marilyn live in Littleton, CO, and have three sons; Daniel, Thomas and Jonathan.

JAMES G. DONOVAN, "BUD" Colonel, (Ret.), received his commission through the AFROTC in 1960. He entered navigator training at Waco, TX, and remained as an instructor. In 1965, he was assigned to Vietnam with the 309th Air Commando Squadron flying in C-123s.

In 1971, he moved to Charleston AFB, SC, and served there as chief of Wing Combat Tactics in the C-141. After an AFIT assignment he returned to MAC at Albuquerque, NM. In 1980, he moved to MAC Headquarters and worked in the Manpower and Organization Directorate, DCS Plans. In 1983, he moved to Lajes AB, Azores and became Commander, 1605 Air Base Group. In 1985, he returned to the University of Arkansas as the PAS. He retired in 1988.

Bud was a master navigator with over 5,000 hours. His decorations include the Legion of Merit and Distinguished Flying Cross. He and his wife Joyce reside in San Antonio.

GEOFFREY R. DOUGLAS, Captain, USAF, born March 15, 1961, Austin, TX. He received his BA degree in economics from the University of Texas at Austin, in 1984. He has been active duty since Aug. 20, 1986. He has served OTS class 87-03, UPT class 88-01, Reese AFB, TX; C-12F, 1402 MAS, Andrews AFB, MD, February 1988- May 1991; C-5 Galaxy, 9 MAS (now 9As), Dover AFB, DE, May 1991 to present.

He flew C-12s at Andrews, carrying VIPs all over the U.S., for three year and enjoyed it immensely. At Andrews he saw a lot of history, met interesting people, and enjoyed Washington DC. That's also where he met his wife. At the beginning of Desert Storm, they took the seats out of a C-12, positioned it at Dover, and flew a MICAP express service between stateside 21AF bases.

He ended up at Altus for C-5 initial right in the middle of the war, and found it nearly deserted when he arrived. There was one C-5 on the flightline. He checked into the 9th in May and spent the summer and fall of '91 hauling everything home. The trips were round-robins from Dover - Rhein Main or Ramstein - Daharan or Riyad - Torrejon - USA.

More recently, he's flown 30+ missions for Restore Hope, many with multiple ARs. He's also gained an extraordinary familiarity with Cairo West and the Pyramids.

He was an IP in the C-12, and is now an AR-qualified IP in the C-5. He has 2,900 hours military flying time plus about 400 of civilian time.

JAMES F. DOWNING, born June 24, 1925, in Buffalo, NY. He graduated from Canisius High School, in Buffalo, in 1942, and from Canisius Collage, also in Buffalo, with a BS degree in physics in 1948.

Downing retired from the Sierra Research Division of LTV Missiles and Electronics Corp. In May 1988, where among other things he initiated the Sierra Stationkeeping Program. Prior to joining Sierra he was the Washington representative for the Avionics Division of Bell Aerospace Corp.

Downing was active in the founding of the Airlift Association having served as industry coordinator for a number of the 834th Air Division Reunions and later as president of the Airlift Association. He is currently a member of the board of advisors.

Downing and his wife Mary reside in Buffalo, NY. They have nine children and 13 grandchildren.

BENNIE G. DRAKE, Chief Master Sergeant, born Dec. 29, 1938, in Wichita Falls, TX. He graduated from Wichita Falls Senior High School in Wichita Falls, TX, and then entered the United States Air Force on Aug. 1, 1957. He subsequently received a bachelor of science degree in occupational education from Southern Illinois University at Carbondale, IL, in 1977. He is a graduate of the SAC Leadership School, MAC NCO Academy and the Fifth Anniversary Class of the USAF Senior NCO Academy.

Upon completion of basic training at Lackland AFB, TX, in 1957, Chief Drake was assigned as a security policeman in the 818th Air Police Squadron (SAC), at Lincoln AFB, NE. His first overseas assignment was with the 3926th Air Police Squadron, Ben Guiere, Morroco. He was assigned to the 4123rd Combat Defense Squadron (SAC), Clinton-Sherman AFB, OK, where in July 1961, Chief Drake retrained as a legal specialist with Headquarters 4123rd Combat Support Group (SAC), Clinton-Sherman AFB, OK. He was reassigned to Headquarters 3970th Wing (SAC), Torrejon Air Base, Spain, from 1963 to July 1966. From 1966 to August 1968, he was assigned to Headquarters 92nd Combat support Group, Fairchild AFB, Washington, as claims clerk and eventually noncommissioned officer in charge of the Staff Judge Advocate Office. The Chief retrained to Aircraft Loadmaster in August 1968, and after completing the Aircraft Loadmaster School, Sheppard AFB, TX, he was assigned to the 3rd Military Airlift Squadron, Charleston AFB, South Carolina until September 1969. He served as a loadmaster with the 8th Aerial Port Squadron, Tan Son Nhut Air Base, Republic of Vietnam from September 1969 to September 1970. From September 1970 until 1976, he served as a C-141 instructor loadmaster with the 30th Military Airlift Squadron, McGuire AFB, NJ. He was assigned to the 443rd Technical Training Squadron in April 1976, where he served as an instructor, C-141 loadmaster superintendent and chief of C-5 and C-141 loadmaster training. He served as chief, loadmaster training analyst and C-141 flight examiner until his selection as senior enlisted advisor for the 340th Air Refueling Wing (SAC) on Oct. 15, 1981.

Chief Master Sergeant Drake flew over 100 combat missions in Southeast Asia on C-123K and C-130 aircraft, as well as various helicopters. He has accumulated over 4,000 hours in the C-141A and C-141B aircraft. His decorations include the Bronze Star, Meritorious Service Medal with oak leaf cluster, Air Medal with oak leaf cluster, Air Force Commendation Medal with oak leaf cluster, Combat Readiness Medal with two oak leaf clusters, Republic of Vietnam Service Medal with three service stars and Republic of Vietnam Cross of Gallantry with palm.

Chief Drake retired in 1985, with 28 years in the USAF and a total of 30 years military service. He is presently employed by The National Bank of Commerce Altus, OK, as vice president and branch manager. He is married to the former Joan R. Heise of Bancroft, NE. They have two daughters, Valynda and Kathleen. They have three grandchildren Bennie, Kaylinn and Keira. Chief Drake and his wife presently reside in Blair OK.

MARC L. DRINKHAHN, Colonel, Commander and Professor of Aerospace Studies, Air Force ROTC Detachment 770, Clemson University, Clemson, SC.

Col. Drinkhahn was born Dec. 23, 1945, in Monroe, MI, and graduated from Monroe High School in 1964. He received a bachelor of science degree in psychology and biology from New Mexico Highlands University in 1968, and a master's degree in administration from Georgia College in 1977. He completed Squadron Officer School in 1976, graduated from the Air Command and Staff College in 1983, and the Air War College in 1987.

After earning his commission as a second lieutenant

from Officer Training School as a distinguished graduate in 1969, he entered undergraduate pilot training at Williams Air Force Base, AZ. His first assignment took him to the 537th Tactical Airlift Squadron at Phu Cat Air Base, Republic of Vietnam, in February 1971. There, he served as a co-pilot, pilot and an instructor pilot in C-7A Caribou aircraft, flying over 100 missions totaling 650 combat hours.

In December 1971, Col. Drinkhahn was assigned to the 19th Bombardment Wing, Robins, Air Force Base, GA, as a co-pilot, pilot and instructor pilot in KC-135 Stratotankers. He was also assigned as the chief of tanker scheduling, chief of quality control, the officer in charge of the Bomber Branch and maintenance supervisor with the 19th Organizational Maintenance Squadron. Col. Drinkhahn moved to become the aide-de-camp to the Vice Commander, Headquarters Air Force Reserve at Robins in July 1977.

Col. Drinkhahn was assigned in August 1979 to the 379th Bombardment Wing, Wurtsmith Air Force Base, MI, where he performed duties in tanker scheduling, and standardization and evaluation. He was assigned as an instructor pilot and chief of Tanker Branch in the Wing Standardization Division.

In June 1983, he was assigned to Headquarters Strategic Air Command, Offutt Air Force Base, NE, as chief of Applications Division in the Directorate of Tanker Operations.

Col. Drinkhahn arrived at Grissom Air Force Base, IN, in March 1985, as the commander of the 305th Air Refueling Squadron. In July 1987, he moved to Carswell Air Force Base, TX, as the deputy commander of the 7th Combat Support Group and became commander on April 27, 1989. He was assigned as the Commander, 416th Operations Group, Griffiss Air Force Base, NY, on Dec. 18, 1989. He was assigned as the Commander, 305th Air Refueling Wing, Grissom AFB, IN, on Jan. 31, 1992.

Col. Drinkhahn served as the 1702nd Air Refueling Wing (Provisional) Commander during Operations Desert Shield/Desert Storm. The 1702 AREFW was headquartered at Camp Mirage Nacirema, in Seeb, Oman, from Aug. 26, 1990, through April 11, 1991. The 1702 AREFW was a composite wing of 15 KC-135R and 10 KC-10 aircraft.

Col. Drinkhahn has accumulated more than 3,000 flying hours in the KC-135, C-7A, T-37, T-38 and the B-52. His military awards and decorations include the Legion of Merit with one oak leaf cluster, the Distinguished Flying Cross, the Meritorious Service Medal with three oak leaf cluster, the Air Medal with two oak leaf clusters, the Air Force Commendation Medal, the Air Force Outstanding Unit Award with Valor Device with one oak leaf cluster, the Air Force Organizational Excellence Award, the National Defense Service Medal, the Vietnam Service Medal with three Bronze Stars, the Republic Gallantry Cross with Palm and the Republic of Vietnam Campaign Medal. He was awarded the Legion of Merit for his service in Operations Desert Shield/Desert Storm.

He was promoted to colonel on Aug. 1, 1989, with the same date of rank.

Col. Drinkhahn is married to the former Jo Hewette of Sandersville, GA. They have one son, Matthew.

WILLIAM E. DRIVER, Lieutenant Colonel, born Nov. 29, 1951, in Carrollton, GA. He joined the USAF (Vietnamese Linguist, 1974-75); commissioned in 1975 as second lieutenant Medical Service Corps.

He served as assistant administrator USAF Hospital, George AFB, CA, 1976-78; UPT-Williams AFB, AZ, 1978-79; C-130E AWNS Pilot Rhein-Main AB, Germany, 1979-83; C-130E/H IP Little Rock AFB, AR, 1983-85; transfer to USAFR Jan. 18, 1985, 327TAS Willow Grove, ARF, PA; transferred to 459TAW Andrews AFB, MD; C-141B TW Altus AFB, OK, 1985; continuous service in AFRES in C-141B TIV Andrews AFB, MD (459AW) and 756 AW through 1992.

Awards: Aerial Achievement Medal, Southwest Asia Campaign Medal, Kuwait Liberation, Air Force Commendation Medal, Outstanding Unit, Combat Readiness, etc.

He and his wife, Rosemarie, have one son, Matthew. Today he is operations officer with 756 ALS Andrews AFB.

JOSEPH M. DROBEZKO, Lieutenant Colonel (S), is the 31 ALS Operations Officer, Dover AFB, DE. Born Jan. 23, 1954, in Butler, PA, he graduated from Slippery Rock Area High School in 1971 and entered the USAF in 1974. He received a BS degree from Slippery Rock State College in 1977, and a master's degree from Webster University in 1989. His military education includes Squadron Officers School, Airlift Operations School and Air Command and Staff College. Commissioned in 1977, he completed pilot training with Class 79-05, Williams AFB, AZ. His assignments include KC-135 aircraft commander, 11 AREFS (Hvy), Altus AFB, OK, 1979-1984; C-141B instructor pilot, 6 MAS, McGuire AFB, NJ, 1984-1987; staff officer, AFMPC, Randolph AFB, TX, 1987-1991; C-5 aircraft commander, 9 MAS, 1991-1992; and assumed his present position October 1992. He is a senior pilot with over 3,300 hours in the T-37, KC-135, C-141B, and C-5. His military decorations include the Meritorious Service Medal with one oak leaf cluster, the Air Force Commendation Medal, the Combat Crew Readiness Medal with one oak leaf cluster, and the Southwest Asia Service Medal. He is married to the former Elizabeth Denise Price of Altus, OK. They have two children, Joseph Michael II and Molly Elizabeth.

JOHN R. DUMBROSKI, Colonel, was born in Grove City, PA, on Aug. 5, 1942. He graduated from Grove City College in 1965, with a bachelor's of arts degree in economics. After UNT at Mather AFB, Dumbroski received initial C-130 training at Pope AFB. He was transferred to Mactan AB, Philippines where for 14 months he flew extensive combat missions in South East Asia. Subsequent C-130 assignments were with the 39 TAS Lockborne, AFB, and 7406 SSq., Rhein Main AB, GE. After serving as 322 TAW executive officer, he was assigned as chief SAAM requirements division, HQ MAC. Following an assignment to HQ EUCOM, Dumbroski returned to HQ MAC as director of Cargo and Requirements and then served as the first director of the DOD Air Carrier Survey and Analysis Office. He then was selected to become director of Logistics, Joint Special Operations Command, Ft. Bragg, NC.

Dumbroski received a master's in international relations from the University of Arkansas in 1975. He completed National Security Management and Air War College.

He retired in July 1992. Among his many awards and decorations are: Defense Superior Service Medal, Legion of Merit, Defense Meritorious Service Medal, Meritorious Service Medal with one oak leaf cluster, Air Medal with 11 oak leaf clusters, Armed Forces Expeditionary Medal, and Vietnam Service Medal.

Dumbroski and his wife, Nancy, reside in Belleville, IL. They have three sons: Michael, David and Todd.

GUY B. DUNN JR., Lieutenant Colonel, born Dec. 12, 1922, in Bland, VA. He left Virginia Tech in December 1942 for the Army Aviation Cadet Program. He finished pilot training from Gulf Coast Training Command, Altus AB, in January 1944, then went to RTU at St. Joseph, MO, then to ETO as C-47 pilot.

Dunn stayed in ETO until November 1945, then was sent to Nashville, TN for six months, followed by assignment to Mannison Field, FL to fly C-54 to India. While there he was given the job of base chief pilot. He was then transferred to Brookley Field, Mobile, AL; spent eight months of TDY on the Berlin Air Lift flying as a check pilot in C-54s.

He returned to Brookley and spent five years flying the C-74. Besides being the squadron maintenance officer, he taught Ground School. From Brookley he went to Guam with the 54th Weather Recon Sqdn. where he was the squadron maintenance officer and WB-29 aircraft commander.

Dunn came back to the States to Otis AFB and was assigned to the 551st AEW & C Wing in July 1955 and worked as an aircraft commander in RC-121D and as a maintenance officer. After four years he returned to Air Training Command at Randolph AFB, TX as the assistant director of maintenance plus flying C-54s and T-29s as a VIP pilot.

Went to the 314th Troop Carrier Wing as the chief of maintenance with additional duty as aircraft commander. During his stay at Ching Chuan Kang, Taiwan, he spent two months at Cam Ranh Bay, RVN as Det. 2 Commander, followed by assignment to 9th AF HQ as the director of maintenance.

Dunn retired from the Air Force June 1, 1972 after 29 and a half years of service. He is most proud of the 12,000 hours he logged in heavy transport aircraft.

He is married to Eleanor H. Dunn and they have two daughters and one son, who is a USAF major.

DWIGHT R. DINSMORE, Colonel, born on June 14, 1914, in Fremont, IA. He graduated with honors, from Hedrick, IA, high school in 1931, and from Iowa Wesleyan College, IA in 1935.

He served as a flying instructor with Brayton Flying School at Cueno, TX, training Army Air Corps Cadets.

In July 1943, he received a direct commission as a first lieutenant on the Army Air Corps. He served the duration of the war as a basic and advanced flying instructor. He was discharged as a captain in July 1945, and flew as a pilot on TWA until 1947, when he received a regular commission in the Army Air Corps.

During the balance of this career he served overseas tours in Iceland and Wiesbaden, Germany and two tours in the Office of the Secretary of the Air Force in the Pentagon.

He flew C-54s in the 60th Troop Carrier Group, flying coal and food from Wiesbaden to Berlin during the Berlin Airlift.

Retiring in 1966, with a rank of colonel he joined the LTV Corporation as the East Coast director of public relations, stationed in Washington, DC. He retired from LTV in 1977, and moved to San Antonio, TX, where he now resides.

He has been married to Virginia Warren for 52 years. They have two grown children, Diana and Charles.

DAVID D. DYCHE, graduated from Fremont High School in Sunnyvale, CA, in 1969. He received a bachelor of science degree in physics as a distinguished graduate from the U.S. Air Force Academy in 1975.

In 1976, he was the ATC Commander's Trophy winner as the top graduate in his pilot training class at Reese AFB, TX. He has completed tours as a C-130 instructor pilot at McChord AFB, WA, and Clark AB, Philippines His last flying assignment was in C-12s as the commander of Det. 3, 19th Airlift Squadron, Osan AB, Korea, from 1990 to 1992.

He earned a master's of science degree in optical sciences from the University of Arizona, graduating Magna Cum Laude in 1983. He was a distinguished graduate from Squadron Officers School in 1981, and attended the U.S. Marine Corps Command and Staff College.

He is currently the Air Mobility Command Research Fellow at Maxwell AFB, AL, and is also attending the Air War College.

His decorations include the Meritorious Service Medal, Joint Service Commendation Medal, and Air Force Commendation Medal. He and his wife Yolanda have two sons, Dana Christopher and Robert Scott.

WILLIAM R. EDGAR, Colonel, 4500th Air Base Wing (TAC), Langley Air Force Base, VA.

Col. Edgar was born in North Braddock, PA, on April 15, 1924. A graduate of Swissvale High School, Swissvale, PA, he received his pilot wings and a commission as a second lieutenant in March 1944.

The colonel served in numerous capacities prior to the end of WWII. His primary duties were as pilot and light bombardment instructor. After completion of Air Inspectors School in December 1946, he was assigned to Erding Air Depot, Germany. He returned to the United States late in 1947, for duty with the 313th Troop Carrier Wing (TCW) at Bergstrom AFB, TX.

The 313th TCW deployed to Germany in November 1948, for the Berlin Airlift, where the colonel served as a squadron commander. Early in 1949, he was assigned to the 7520th Air Base Group, Burtonwood, England and assumed duties as satellite base commander and commander, Headquarters and Base Services. In August 1951, the colonel was transferred to the 500th Air Defense Group in Coraopolis, PA, where his duties ranged from squadron operations officer, to squadron commander and group executive officer. While there he attended the University of Pittsburgh under Operation Bootstrap.

Col. Edgar was assigned to the Air Force Institute of Technology as a student officer from August 1955, until June 1957, when he graduated from Harvard University with a master's degree in business administration.

For the next three years he was assigned to Headquarters USAF in Washington, DC. He served as a staff officer in Personnel Retention, Directorate of Procurement, Training and Retention, Deputy Chief of Staff for Personnel. In February 1959, he was assigned to the Selection Board Secretariat, Directorate of Military Personnel.

In June 1960, Col. Edgar was assigned to Allied Air Forces, Southern Europe as Chief, Operational Readiness Evaluation Branch. He was instrumental in establishing a Tactical Evaluation Program for the air forces of Italy, Greece and Turkey, integrating them into the NATO organizational structure.

When he returned from Europe in August 1963, Col. Edgar was assigned to the Directorate of Legislative Liaison, Secretary of the Air Force. He served as assistant chief, then chief, of legislative sections within the Congressional Inquiry Division and subsequently became an Air Force Congressional Liaison Officer to the House of Representatives. In April and May of 1966, the colonel was a military assistant to Vice President Hubert H. Humphrey.

Col. Edgar was assigned to Headquarters TAC in July 1967, as Deputy Director of Information. He became the Director of Information in September 1968, and remained in that capacity until May 31, 1971. Subsequently, after serving as Director of Information he took Command of TAC's 4500th Air Base Wing on June 1, 1971.

The colonel is a Command pilot with nearly 4,000 hours of flying time. He is a graduate of the Industrial College of the Armed Forces (ICAF). His military decorations include the Legion of Merit, Air Force Commendation Medal, Army Commendation Medal, American Campaign Medal, World War II Victory Medal, Medal for Humane Action and the Army of Occupation Medal with Berlin Airlift Device.

Col. Edgar is married to the former Lois Jean Price of Swissvale, PA. They have two children, Lori Jeanne and James Curtis.

DARRELL E. EICKHOFF, Colonel, USAF, born April 25, 1947, in Evansville, IN. He joined the U.S. Air Force June 8, 1969, serving as medical service corps officer. Military locations at which he served were Pope AFB, NC; Clark AB,; RPI; Andrews AFB, DC; Bolling AFB, DC; Wiesbaden, Germany; Fweibrucken AB, Germany; and Scott AFB, IL.

His memorable experiences include providing aeromedical evacuation support for repatriated POWs during Operation Homecoming, 1973; planning AE support for Operation Desert Storm, 1990.

Awards/Medals: Air Force Meritorious Service Medal with four oak leaf clusters.

He is married to Ann. Today he is Chief, Medical Manpower Division, Office of Air Force Surgeon General, Bolling AFB, Washington, DC.

ROBERT F. ELLINGTON, born May 1, 1932, in Ellijay, GA. He attended Ellijay High School and North Georgia College and completed Squadron Officers School, Armed Forces Staff College and Industrial College of the Armed Forces. Aeronautical ratings include Command pilot, navigator, aircraft performance engineer and parachutist.

Ellington retired as a colonel Sept. 1, 1980. During his last assignment he served as the Director of Command and Control, Headquarters Military Airlift Command, Scott AFB, IL. Prior to that he served in various flying, operational and staff assignments primarily in airlift.

He is one of the founders of the Airlift Association serving as national president from 1981-1983, and chairman from 1983-1985. Among his awards are the Legion of Merit, Bronze Star, Air Medal and the Army Commendation Medal. Ellington and his wife, Pat, reside in Seffner, FL. They have six children: Mike, Steve, Ann, Lisa, Charlene and Shannon.

JIM ENGELKER, born June 4, 1946, in Ft. Collins, CO. Dates of service are November 1965 to September 1988. He served the U.S. Air Force in the following units: 47th TAS, Forbes AFB, KS, 1966-68, C-130B and C-130E basic and instructor loadmaster; 313th TAW Forbes AFB, KS, 1968-69, C-130E flight examiner loadmaster; 7th Aerial Port Squadron, Naha AB, Okinawa, 1969-71, C-130A instructor loadmaster, assistant NCOIC of Mobility Section; 15th MAS, Norton AFB, CA, 1971-72, instructor C-141A loadmaster; 53rd MAS, Norton AFB, CA, 1972-78, flight examiner loadmaster, assistant standardization loadmaster; 63rd MAW, Norton AFB, CA, 1978-79, assistant standardization loadmaster; HQ Military Airlift Command, 1979-88, assistant C-141A/B standardization loadmaster, chief C-141B standardization loadmaster, first HQ MAC/DOV C-17 loadmaster, participated in C-X source selection, worked with McDonnell-Douglas on design of C-17. Instrumental in reducing the forward G requirement from 8 to 3. Instrumental in the redesign of the current DD Form 365F. Participated in Operation Nickel Grass, Vietnam and Grenada.

He is married to the former Jean Tatum and has two sons, Jerry and Jeffrey.

Today he is the superintendent's assistant for Revere Junior-Senior High School and the district guidance counselor for Platte Valley School District, both located in Ovid, CO.

HOWELL MARION ESTES JR., former Air Force officer, airline executive. He was born in Ft. Oglethorpe, GA, Sept. 18, 1914, the son of Col. Howell Marion and Juanita Dickson E. Estes. He received his BS from the United States Military Academy in 1936. Graduated Flying Schools in 1940 and Air War College in 1949. He married Annah Verbeck, on March 8, 1941, their children: Howell Marion III, Michael Summer, Charles Dickson. Command, second lieutenant Cavalry, U.S. Army, 1936, USAAF, 1940-69, advanced through grades to general. USAF, 1964; assigned 7th Cav., Ft. Bliss, TX, 1936-37; acting a.d.c. to commander 1st Cav. Div., 1937-39; flight instructor, Brooks Field, 1940, comdt. cadets Advanced Flying School, 1940-42, director flying, 1942-43, director tng., 1943-44; commanding officer Lackland Air Base, Waco, TX, 1944; Commander, Lubbock Army Air Field, 1944-45; Chief, Plan and Policy Branch, Operations Division USAAF Europe, Wiesbaden, Germany, 1946, Chief of Staff for Operations, 1946-47, Assistant Chief of Staff for Plans, 1947; assigned 22nd Bomb Wing, March Air Base, CA, 1949, Commander, 1st Air Base Group, 1949; with 22nd Bomb Group, U.K., 1949-50.

Deputy Commander, Chief of Staff 22nd Bomb Wing, 1950-51; Commander, 44th Bomb Wing, March Air Base, 1951; Vice Commander, Far East Air Forces Bomber Command, 1951; Commander, 320th Bomb Wing, March Air Base, 1951-52, 12th Air Div., 1952-53, Air Task Group 7,4, 1952-54; director weapon systems Operations Wright Air Development Center, Air Research and Development Command, Wright-Patterson AFB, OH, 1954-55; Assistant Chief of Staff for Weapon Systems, Commander, Detachment #1, HQ Air Research and Development Command, Director of Systems Management 1955-57; Assistant Chief of Staff for Air Defense Systems HQ USAF, Washington, 1957-58, Assistant Deputy Chief of Staff, Operations, 1958-61; Deputy Commander, aerospace systems Air Force Systems Command, Los Angeles, 1961-62; Vice Commander, Air Force systems Command, Andrews AFB, 1962-64; Commander, Military Airlift Command, 1964-69; retired 1969.

Senior vice president World Airways, Inc., Oakland, CA, 1969-71, executive vice president, 1971, president, 1971-74, also director; chairman of the board, chief executive officer, director Federal Express Corp., Memphis, 1974-75; chairman of the board Vanguard International Aircargo Inc., 1975.

Decorated Legion of Merit with two clusters, Distinguished Flying Cross, Air Medal with two clusters, Army Distinguished Service Medal, U.S. Air Force Distinguished Service Medal, War Cross (Czechoslovakia). Recipient Gen. H.H. Arnold trophy. Member of the National Defense Transportation Association (def. adv. council).

RONALD C. EUSCHER, was born March 16, 1936, in Jamaica Plains, MA. He graduated from Josephine F. Wilber High School, Little Compton, RI, in 1954. he entered the U.S. Air Force in 1955, and was assigned to the 1st Aerial Port Squadron, Donaldson AFB, SC. He participated in Operation Deep Freeze in Antarctica from 1956 to 1962, and has 62 airdrop missions to every part of the Antarctic continent. He has flown on the C-124, C-54, C-119, C-123 and was initial cadre on C-141 and C-5 aircraft. He had a tour of duty in Vietnam on the C-123 at Da Nang AB, Republic of Vietnam. He preplanned the cargo load for the first operational C-5 mission in 1970, and flew on the first operational C-5 to be delivered to Charleston AFB, SC.

He was the primary loadmaster on the U.S. Air Force Mackay Trophy winner for 1977, delivering a 40 ton superconducting electromagnet to Moscow, U.S.S.R., thus accomplishing the longest flight in aviation history for a payload of the size and weight of the magnet (125,000 lbs. total). He has over 11,850 flying hours. He is a graduate of the NCO Academy and the USAF Senior NCO Academy. He received a bachelor' degree in business management in 1989. Chief Master Sergeant Euscher retired from the U.S. Air Force in December 1985. During his last assignment, he served as the 436th Military Airlift Wing, Dover AFB, DE, aircrew standardization loadmaster. His decorations include the Distinguished Flying Cross, the Meritorious Service Medal with one oak leaf cluster, the Air Medal with 23 oak leaf clusters, the Air Force Commendation Medal, the Army/Air Force Commendation Medal, the Air Force Outstanding Unit Award with 10 oak leaf clusters, and a V, the Combat Readiness Medal, the Air Force Good Conduct Medal with six oak leaf clusters, the Army Good Conduct Medal with two loops, the National Defense Service Medal, the Antarctic Service Medal, the Expeditionary Medal, the Republic of Vietnam Service Medal, the Air Force Longevity Service Ribbon with six oak leaf clusters, the Professional Military Education Ribbon with one oak leaf cluster, the Small Arms Marksmanship Ribbon, the Republic of Vietnam Gallantry Cross with Palm and the Republic of Vietnam Campaign Medal. He and his wife, Pauline, have one daughter, Sharon.

JACK D. FELLMAN, born Oct. 22, 1928, in Williston, ND. He graduated high school in Alexandria, MN, in 1946, and attended the University of Minnesota for two years and the University of Washington for two years.

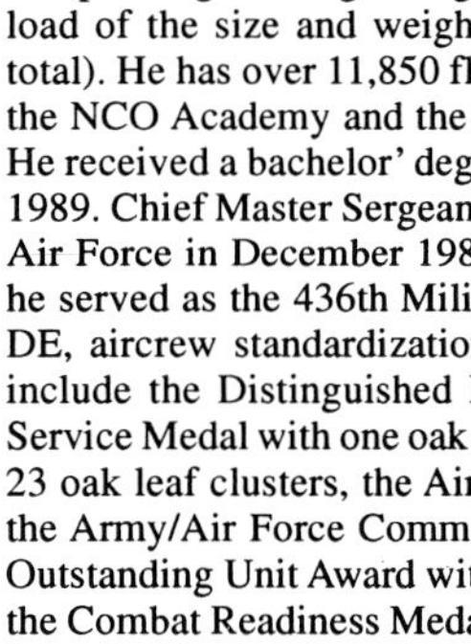

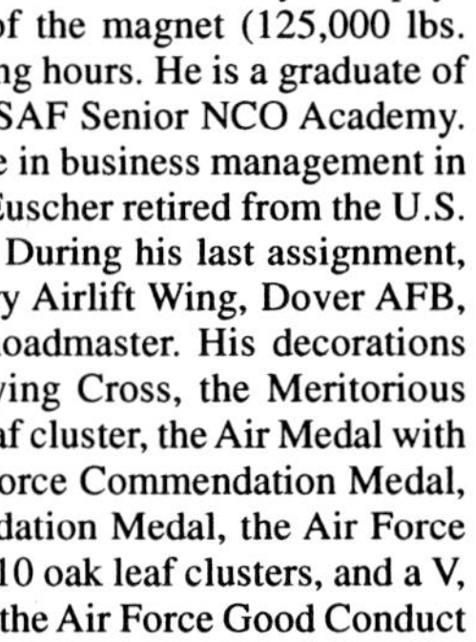

Joined the U.S.A.F. on April 7, 1948, serving with the 1946th AACS, Berlin; 1968th AACS, Frankfurt; 1917th AACS, Westover. Military locations at which he served include Berlin, Frankfurt, Germany, and Westover AFB, MA.

His memorable experiences include starting airlift at Tegel Air Base in Berlin in a mobile control tower (hot dog tower) and becoming a radar air traffic controller at Templehof, Berlin, during the Berlin Airlift of 1948-1949.

He was discharged April 7, 1952, with the rank of technical sergeant.

He married Donna Bohlin July 12, 1956, in Seattle, WA. They have a son and a daughter, and four grandchildren, two boys and two girls. As a civilian he was employed with Boeing Airplane Company, as procurement manager. He retired April 1, 1988, and now lives on an island in Puget Sound, (Hartstene Island). He is a member of Berlin Airlift Veterans Association and the Airlift/Tanker Association.

GARY C. FERGUSON, born April 14, 1938, in Fairport, NY. He enlisted in the USAF, June 6, 1960, through ROTC at the University of Rochester. He entered navigator training December 1960, at James Connally AFB, Waco, TX. Naval Bomb Training, 1961-62, at Mather AFB, CA. He was assigned 41st MAS, Charleston AFB, SC, 1962-65, as navigator C-130Es. He was assigned 79th ARRS, Anderson AFB, Guam, 1965-67, in C-54 and C-130H. Helped organize and participated in Crown/King rescue mission over Laos and North Vietnam until permanent squadrons set-up in RVN and Thailand. Assigned C-141, 20th MAS, Donu AFB, 1967-70. Assigned 9th SOS, Phan Rang AB, and Ta Son Nhut AB, RVN in C-47s. He was chief navigator and assistant operations officer, 1970-71. Assigned 20th MAS, Dover AFB, 1971-73, and Charleston AFB, SC, 1973-75, in C141. Chief mission scheduler, 437th MAW Current Operations. Assigned HQ MAC, Operations Plans, 1975-1981, was director readiness and tactic development and Chief, HQ MBC Readiness Initiative Group. Responsible for preparation MAC JCS Sckiannual Situation Report and C-5, C-141, C-130 camouflage paint schemes. Assigned McGuire AFB, NJ, in C-141s, 1981-88, 18th MAS chief navigator and operations officer, 437th MAW, Chief Current Operations, Chief Combat Plans, and executive officer to Deputy Commander for Operations. He has over 7,300 flying hours.

Awards/Medals: Distinguished Flying Cross, Bronze Star, Defense Meritorious Service Medal, Meritorious Service Medal, Air Medal, Air Force Commendation Medal, Air Force Achievement Medal, plus other awards.

He was discharged May 1, 1988, with the rank of lieutenant colonel.

He and his spouse, Helen, reside in Mt. Lauchel, NJ. He is a registered representative with USPA and IRA.

ED FIELD, is a native of McGregor, MN. He was commissioned in the Air Force in 1956, through the Aviation Cadet Program. His education includes a bachelor's degree in mathematics, University of Omaha; MBA, Kent University; Air War College; and the Executive Management Program, Northwestern University.

He retired from the Air Force in July 1986. During his career he commanded a C-141 Airlift Group, Air Rescue Wing, C-130 Tactical Airlift Wing and the USAF Airlift Center. He also served as the deputy commander of the Joint Special Operations Command. He has over 10,000 flying hours primarily in Strategic and Tactical Airlift, Tanker, Rescue and Special Operations aircraft. He participated in operations such as the Congo, Vietnam and Grenada. He is presently employed by the CAE-Link Corporation as vice president, Training Operations.

Field and his wife, Elsie, reside in Endicott, NY. They have four children: David, Catherine, Elizabeth and Susan.

ANNE MARTIN FLETCHER, Major, USAF, born in 1958, in Albuquerque, NM. Entered the USAF Academy in 1976, in 1980 graduated with commission. Pilot training graduate in 1981, Reese AFB, Lubbock, TX. C-141 school, distinguished graduate, Altus AFB, OK. C-141 pilot/flight examiner, 1981-1986, Charleston AFB, SC; Air Force Institute of Technology School of Engineering, MS, in strategic and tactical science, 1988, Dayton, OH. Strategic Planning Research Analyst, Joint Strategic Target Planning Staff, Offutt AFB, NE, 1988-1991. C-12 and C-137 pilot, 1st ALS, 89th Airlift Wing, Andrews AFB, MD, 1991-present. Graduated with first class containing women from USAF Academy; flew contingency missions to Grenada and Africa; first women to graduate from AFIT Strategic and Tactical Science Program; helped make nuclear planning more responsive to changing threat in "New World Order," first woman to fly C-137 at Andrews, AFB.

She is married to Air Force officer, Lt. Col. Charles "Chuck" Fletcher. Today she is flying C-137s for 89th Airlift Wing.

Maj. Anne Martin Fletcher's career has included a number of "firsts," beginning when she graduated in the first class at the U.S. Air Force Academy to include women. She was a Distinguished Graduate from Initial C-141 Training and went on to gain qualification as an evaluator and air-refueling aircraft commander (the most qualifications available at that time for female pilots). She flew the first U.S. airplane into Kenya following a coup attempt and commanded successful missions into Grenada during the 1983 Rescue Operation. After earning a master's degree in operations research with a thesis in Artificial Intelligence she served on the Joint Strategic Target Planning Staff. There she developed a model of the complex nuclear planning process and used it to test the feasibility of numbers and implementation dates being proposed in Strategic Arms Reductions talks. She has flown two Vice Presidents and accumulated over 3,000 hours in C-141, C-12C, C-137, and C-21 aircraft. She currently is a flight commander with the 457th Airlift Squadron at Andrews AFB, MD.

JOHN C. FLOURNOY, Colonel, USAF, (Ret.), born Nov. 30, 1936, in Florala, AL. He graduated from Vigor High School in Prichard, AL, in 1955, and attended Auburn University (API), where he received a BS degree in BA in 1959. He is a graduate of SOS, ACSC, and ICAF. He enlisted in the Alabama National Guard in 1954, and was commissioned a second lieutenant in the USAF in 1959, where he phased through UPT at Lackland AFB, TX, Graham AB, FL, and Larado AFB, TX, graduating in 1961. He entered the Air Rescue Service (MATS) following helicopter training at Stead AFB, NV, in 1962.

Col. Flournoy retired from the USAF in October 1988. During his last assignment, he served as DCS Ops., 23rd AF (AFSOC), Hurlburt Field, FL. During his tenure as 23rd AF/DO he directed 326 aircraft and their crews assigned to the varied missions of the NAF.

Among his many awards and decorations are: Legion of Merit, Meritorious Service Medal with three oak leaf clusters, Air Force Commendation Medal with four oak leaf clusters, Air Force Outstanding Unit Award with four oak leaf clusters, Air Force Combat Readiness Medal with four oak leaf clusters, Vietnam Services Medal, and the German Medal of Gratitude. He is a recipient of the MAC 5,000 hour accident free flying award and the Sikorsky and Kaman Aircraft companies' individual lifesaving awards.

He is currently the site manager, CAE-Link Corp., Kirtland AFB, NM, HC-130 flight simulator. Flournoy and his wife, Charlene, reside in Albuquerque, NM. They have three children: Jamie, John Jr. "Jay" and Jeffrey.

W.J. "JIM" FORD, Colonel, USAF, (Ret.), born in Cedar Rapids, IA, on March 18, 1931. He graduated from Waller High School in Chicago, IL, in 1949, and attended Coe College, IA. He received a bachelor's degree in physical education and master's degree in secondary school administration from Hampton Institute, VA. Ford retired from the U.S. Air Force in February 1981. His last two duty assignment included wing commander of the 513 TAW Mildenhall, England and base commander of Lackland AFB, TX. Awards include the Legion of Merit w/one oak leaf cluster, the Bronze Star, the Meritorious Service Medal, and the Air Medal with one oak leaf cluster. He is also the recipient of the Freedoms Foundation Honor Certificate.

Jim is currently the owner of four carpet, upholstery and drapery cleaning franchisees located in San Antonio and Austin, TX. He is a life member of Daedalians, Air Force Association and the Airlift Association. He has been member of Rotary International for the past 12 year and a district governor for the Rotary year 1993-1994.

Jim and his wife, Bernie live in Fair Oaks Ranch, TX. They have seven children and 14 grandchildren.

JOHN E. FORD III, born Dec. 12, 1937, in Englewood, NJ. He received a bachelor's degree in cinematography from UCLA in 1959, and a master's degree in education from SIUE in 1974. He completed Air Command and Staff College in 1974 and Air War College in 1984.

Col. Ford retired from the U.S. Air Force in September 1992. In his last assignment he served as Vice Commander, Air Combat Camera Service. While here, he deployed to Desert Shield/Desert Storm. He served on Gen. Schwarzkopf's J-3 staff and was commander of all combat camera forces in theater. 11.5% of his career was in combat. He has 124 combat missions from Southeast Asia and the Persian Gulf region. His decorations include the Legion of Merit with one oak leaf cluster, the Bronze Star, Meritorious Service Medal with four oak leaf clusters, Air Medal with five oak leaf clusters, Joint Service Commendation Medal, Air Force Commendation Medal with two oak leaf clusters, and the Republic of Vietnam Medal of Honor, First Class.

Ford and his wife, Sherry, reside in Redlands, CA. They have one daughter, Shonna.

STEPHEN C. FORGIEL, Lieutenant Colonel, USAF, (Ret.), received his commission through the Officer Training School program in November 1971, and graduated from Undergraduate Navigator Training at Mather AFB, in August 1972. From October 1972 to June 1979, he was assigned to the 41st Tactical Airlift Squadron and the 317th Airlift Control Element at Pope Air Force Base, NC. Following an assignment to Hanscom AFB from 1979 to 1982, Col. Forgiel returned to operations with the 62nd Tactical Airlift Squadron at Little Rock AFB.

In July 1986, after graduating from the Army Command and General Staff College at Ft. Leavenworth, KS, Col. Forgiel was assigned to the 322nd Airlift Division at Ramstein AB, Germany. He returned to Hanscom AFB, in August 1989.

Col. Forgiel was a master navigator with over 3,500 flying hours in the C-130 aircraft. He is married to the former Karla Griffin of Laurinburg, NC. They have one son, Stephen. Col. Forgiel retired from active duty in September 1993.

LEWIS H. FOUNTAIN, was born Dec. 13, 1935, in Wilmington, DE. Education received: MS degree in education; BSC, AA with honors; CCAF-62 credits, certified teacher.

Fountain served 20 years with the U.S. Air Force, he

retired in 1973. Under a Secretary of the U.S. Air Force Waiver, a special Air Force program for retired flight engineers, Fountain served 19 years in the Air Force Reserves, in the 335th Squadron, under the 514th Wg at McGuire AFB, NJ.

Children: Linda and Steven. One grandson, David.

KAROL E. FRANZYSHEN, born June 18, 1940, in South Bound Brook, NJ. He graduated from Bound Brook High School and entered the U.S. Air Force Aviation Cadet Program in 1959. After completing B-47E navigator training, he served tours at Little Rock AFB, AR and Lincoln AFB, NB, until June 1966, when he entered Air Force pilot training at Laredo AFB, TX. Following UPT, he was assigned to C-141s at McGuire AFB, NJ. He served as a 0-2A Forward Air Controller at Ban Me Thuot, RVN in 1969-70. He returned to McGuire.

He was assigned to the 89th MAW, Andrews AFB, MD, in 1972. In 1977, he became the first commander of the 58th MAS, Ramstein AB, Germany. Upon returning from Germany, he served as Chief, Logistics Plans, HQ21AF, McGuire AFB, NJ, until his retirement in November 1988.

Among his many awards and decorations are: Distinguished Flying Cross, Meritorious Service Medal with two oak leaf cluster, Air Medal with nine oak leaf clusters, Distinguished Presidential Unit Citation, Vietnam Cross of Gallantry, and the Air Force Longevity Award with seven oak leaf clusters.

Franzyshen resides in Roebling, NJ. He has four children, Karolyn, Wendi and twins Scott and Jeffrey.

MATTHEW J. FREDA, "JOE", Colonel, USAF, (Ret.), received his BS degree from Cornell, and entered the Army Air Corps in January 1942. He completed Air Command and Staff College in 1954.

He retired in April 1970, from the Pentagon, as Director, Tactical Airlift Division, Studies/Analysis. Prior assignments were Commander, 603 MASSQ, Kadena, Okinawa, Director, Plans and Programs Division, DCS Plans, HQ MAC, and various administrative duties in MATS flying airlift planes: C-141, C-124, C-133, C-54, C-46, C-47, accumulating 19,000 pilot hours.

He flew the Hump in CBI in WWII, C-124, C-74 and B-26, during Korea, and C-124s from Japan to Vietnam in that war.

His decorations include the Distinguished Flying Cross, Meritorious Service Medal with two oak leaf clusters, Air Medal with oak leaf cluster, U.S. Air Force Commendation Medal with two oak leaf clusters, two Presidential Unit Citations.

He resides with his wife, Betty, in Galille, PA. They have five children.

CHARLES G. GALEMORE, "CHUCK", born May 4, 1933, in Sucess, AR. He enlisted in the U.S. Air Force on May 25, 1951, serving for 10 years. Also served the Tennessee Air National Guard for 28 years. Military locations in which he served include Randolph AFB; Mt. Home, Goose Bay, Nashville Metro Airport (with Tennessee Air National Guard).

His memorable experiences include numerous mis-

sions to Vietnam, participating in Red Flag, Valiant Oak in Panama Canal.

He is the recipient of Meritorious Service Medal and other awards.

He was discharged Dec. 31, 1988, with the rank of CMSgt.

He and his wife, Carlene have four sons and two grandchildren. Today they are traveling every chance they get.

EDWARD A. GALLAGHER, Lieutenant Colonel, USAF, (Active) born March 29, 1954, in San Antonio, TX. Military locations at which he served include Dyess AFB, TX, 1992-present; Pentagon, 1988-92; Pope AFB, NC, 1985-88; Berlin 1983-85; Rhein Main AB, 1979-1983; and Little Rock AFB, 1977-79.

His memorable experiences include being liaison officer to Soviet Army, 1983-85.

Awards/Medals: Defense Meritorious Service Medal, Service Medal, Meritorious Service Medal, Commendation Medal, and Air Medal.

Today he is operations officer, 772 Airlift SQ.

MICHAEL R. GALLAGHER, Colonel, USAF, is the Assistant Deputy Chief of Staff for requirements, Headquarters Military Airlift Command, Scott Air Force Base, IL.

Col. Gallagher was born Dec. 20, 1944, in Sacramento, CA. He received a bachelor of arts degree in public administration from California State University at San Francisco and a master of arts degree in public administration from Golden Gate University. He is a distinguished graduate of the Reserve Officer Training Corps and Squadron Officer School.

The colonel received his commission and entered active duty in June 1967. He was a distinguished graduate from undergraduate pilot training at Webb Air Force Base, TX.

From 1968 through 1970, Col. Gallagher flew C-130E aircraft for the 345th Tactical Airlift Squadron at Ching Chuan Kang Air Base, Taiwan, and accumulated 732 combat hours in Southeast Asia. He returned to Webb AFB, in 1970, where he served as a T-37, instructor pilot and later as wing safety officer for the 78th Flying Training Wing.

Col. Gallagher was assigned to Mather AFB, CA, from 1974, until 1976. He was a member of the initial cadre for the T-37 Undergraduate Navigator Training Program; an assistant operations officer for the 455th Flying Training Squadron; and airspace planning officer for the 323rd Flying Training Wing. In 1976, he was assigned to Maxwell AFB, AL, first as section commander and manager of communication skills instruction at Squadron Officer School, and later as a student at Air Command and Staff College.

He was assigned to McGuire AFB, NJ, in 1979 as a C-141 instructor pilot and Command post chief with the 438th Military Airlift Wing, and then assistant operations officer for the 18th Military Airlift Squadron. From 1982 until 1985, the colonel was assigned to the Pentagon as air staff planner in the Airlift Forces Division, where he worked C-17 force structure issues.

From 1985 until 1987, Col. Gallagher served at Travis AFB, CA, as commander of the 7th Military Airlift Squadron, and then as assistant deputy commander for Operations, 60th Military Airlift Wing. Following this tour, he attended Air War College. Upon graduation, he was assigned to McChord AFB as deputy commander for Operations, 62nd Military Airlift Wing. While at McChord, he maintained currency in the C-130E and C-141B and flew in the airborne assault on Panama during Operation Just Cause.

In 1990, Col. Gallagher became vice wing commander of the 63rd Military Airlift Wing, Norton AFB, CA. He assumed his present position in July 1991.

Col. Gallagher is a Command pilot with more than 6,100 flying hours in T-37, T-38, C-130E, T-29, and C-141 aircraft. His military awards and decorations include the Distinguished Flying Cross, Meritorious Service Medal with four oak leaf clusters, Air Medal with five oak leaf clusters, Air Force Commendation Medal, Combat Readiness Medal, Vietnam Service Medal with six Bronze Service Stars, Southwest Asia Service Medal with two Bronze Service Stars, and Humanitarian Service Medal.

He is married to the former Jayne Walker of Whittier, CA. They have three children: Robert (serving in the U.S. Marine Corps), Peter (serving in the U.S. Army), and Katharine.

ANTHONY PAGLINAWAN GALVEZ, SMSgt, USAF, born Oct. 18, 1951, in Honolulu, HI. He joined the U.S. Air Force on July 7, 1970, at Ft. De Russy, HI, serving as a fuels specialist, 1970-74; cross-trained into aircraft loadmaster, 1974 to present. Military locations at which he served include RTNAB Utapao, Thailand during the Vietnam War, 1972-73, as a fuel specialist. Deployed to Jeddah, Saudi Arabia and served as the superintendent, Air Terminal Operations in support of Operation Desert Shield and Desert Storm.

Memorable experiences include: being on the crew of the first American C-130 of the 374th Tactical Airlift Wing, into Loakan Airport, Baguio City, Philippines, July 1990 after a 7.8 earthquake struck central Luzon. Secondly, again flying with the 374th TAW on Operation Sea Angel, Bangladesh May 1991, a humanitarian airlift mission after a cyclone ravaged that country. Lastly, supported the evacuation of U.S. citizens and dependents from the Philippines during Operation Fiery Vigil, June 1991 due to the eruption of Mount Pinatubo.

Awards/Medals: Bronze Star Medal, Meritorious Service Medal with one oak leaf cluster, Air Force Commendation Medal with two oak leaf cluster, Air Force Achievement Medal, 1984 Young Leadership Award, 22AF 1991 Lance P. Sijan Award, 834ALD 1991 Outstanding Transportation Unit Support, NCO of the Year.

He is married to the former Erlinda Aspa Pagatpat, and has five children: Gina, Maria, Jordon, Jason and Jeremy.

Today he is currently superintendent, Loadmasters, 345th Airlift Squadron, 374th Operations Group, 374th Airlift Wing, Yokota Air Base, Japan.

GEORGE G. GANJON, born June 1, 1929, in Baltimore, MD. He joined the U.S. Air Force on Nov. 22, 1950, serving as a gunner and Command pilot in KB-29P, BK50K tankers, C-47, C-141, and C-9s. He retired on Nov. 30, 1970, with the rank of major.

His memorable experience include TDY Thulz Greenland flying over the pole getting radar date for Pinetree Radar Line. Flying in and out of Peshawar Pakistan moving the wounded from Vietnam to hospitals. Hauling jet engines in C-47 engines dolly wheels sometimes went through the floor in rough weather.

Awards: Bronze Star and Air Force Commendation Medal.

He is married to the former Alvina Lusie and has a daughter Sara Lynn. Since retiring and selling his business he helps on the family farm.

GAROLD L. GARBER, "GARY" Colonel, was born in Dayton, OH, on May 30, 1948. He grew up in Englewood, OH, graduating from Northmont High School in 1966. He graduated from Grace College in 1970 with a degree in mathematics. He earned his commission at OTS and gradated from pilot training at Columbus AFB, MS. His first assignment was to NKP, Thailand, where he flew EC-47s. He has flown the C-118, T-39, and the C-5A/B. He commanded the 3rd Military Airlift Squadron at Dover AFB while it performed in both Operations Just Cause in Panama and Desert Shield/Storm in the Middle East. He is an in-residence graduate of Squadron Officers' School, the Army Command and General Staff College, and Air War College.

He is married to the former Barbara Ann Fife of Trotwood, OH. They have two children, Michelle and Greg.

JOHN S. GIBSON, "JACK", graduated from high school in Deland, FL. He entered Stetson University, also in Deland, and graduated in 1956, with a BS degree and a commission as an Army second lieutenant. He served on active duty at Redstone Arsenal, AL, as a research program officer.

In 1959, Gibson started a long career in military airlift at Lockheed, Marietta, GA. Starting in engineering analysis, design and test on the C-130 and JetStar, he then worked on the development of the C-141. Later he participated in the proposal program for the C-5A and was an engineering group manager on the C-5A for several years. Gibson was then a manager for airlift, related R&D, derivative aircraft, and international airlift development programs. In 1986, he was elected a Fellow of the American Institute of Aeronautics and Astronautics. Currently he is a program manager in Lockheeds' Advanced Systems organization.

Gibson and his wife, Diane, reside in Smyrna, GA. They have one daughter, Lisbeth and one granddaughter, Brittany.

FRANK E. GILK, Colonel, USAF, (Ret.), entered the U.S. Air Force, August 1952. Attended Navigation and Advanced Electronics Countermeasures Training, graduating from the class 53-06 (ECM) in December 1953. 1st Communications Squadron, Nagoya, Japan, until entering pilot training Class 57 Hotel, graduating February 1957. Assigned 30th Air Transport Squadron (MATS), McGuire AFB, NJ, 1957-1961. Upgraded to MATS Aircraft Commander. Voluntary assignment to Base Airlift Command Post, flying the line as an additional duty. Assigned to EAATCC (MATS), South Ruislip, England, 1961-1964. Command and Staff Class 1965. Assigned to the Directorate of Current Operations, MAC Headquarters, 1965-1970. Squadron Commander 618 MASSQ (MAC), Utapao, Thailand, 1970-1971. Assigned to Directorate of Plans, Headquarters Air Force, 1971-1977. MAC Liaison Officer to the Joint Staff, 1977-1981.

As Chief, Airlift Forces Division, Directorate of Plans, HQ USAF, developed the strategic and tactical airlift force structure (numbers and types of passenger and cargo aircraft) in the future airlift force. With a member of the OSD/PA&E staff (L/C Pat Renehan) and a member of the Lockheed, GA, Aircraft Corporation (Jim May), resurrected the concept of the C-141B modification that increased the volume of the C-141A by 30 percent, and also added an aerial refueling capability. Briefed the plan for Airlift Enhancement (which included the C-141B) to the Air Force Council. The plan was approved and implemented, increasing the airlift capability of the airlift force by over 50 percent without an increase in the number of airframes.

As the MAC Liaison Officer to the Joint Chiefs of Staff, acted as the eyes and years for CINCMAC, monitoring all of the actions that affected the Command. Coordinated and assisted in the planning that provided the air transportation for every humanitarian crisis and contingency operation during the assignment, including the airlift of military forces to Zaire, the airlift of English forces to Rhodesia, the recovery of remains from Georgetown, Guiana, and the successful positioning of all forces (prior the actual attempt) for the aborted Iranian hostage rescue attempt.

JOHN A. GILPIN, born Dec. 27, 1916, in Astoria, OR. He enlisted in the service Oct. 9, 1941, serving in the Army Air Corps and U.S. Air Force, as aircraft observer bombardment, master navigator.

He is proud to have been assigned to four tactical organizations. The 12Bomb Group, B-25s on the British 8th Army Front in Egypt, Libya and Tunisia, participating in the Battle of El Alamain, October 1942, against Rommel's Africa Corps. The 19th Bomb Group, B-29s in 1947-48, on Guam, navigator on a flight carrying special observer, Charles A. Lindberg to Japan and back. From 1957-60 the 421st Air Refueling Squadron in Japan. This K-B-50 outfit was the only one in PACAF. The 314 TCW, 1960-66, worldwide activities made this a fascinating mission.

His memorable experiences include 1965, while at the Pentagon on a briefing mission, he attempted to pour himself a cup of coffee. The top came off the pot spilling coffee all over the table. As he stood there red-faced, Gen. Ernest C. "Moose" Hardin walked up and laconically said, "We gotta get Gilpin an assignment to the Pentagon so he can learn how to work one of these coffee cups." What a guy, the Moose.

Awards/Medals: Distinguished Service Cross, Purple Heart, Air Medal, Air Force Commendation Medal.

Discharged Oct. 3, 1968, with the rank of lieutenant colonel, regular Air Force. He and his wife, Polly (deceased), have two sons John F. and William A., in Nashville. He is retired and a life member of ALA, AFA, PROA, VFW, MOPH, also 19th Bomb Association, Aerospace Education Force, Silman Field Historic Society, Air Force Historical Society Bombardiers, Inc.

ALDEN G. GLAUCH, born Nov. 14, 1919, Traverse City, MI. Education Traverse City High School, College of Mines, EL Paso, TX, and University of Maryland Europe and Pentagon. Air Command and Staff College, 1968.

Enlisted Army, June 12, 1942, graduated Aviation Cadet, Luke Field 43J flying P-40. Arrived Casablanca Christmas day 1943. North Africa, Sicily (Unit converted to Troop Carrier). Then to CBI, Assam - Imphal Valley. Return to Rome June 1944, Invasion of Southern France, dropped troops then gliders. 149 combat sorties, returned ZI June 1945. Inactive reserve recalled for Berlin Airlift, 123 missions from Rhein Main.

Assigned to MATS Westover 52, McGuire, ARS Cmdr. Thule, Pentagon (forever-C 141 &C-5 programs) Travis, Vietnam, McGuire 21 DCS/O, Rhein Main Wing Commander, Scott AFB MAC DCS/O, Commander 21 AF McGuire. TDRL 1977 retired in 1981. He has 8,000+ flying hours in 39 different aircraft.

Awards and decorations include Distinguished Service Medal (2), Legion of Merit (2), Distinguished Flying Cross, Air Medal, Presidential Unit Citation. Michigan summer, St. Pete winter. Married Rhea Forton in 1940, have three children Alden W., Grant and Cheri. 16 Handicap.

RICHARD M. GLOGOWSKI, "DICK", born Dec. 1, 1930, in Chicago, IL, where he attended Illinois Institute of Technology and entered the Air Force in 1951, as an Armed Forces Security Services Traffic Analyst. He became a navigator in 1953, and by 1956, served as the 40th BW Air Refueling Standardization Navigator with over 1,500 hours in KC-97s before entering pilot training. He flew over 1,700 hours in B-47s at Plattsburg AFB, NY, with reflex to Brize Norton, United Kingdom. In 1967, Dick was reassigned to the 76th MAS at Charleston AFB, SC, where he logged over 2,600 hours in the C-130 and C-141 aircraft.

After a C-121 tour in SEA, he was reassigned to HQ MAC in Cargo Services. He help form HQ MAC Transportation (TR), developed automated cargo and passenger systems. Col. Glogowski became USAFE Director of Transportation from 1977 until 1980, and retired in 1983, after serving as HQ 22AF DCS Transportation. His military decorations include Legion of Merit with oak leaf cluster, Meritorious Service Medal with oak leaf cluster, Air Medal with oak leaf cluster, Commendation Medal and Combat Readiness Medal with oak leaf cluster. Dick currently manages automation projects and resides in O'Fallon, IL. He married the former Miss Donna Mauk of Torrance, CA, in 1957, and has two children, Dayna and David. David is currently in the Air Force as a communications officer at HQ Space Command.

RICHARD GLORIOSO, Colonel, born Nov. 22, 1943, in Danbury, CT. He graduated from high school in 1962; Northeastern University (mathematics), in 1967; and Central Michigan University (master's) in 1974.

Assignment include: computer operations officer at Travis and Francis E. Warren; UPT at Reese in 1972, Distinguished Graduate, receiving Academic Training Award; flew C-141s as instructor pilot at Charleston and Travis and ice qualified at McMurdo Station, Antarctica (was on the first C-141 to land in Israel during the Israeli Airlift); Chief, Airlift control Element and flew CT-39s at Kadena; with 22 AF, served as contingency and exercise plans officer and Chief, Aircraft Control Center Division, directing deployed ALCC and ALCE operations; commanded the 374th Field Maintenance Squadron at Clark and he served as Chief, Standardization and Evaluation Division and director of Logistics Plans at MAC HQ; was 375th Deputy Commander for Maintenance and Logistics Group Command at Scott.

A Command pilot, some of his military decorations include the Meritorious Service Medal, Air Force Commendation Medal, Combat Readiness Medal and Vietnam Service Medal.

He and Judy, have two sons, Richard and Jeffrey.

PAUL A. GODFREY, Lieutenant Colonel, (Ret.), was commissioned as a Distinguished Graduate from the ROTC Program at the University of Missouri, Columbia Campus, in June, 1966. He was assigned to Pilot Training Class 68A at Vance AFB, OK, and received his wings in August 1967. He completed C-141 training at Tinker AFB and was assigned to the 75th MAS at Travis AFB, CA. In 1969, he made the transition to the OV-10, and flew as a forward air controller in the Southeast Asia Conflict.

In 1970, he returned to the states and completed C-5A training at Altus AFB, with assignment to the 3rd MAS at Charleston AFB. In 1973, the squadron made a unit move to Dover AFB, DE, where he was stationed for two years, upgrading to pilot flight examiner. In 1975, was reassigned to the Air Force Academy as a military training instructor for the Commandant of Cadets. Paul was then sent back to the C-5 and the 75th MAS in 1978, with moves to 60th MAW Aircrew Standardization and then Travis AFB Chief of Safety. During this period, he also had the opportunity to serve as the Chief Test Pilot for all three phases of the C-5 "off-pavement operational tests" (Operational Utility Evaluation). In January 1984, he moved to a Secretary of the Air Force staff position as Deputy Chief, Air Force Liaison Office, U.S. House of Representatives, in Washington, DC. He retired from the Air Force in 1987. Paul was a Command pilot with over 6,000 hours of flying. His decorations include Distinguished Flying Cross, Meritorious Service Medal, and Air Medals. Paul and his wife Nancy, reside in New Windsor, NY, where Paul is a site manager for Flight Safety Services Corporation at Stewart ANGB.

GARY P. GOLDSTONE, Captain, born Feb. 4, 1965, in Los Angels, CA. He received his bachelor's degree in biology from California State University, Northridge, in 1988.

Capt. Goldstone earned his commission through OTS in 1989. He completed Undergraduate Pilot Training at Laughlin AFB, TX. He was then assigned to the 773rd TAS, now the 40th AS, at Dyess AFB, TX, where he is currently a C-130H instructor pilot.

Capt. Goldstone represented Dyess at the 1994 Airlift Rodeo, where he and his crew won the Engine Running Onload/Offload competition. He has also participated in numerous Operations (Desert Shield/Storm, Southern Watch, Restore Hope, Provide Relief, Ocean Venture, Flowing Pen, Volant Pine and numerous exercises and worldwide missions. During Desert Storm, he flew combat and combat support missions in support of the coalition forces.

Capt. Goldstone's decorations include the Air Medal, Aerial Achievement Medal, Air Force Outstanding Unit with Valor, Combat Readiness Medal, National Defense Service Medal, Armed Forces Expeditionary Medal, Southwest Asia Service Medal, and the Kuwait Liberation Medal.

Capt. Goldstone is married to the former Mary E. Hass of Westlake Village, CA. They have one son, Garrett Edward.

FRANK S. GOODELL, Brigadier General, USAF, (Ret.),born Sept. 28, 1933, in New Carlisle, OH, and graduated from Withrow High School, Cincinnati, in 1951. He earned a bachelor of science degree from Ohio State University in 1955, and a master of business administration degree from Auburn University in 1971. He graduated from Air Command and Staff College in 1971 and the Naval War College in 1976.

Goodell retired from the United States Air Force in January 1989. During his last assignment he was the Air Force Special Assistant for Reliability and Maintainability. In this position he developed the R&M 2,000 Variability Reduction Process, which yields high quality products at affordable costs. For his efforts he was presented several awards including the Society of Logistics Engineers Founders Medal and the American Defense Preparedness Association Simon Award.

Goodell and his wife Wanda reside in Kent, WA. They have three children, Frank JR., Scott and Karen.

TERRY A. GRAHAM, Lockheed Aeronautical Systems Company's vice president for Business Development, is responsible for overseeing the firm's activities relating to new business from U.S. and foreign governments, and for developing closely related commercial markets for Lockheed's aeronautical products. He is also responsible for the company's product support branch, which includes integrated logistics support, technical publications and field service. Prior to this position, he served six years as director of Hercules Programs, responsible for total program management of Lockheed's C-130 activities.

He was born on Sept. 19, 1945, in Memphis, TN. In 1969, he received a bachelor's degree in industrial management from the Georgia Institute of Technology. In addition, he has completed several Lockheed sponsored executive programs.

He is a pilot and a member of the National Management Association, the Airlift Association, Kappa Sigma Alumni, Atlanta Steeple Chase Board of Stewards, U.S. Coast Guard Auxiliary, Columbia Alumni Association, and Georgia Tech Alumni Association.

CHARLES W. GRAY, "CHARLIE" Lieutenant Colonel, USAF, (Ret.), an Aerospace Marketing Consultant residing in Burke, VA, near Washington, DC. Born in Philadelphia, PA, on July 2, 1935, he was raised in New Jersey and Washington, DC. After attending ST. John's College High School in Washington and the University of Maryland, he entered the Aviation Cadet program in 1955. Upon commissioning and graduation from the Primary Basic Observer Course (PBOC) at Ellington AFB, TX, in 1956, he reported to Westover AFB, MA, for duty as a KC-97 navigator with the 384th Air Refueling Squadron. He completed pilot training at Williams AFB, AZ, in 1962, and was assigned to the 41st Air Transport Squadron, Charleston AFB, SC, as the unit was completing transition from C-121 to C-130E aircraft. He next reported to the 1st Air Commando Wing at Hurlburt Field, FL, in early 1964, to enter the Farm Gate program. After completion of training in the C-47, he reported to Bien Hoa AB, Republic of Vietnam to join the 1st Air Commando Squadron, 34th Tactical Group in June 1964. During the next 12 months he flew over 150 combat missions in C/AC-47 aircraft, mostly in support of U.S. Army Special Forces.

Returning to the CONUS in 1965, he was assigned to the 29th Military Airlift Squadron at McGuire AFB, NJ, in C-130Es. While at McGuire AFB, he was directed TDY to Hurlburt Field as an instructor/flight examiner in the C/AC-47. He attended the U.S. Army Command and General Staff College, graduating in 1968. He served from 1968 to 1970, as the exercise/contingency plans branch chief at 22nd Air Force Headquarters, Travis AFB, CA, while flying C-141s in attached status. During this period, he participated in the formulation of the first Reforger/Crested Cap exercises, as well as being the key planner for several Garden Plot operations. In 1971, he reported to the 65th Military Airlift Support Group, Yokota AB, Japan as Chief of Plans and Programs. In 1972, he organized the 22nd Air Force Airlift Operations Center at Yokota which was responsible for the management of up to 200+ aircraft operating in the Western Pacific region. Returning to 22nd Air Force Headquarters in 1973, as the Chief, Airlift Management Division, he was assigned as Deputy Director, Command and Control in 1974. In 1975, he moved to the Pentagon as the Airlift Deputy Branch Chief, Directorate of Programs, Deputy Chief of Staff, Programs and Resources, Headquarters, United States Air Force. In that position, he managed several key airlift/mobility programs, including the C-5A H wing Modification, C-141 Stretch (C-141B), and the Advanced Tanker Cargo Aircraft (ATCA), which resulted in the acquisition of the KC-10 fleet. He retired from the Air Force in 1979, accepting a position with Canadair Challenger Incorporated as director, (later vice-president) of Government Relations. He left Canadair in early 1989 to form his consulting firm which is serving a variety of aerospace and defense related clients. He has been married to the former Margaret Frances "Peggy" Green for 37 years. They have a daughter, Julie Anne, born in 1974.

EVERETT F. GREEN, born March 3, 1925, in Fairburn, SD. He served the U.S. Navy 1943-46 and the U.S. Air Force 1946-63, with the MATS 1946-48, TC 1948-51, and SAC 1951-63. Military locations at which he served include USN - South Pacific and USAF - world wide.

He was at Kwajalein Island from March 1948 to June 1948 for the atomic bomb tests on the island of Eniwetok.

He returned to the states in June 1948 and then was sent to Germany for the Berlin Airlift. He was stationed at Fassberg in the British Zone where he flew 234 times to Berlin hauling coal. On one occasion the base commander, Col. Theron J. Coulters' wife, who was Constance Bennett the movie star, boarded his airplane for a trip to Berlin, quite exciting.

He came back to the States in 1951, and attended Flight Engineers School at Chanute Field, IL. He was assigned to SAC, where he spent the rest of his career. In January 1957, he was sent to the Philippines, Clark Field, to participate in the refueling of the B-52s that flew around the world non-stop. The airplanes associated with in his career were C-47, C-54, B-29, KB-29 and KC97.

He retired from the U.S. Air Force at Lockbourne AFB, OH, on June 30, 1963, with the rank of master sergeant. His career spanned 20 years and was quite memorable.

Today he lives in California City, CA, close to Edwards AFB, CA. He has attended two Berlin Airlift Reunions in Germany and if able, will attend the next one in 1999.

FELIX M. GRIEDER, Lieutenant Colonel, USAF, currently Commander, 75th Airlift Squadron, Travis AFB, CA. He was born on Aug. 30, 1953, in Zurich, Switzerland. In 1955, the family moved to Chicago, IL. Lt. Col. Grieder graduated from Lyons Township High School, La Grange, IL, in 1971. He attended the U.S. Air Force Academy and graduated in 1975, with a bachelor of science degree in international relations. Lt. Col. Grieder then entered pilot training at Columbus AFB, MS. In August 1976, he reported to Scott AFB, IL, for his first operational flying assignment as a C-9 pilot. He completed his tour as a squadron C-9 flight examiner in 1981. Lt. Col. Grieder then attended language training at the Defense Language Institute at Monterey, CA, prior to reporting in May 1982 to the Ecole de L'Air, Salon de Provence, France, as the USAF Exchange Officer to the French Air Force Academy. In July 1984, Lt. Col. Grieder was assigned to the 3rd Military Airlift Squadron, Dover AFB, Delaware, to fly the C-5 Galaxy aircraft. While at Dover, he served as flight commander, wing plans officer, and wing executive officer. In July 1987, he reported to the Naval Command and Staff College (NCSC), Newport, RI. Following graduation from NCSC in June 1988, Lt. Col. Grieder was assigned to the Air Force Military Personnel Center (AFMPC), Randolph AFB, TX. While at AFMPC, Lt. Col. Grieder served as Chief, Plans and Operations Division, Colonels' Group, and as Chief, Airlift/Tanker/Helicopter Assignments Branch, Operations Officer Assignments Division. Lt. Col. Grieder is a Command pilot with over 4,300 flying hours. His military decorations include the Meritorious Service Medal with two oak leaf clusters.

Lt. Col. Grieder is married to the former Donna Marie Dalton of Jersey City, NJ. They have five children, Daniel, Erica, John, Mark and David.

JOHN E. GRIFFITH, born in Kansas City, KS, Dec. 4, 1930, and entered the USAF in 1948. He was a NCO for nine years before commissioning in 1957. He holds a BS degree from Troy State University, graduated from Armed Forces Staff College and is a distinguished graduate of the Air War College.

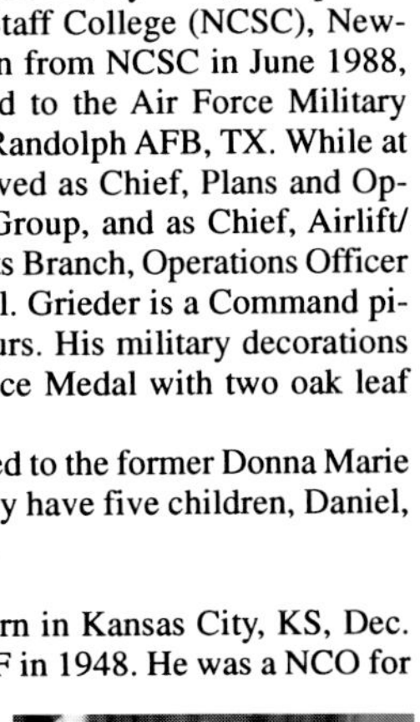

Major General Griffith retired in June 1990, after over 41 years active duty. He was the first Director of Operations and Logistics of USTRANSCOM. Prior general officer assignments include: Commander, Defense Fuel Supply Center, Director of Transportation U.S. Air Force and DCS Air Transportation, HQ MAC. Griffith served 15 years in MAC and five years in SAC air refueling squadrons. His awards include the Defense and Air Force Distinguished Service Medal with oak leaf clusters, the Legion of Merit with two oak leaf clusters, Meritorious Service Medal and Commendation Medal.

Jack and Bonnie Griffith reside near Belleville, IL. They have five children and six grandchildren.

AL GUIDOTTI, a Californian, who graduated from Castlemont High School in Oakland, CA, and attended Oregon State University on a baseball scholarship where he was a leader in student government and graduated in 1956. He was commissioned into the Air Force in 1956, and completed SOS in 1965 and Air War College in 1972.

Al served 24 of his 30 years in the Air Force in the airlift business with MATS/MAC. His duties and responsibilities touched all aspects of airlift from buck pilot to flight examiner, scheduler to Chief of Current Operations, Squadron Commander to Wing Commander, Headquarters Staff Officer to Division Command and Vice Commander 21AF.

Highlights of his career include serving as commander of the 436 MAW Eagle Wing, one-third of his career with the 89th MAW Presidential Support Wing, in the SEA Rescue Coordination Center in Vietnam, flying critical missions to world hot-sports and having the honor of serving our great nation with the best of the best. Al and Pat live in Huntington Beach, CA.

WILLIAM F. GUNKEL, Major, (Ret.), born in Muskogee, OK, on March 14, 1943. He graduated from Lawton High School, OK, in 1961, and Arizona State University in Tempe, AZ, in 1966. After pilot training at Williams AFB, he flew 202 C-130 combat missions in South Vietnam, and was awarded the Distinguished Flying Cross and six Air Medals. In 1970, at Little Rock AFB, AR, he was involved with the C-130 Replacement Training Unit and Instructor School. He then was an air liaison officer with the 9th Infantry Division at Ft. Lewis, WA. Major Gunkel completed his career with assignments at Bergstrom AFB, TX, Rhein-Main AB, Germany, and McClellan AFB, CA. While at Rhein-Main, Major Gunkel was awarded an Air Medal in the rescue of a capsized sailor in the middle of the Atlantic. Major Gunkel's other decorations include the Air Force Commendation Medal with one oak leaf cluster and the Army Commendation Medal.

GAIL S. HALVORSEN, born Oct. 10, 1920, in Salt Lake City, UT. He graduated from Bear River High School in Garland, UT, in 1939. He received a bachelor's degree in aeronautical engineering from the University of Florida in 1951, a master's degree in 1952, and a master's degree in counseling and guidance from Wayne State in 1972. He completed Air Tactical School at Tyndall AB in 1947, and Air Command and Staff in 1958.

Halvorsen signed as an aviation cadet in 1942, received his wings with the RAF and Air Corps in Class 44F. He was an airlift pilot from 1944 to 1949. He received the Cheney Award (1948-49) for humanitarian service in his role as the Candy Bomber of the Berlin Airlift. He was in research and development of aircraft and space systems (AFSC) from 1952-1970. He had a key role in the development of the Titan 111 launch vehicle. He returned to Berlin as the commander of Tempelhof from 1970-74. On March 15, 1994, Col. Halvorsen flew a drop mission over Bosnia, dropping parachutes with candy bars to Bosnian children. Here he was awarded the German Service Cross to the Order of Merit. He retired from the U.S. Air Force as a colonel on Aug. 31, 1974.

His several awards and decorations include the Legion of Merit.

Halvorsen and wife Alta are cattle ranchers and reside in Provo, UT. They have five children, Brad, Denise, Marilyn, Robert, Mike and 22 grandchildren.

ALFRED G. HANSEN, General USAF (Ret.),born April 13, 1933, in New York City. He graduated from Sewanhaka High School in Floral Park, Long Island, NY in 1951 and attended Hofstra College. He received a bachelor's degree in business and administration and management from Troy State University in 1972. He completed Air Command and Staff College in 1966, and Air War College in 1972.

Hansen retired from the U.S. Air Force in October 1989. He had a major impact on the air mobility forces of the United States. During his career he lead the team that finalized the consolidation of C-130s into the Military Airlift Command in the early 1980's. He was also the Joint Chiefs of Staff Study Director for the project that created U.S. Transcom. The potential and success of this command was manifested in the Desert Storm victory, considered one of the most massive logistics undertakings in history. During his last assignment, he served as commander of the Air Force Logistics Command, headquartered at Wright-Patterson AFB. During his tenure as commander of the Logistics Command, he started the first total quality effort in the Air Force. For his efforts in this area, he was awarded the Eugene Zuckert Award for Management Excellence. He is currently the president of The Airlift Association.

Among his many awards and decoration are: Defense Distinguished Service Medal, Legion of Merit with two oak leaf clusters and the Air Force Longevity Service Award Ribbon with eight oak leaf clusters.

Hansen and his wife, Donna, reside in Marietta, GA. They have four children: Rick, Gary, Debbie and Tim.

O. THOMAS HANSEN, born Sept. 17, 1939, Marblehead, MA. Graduated Marblehead High School in 1957, enlisted USAF in 1959, Officer Candidate School Distinguished Graduate, Commissioned December 1962. Completed Squadron Officers School April 1967; received BA, political science, Maryland University 1975; MA, management/supervision from Central Michigan University 1976. Distinguished Graduate, Air Command and Staff College June 1977; completed National Security Management Course April 1990.

Col. Hansen retired USAF, December 1992. Last assignment, Commander, Detachment 2, Air Mobility Combat Operations Staff, operating AMC Combat Aircrew Training School and Red Flag Tanker Airlift Control Center. Cited for leadership regarding anti-terrorism and air base survivability as Base Commander, Rhein-Main AB. Also commanded 36th Tactical Airlift Squadron. Past Luftbrucke chapter president, is currently the Tunner chapter president.

Awards/Decorations include: Legion of Merit, with one cluster; Distinguished Flying Cross, Air Medal with three clusters.

Col. Hansen and his wife Peggy, have two children, Lisa and Darren.

12th Annual Airlift convention Antedote from 1980:

After the banquet in the Grande Ballroom of the Maxwell House Hotel in Nashville, TN, the following activity level in the hospitality room was very high even as the one o'clock hour arrived. Due to noise complaints from other guests on that floor, the hotel requested that we close down that operating location for the night. A rather large hard-core group decided that the party should continue elsewhere. So scouts were sent out to find an alternate site, while the logistics team prepared the supplies for relocation. At the top of the hotel was a lovely dining room and lounge, that apparently had been left unsecured when they closed. Within minutes the operation had been shifted to the new location by opportune airlift (elevators) and the party continued unabated into the morning hours. For some strange reason there was a steady stream of traffic between the new, but splendid, hospitality room and the Jacuzzi/sauna on the basement floor. About 4 a.m., two hotel employees arrived to set-up the facility for breakfast. They were given several free drinks, and retired to a table in the back of the restaurant to stay out of our way. When the party finally ran out of steam approaching 6 o'clock, they were still sitting at the table laughing, having a good time and thanking the Airlift Association for its hospitality!

ROGER HANSEN, Colonel, born in Rochester, NY, on Jan. 12, 1945. He graduated from Maryland University in 1968 and was commissioned in September.

He graduated from UPT at Moody in October 1969. His assignments were: 56 WRS-Yokota, 362 TEWS-DaNang, 3 MAS-Dover, 56 MAS-Altus, ACSC-Maxwell, and then Army War College at Carlisle.

Assigned to the Pentagon in 1988 he worked in the air staff and the office of the deputy assistant secretary of Defense-Installations. In 1991 he became the commander of the 437 CSG at Charleston. In 1992, he became the commander of the 362 Airlift Support Group.

He is a Command pilot with 5,200 flying hours. His decorations include: Defense Superior Service Award, Distinguished Flying Cross, Meritorious Service Medal with four oak leaf clusters, and the Air Medal with two oak leaf clusters. He and his wife Carol have two children, Christie and Randy.

JOHN H. HAPP JR., Colonel, born Aug. 4, 1945, in Austin, TX, and graduated from Medina High School, OH, in 1963. He earned a bachelor of science degree in agriculture from Texas A&M University and a master's degree in human relations/management from Webster College in 1979. He is a graduate of Squadron Officer School, Armed Forces Staff College and Air War College.

Col. Happ received his pilot wings at Randolph AFB, TX, in June 1969. After completing the C-130 training, he was assigned to Forbes AFB, KS, until January 1970, when he joined E-Flight at CCK AB, Taiwan, flying the C-130 throughout Southeast Asia. In 1972, he returned to Forbes AFB until it closed in 1973. After moving to Dyess AFB, TX, he attended Instructor Pilots' Instrument School and Maintenance Officers' School. In 1976, he was selected to be the C-130 exchange pilot to the Royal New Zealand Air Force, Whenupia AB, New Zealand.

Returning from New Zealand in 1979, Col. Happ was assigned to the HQ MAC Inspector General's office and attended the Safety Officer/Accident Investigation Course at the University of Southern California. During this assignment, he was selected to be a member of the CINCMAC-directed Close Look II Project which examined airlift tactics and chosen as one of the 11 source selection item evaluators for the C-17 military airlifter.

Following intermediate service school in 1982, Col. Happ was assigned to the Military Personnel Center, Randolph AFB, TX. In 1984, he returned to Dyess AFB, TX, as commander of the 463rd Field Maintenance Squadron. He graduated from Air War College in 1987 and reported to Headquarters Air Force Logistics Command, Wright-Patterson AFB, OH, as the program management division chief, Deputy Chief of Staff/Plans and Programs.

In August 1990, he became the director of systems requirements, Deputy Chief of Staff, Requirements, Headquarters Military Airlift Command, Scott AFB, IL.

Col. Happ is a Command pilot with 4,000 flying hours.

His military decorations include the Meritorious Service Medal with four oak leaf clusters, Air Medal with one oak leaf cluster, Air Force Commendation Medal, and the Antarctica Service Medal.

Col. Happ is married to the former Carol Ann Lucas of Medina, OH. They have one son, John III.

JOHN DOUGLAS HARRINGTON, Major, (Active Duty), born Jan. 3, 1958, in Tokyo, Japan. He was commissioned May 28, 1980, USAF Academy. He served with the 89th Airlift Wing, Andrews AFB MD, VC-9 Aircraft Commander, Wing Exec. Military locations he served were 1980-81 UPT, Reese AFB, TX, 1981-86; C141 flight examiner, McGuire AFB, NJ, 1986-88, AFIT student, University of Pennsylvania, 1988-91, assistant professor of political science, U.S. Air Force Academy.

His memorable experiences include flying C-141 into Beirut the day the bomb exploded the Marine barracks.

Awards/Medals: Meritorious Service Medal, Air Force Commendation Medal with two oak leaf clusters, McGuire AFB Junior Officer of Quarter October - December 1985, Political Science Junior Officer of the Year 1990.

He is married to Donna de Pasquale, and has children: Katherine Ann, 5; and John Douglas Jr., 4.

Today he is C-9 Aircraft Commander, Wing Exec., 89th Airlift Wing, Andrews AFB, MD.

WOODROW W. HARTSOCK, born Nov. 21, 1916, in Gary, WV. He received his AB at the University of South Carolina, MA from SMU, in EdD from West Virginia University.

He joined the U.S. Air Force on July 10, 1941, serving with the Bomb Wings: 2nd, 98th, 310th, 306th, 313th TCW. He was stationed in Italy; Yokota; Salina, KS; Orlando, FL; and Fassberg, Germany.

His time was spent mostly in bombers. His only association with airlift was on the Berlin Airlift September 1948 to September 1949, at Fassberg, up in the British Zone of Germany. He was the provost marshal for awhile (what a snafu that was), and special services officer.

One interesting story concerned a girl who was turned over to him for custody during his time as provost marshal. She had been apprehended in Berlin in a coal sack on one of our C-54s carrying coal into Berlin from Fassberg, who was disguised as a sack of coal. Her national identity was hard to determine, although it was probably Belgian. She had been in France just prior to turning up at Fassberg. She had some ties to Germany, although it was difficult to determine just what they were. Now she was in German's British Zone, a Belgian, taken from an American airplane, now who has custody? Fortunately for him, he managed to get out from under the provost marshal mantle about that time, and he understood the captain who was flying the C-54 from which she was taken was called on the carpet at Wiesbaden, where the girl was taken. He doesn't know what happened to him, or to her. Providing 24-hour protection to her, meals, sleeping quarters, etc., while awaiting disposition of her case was quite a problem. He was glad to leave it to somebody else.

His best memory was flying 81 combat missions with out ever getting a scratch, or anybody on his airplane getting hurt, or even being in serious trouble. Just scared half-to-death.

He was discharged Aug. 31, 1964, with the rank of lieutenant colonel. As a civilian he was employed as director of admissions, BSC, Bluefield, WV. He retired from his last formal job, June 30, 1971.

PAUL R. HARWOOD, Colonel, born in 1945, in Phoenix, AZ. He earned a degree and ROTC commission at Arizona State University in 1967 and a master's from Central Michigan University. He attended pilot training at Williams AFB, AZ, in 1968.

A Command pilot with 5,200 flying hours, including 575 combat hours in SE Asia, primarily in the C-130E/H and C-21A, he flew with the 776 TAS, the 347 TAS, the 772 TAS and the 1401 MAS. He also served on the 21st Air Force and HQ Military Airlift Command Staffs, commanded the 317th Combat Support Group at Pope AFB, NC, where he deployed to the Persian Gulf during Desert Storm, and served at Incirlik AB, Turkey.

His military decorations include the Distinguished Flying Cross, Defense Meritorious Service Medal, the Air Force Meritorious Service Medal, the Air Force and Army Commendation Medals, and the Legion of Merit. He was promoted to colonel in 1989. He retired Aug. 1, 1993, with 26 years of service to Phoenix, AZ. Harwood married the former Barbara Beard of Albuquerque, NM. They have two children, Stephanie and Donald.

WILLIAM DECKER HATCH, Lieutenant Colonel, a Berlin Airlift veteran and a lifetime member of the Airlift Association, died on Sept. 12, 1990. William Decker Hatch, June 26, 1917-Sept. 12, 1990, born and raised in Mill Valley, CA. Hatch joined the Navy in 1935 at the age of 17.

In 1940 he met his wife. Those in the know say it was love at first sight. They courted and on May 3, 1941, eloped to Carson City, NV, to be married with their good friends Marge and Brad. This next May would have been their 50th Wedding Anniversary.

With the war raging in Europe and the U.S. not yet involved, he volunteered to fly for the Canadian Royal Air Force. Pearl Harbor was attacked just before he left for Canada, so he applied for a commission in the Army Air-Corps. The war years were spent at Cadet school for officer training, then flight training, then he became a B-12 flight instructor. He remained in the Reserves after the war and then went back on active duty status to participate in the Berlin Airlift, where he flew precious cargo into the desperate city of Berlin.

Upon return to the States, there were assignments to Treasure Island, CA; Vancouver, WA; a year in Korea without his family. He relished his assignments at Castle AFB, Atwater, CA, and Griffiths AFB, Rome, NY. He flew KC 135 refuelers.

Hatch military career ended with retirement in 1962. They moved to Sunnyvale where he joined Lockheed Missiles and Space Company. He spent 20 years working on top secret projects that he couldn't talk about, loving every minute of it.

Hatch is a member of: PUFL member of the American Legion, the Air Force Association, The American Aviation Historical Society, The Retired Officers Association (and PROC), MOWW (Military Order of World Wars), and a life member of Veterans of Foreign Wars, a Mason and a Shriner, Mill Valley Chapter of DeMolay and became a third degree Mason at Montgomery Lodge #11, A.F.&A.M., in Alabama in 1943, later followed by York Rite of the Islam Temple of San Francisco. Additionally, he was a member of National Sojourners and the Shriner's $100 Million Club, being a significant contributor to the Shriner's Children's Hospital in San Francisco.

Some of his many other interests included being president of their condominium's Homeowners' Association, a Life Member of the National Rifle Association, a member of SIRS, Toastmasters. The World Proof Numismatic Association, sponsor of the U.S. Olympic Shooting Team, loving diverted by hummingbirds and Kubo, his cat.

His void can never be replaced, but he will be forever remembered.

RONALD F. HENDERSON, Colonel, (Ret.), born in St. Clairsville, PA, on June 18, 1934. He graduated from the PennState University in June 1957, and graduated from the USAF Jet Flying Training Program in 1958. He also graduated from the George Washington University USAF Advanced Management MBA Program in 1966.

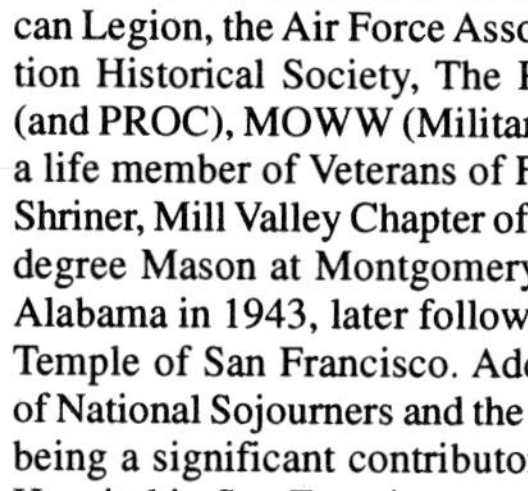

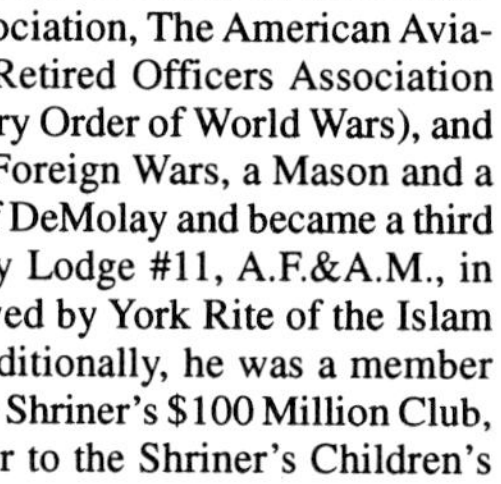

Ron is now a retired U.S. Air Force Colonel and Command Pilot with over 4,000 flying hours. In this capacity he performed duty as a Combat Ready B-47 aircraft commander, and served as C-141 Command pilot and squadron commander responsible for Special Operations Low Level Mission Operations. He also participated in Red Flag Low Level Airlift Missions with C-141/C-130 aircraft, was the first commandant and founder of the Mobility Airlift Command Airlift Operations School. He also served as director of Current Operations 21st Air Force, and served as the program element monitor (PEM) for the C-5A/C-141 and C-130 during his tour of duty with the Air Staff. He also was assigned to the Joint Chief's of Staff as the USAF project officer for the DOD Joint Strategic Airlift Mobility Capability Study and the SECDEF C-5A Program Reviews. Mr. Henderson also served as director of Training and Combat Airlift Tactics for the CINC, Military Airlift Command where he was responsible for the development and implementation of Airlift and Special Operations Combat Tactics.

Following retirement from the USAF in 1983, Ron was employed by the Lockheed Aeronautical Systems Company, Marietta, GA, in Advanced Airlift programs. During this time he enjoined Gen. Robert Huyser to form the Airlift-Country Chapter of the Airlift Association. Ron is presently director/manager of Special Operations Programs for McDonnell Aerospace (MDA) Company. Ron resides in Edwardsville, IL, with his wife Peg. Both Ron and Peg are lifetime members of the Airlift Tanker Association. Ron and Peg have four boys, Ron Jr., Cody, Dale and Kyle.

EARL H. HEAL, was awarded an ROTC commission and a bachelor of science degree from the University of Illinois in June 1955. He entered the Air Force that September and upgraded to C-124 instructor pilot at Dover AFB before separating from active duty in April 1960. Recalled in April 1964, to McChord AFB he resumed airlift operations including a short tour with Operation Shoehorn. Subsequent assignments were 58 WRS, Kirtland AFB; Rescue Center, Queen, Son Tra AB, RVN; Rescue Center - Jack, Nakhon Phanom RTAFB, Thailand; 437 APS, 76 MAS, and 20 MAS, Charleston AFB; Chief of Transportation Division, Oklahoma City ALC; Director of Air Transportation, 834 ALD/PACAF, Hickam AFB; and Deputy Wing Commander/Air Transportation, Travis AFB. He retired in October 1989 with a career that included 10,000 accident-free flight hours in 13 years of airlift operations and 12 years of air and surface transportation. His awards and decorations include the Legion of Merit, Bronze Star, Meritorious Service Medal with three oak leaf clusters, Air Medal with four oak leaf clusters, Combat Readiness Medal with two oak leaf clusters, Vietnam Service Medal with 11 devices, Presidential Unit Citation with one oak leaf cluster, Air Force Outstanding Unit Award with four oak leaf clusters, and Air Force Excellence Award. Col. Heal was accorded special recognition for development of unique mobility concepts which were implemented in coordination with the 7th Infantry Division and validated impressively with their subsequent emergency deployments to Honduras and Panama in 1989. He and his wife, Niki, reside now in Florence, KY.

ROBERT A. HEINLEIN, Lieutenant Colonel, USAF, (Active Duty), graduated from Chaminade High School in Mineola, NY, in 1971. He earned a bachelor's degree in mechanical engineering from Stevens Institute of Technology in 1975, a master of business administration degree from Southern Illinois University in 1983, and a master of science degree in acquisition logistics management from the Air Force Institute of Technology (AFIT) in 1985.

Heinlein served as a navigator aboard the C-141A at McGuire AFB, NJ, and aboard the C-5A at Dover AFB, DE. He was an acquisition program manager at Aeronautical Systems Division, Wright Patterson AFB, OH, on the C-17, C-130 program and the C-27A STOL aircraft program. Heinlein also directed the operational test and evaluation of the C-27A at the USAF Airlift Center at Pope AFB, NC, and is currently the USAF liaison officer at the Naval Training Systems Center and the U.S. Army Simulation Training and Instrumentation Command in Orlando, FL.

His decorations include the Meritorious Service Medal with one oak leaf cluster, and the Air Force Commendation Medal. Heinlein and his wife, Pate, have three children: Rob, Kelly and Jennifer.

PATRICK M. HENRY, Colonel, born Oct. 31, 1944, in El Paso, TX. He earned a bachelor's degree from the University of Colorado in 1967 and a master's degree from Troy State University in 1976. He completed Squadron Officer School in 1974, Army Command and General Staff College, in 1981, and Air War College in 1984.

He was commissioned through the Reserve Officer Training Corps as a distinguished graduate in 1967 and earned his pilot wings at Vance AFB, OK. His first operational assignment was in January 1969 to the 772nd Tactical Airlift Squadron, 463rd Tactical Airlift Wing, Clark AB, the Philippines. He was a copilot, then aircraft commander. In January 1971, Col. Henry was reassigned to the 314th Tactical Airlift Wing, Little Rock AFB, AR, as an instructor pilot and wing flight examiner. After a year in the 16th Tactical Airlift Training Squadron, he became part of the initial cadre for the C-130 Instructor School, and later a wing flight examiner.

In August 1974, he was assigned to Squadron Officer School, Maxwell AFB, AL, as an instructor. Following Maxwell AFB, Col. Henry returned to operations in March 1978 to the 21st Tactical Airlift Squadron, 374th Tactical Airlift Wing, Clark AB, the Philippines, as chief of training and chief of standardization and evaluation. After his second tour in the Philippines, he went to Ft. Leavenworth, KS, to the Army Command and General Staff College, graduating as a distinguished graduate in June 1981. From 1981 to 1986, he was assigned to Scott AFB, IL, first in Headquarters Military Airlift Command as action officer and later as division chief in analytical studies and plans division. In 1984 Col. Henry was assigned as squadron commander for Headquarters Military Airlift Command, Headquarters 23rd Air Force, and Headquarters Air Weather Service personnel.

In June 1986, Col. Henry went back to operations as assistant director of operations, 834th Airlift Division, Hickam AFB, HI. In September 1988, he was named the chief of the USAF Retention Division, Directorate of Personnel Program Management, Headquarters Air Force Military Personnel Center, Randolph AFB, TX. In August 1990, he assumed Command of the 34th Tactical Airlift Training Group at Little Rock AFB, AR. In that position he was responsible for all the C-130 formal school training in the Department of Defense. In February 1991 he became vice commander of the 435th Airlift Wing, Rhein-Main AB, Germany. Several achievements while vice commander included supporting Provide Hope and the first Commander of Mobility Forces for Provide Promise, the humanitarian support for Bosnia-Herzegovina. He retired effective Oct. 1, 1993, to Denver, CO.

Col. Henry has more than 4,275 flying hours and 871 combat hours. His military decorations include the Legion of Merit, Distinguished Flying Cross, the Defense Meritorious Service Medal, the Meritorious Service Medal with four oak leaf clusters, the Air Medal with five oak leaf clusters, and the Air Force Commendation Medal.

Col. Henry is married to the former Janet Lucki Heppes of Oak Park, IL. They have two sons, Michael and Robert.

KENNETH HERMAN, Colonel, USAF, (Ret.), born Sept. 4, 1922, in Pleasant Valley, IA. He graduated from Davenport High School, Davenport, IA, in June 1940. He attended Lincoln Aeronautical Institute, Lincoln, NE, 1940-41, majoring in aeronautical engineering. Received his BS degree in aero-engineering, from the University of Illinois, 1957.

He joined the U.S. Air Force on March 28, 1942, serving with the Air Transport Command, Military Air Transport Service: Alaskan, Pacific and Continental Divisions. He served in Canada, Alaska, Hawaii and Germany. Flew many Alaskan, Trans-Pacific and Trans-Atlantic missions to Aleutians, Japan, North Africa, England, Germany, Guam, Enewetak, etc.

His aeronautical ratings and experience is Air Force Command pilot, commercial pilot with single/multi engine and instrument ratings. Qualified in conventual, turbo-prop and jet aircraft and helicopters. He has 15,000 hours flight time in 80 different types of military and civil aircraft.

His most memorable experiences are 190 missions on Berlin Airlift. YC-97 Project service test pilot. Aircraft commander on first C-97 flight Hickam AFB to Tokyo returning with first med-evac of Korean War wounded.

He retired Jan. 31, 1966, with the rank of colonel.

Awards/Medals: Air Medal, Air Force Commendation Medal with oak leaf cluster, Medal for Humane Action, Occupation Medal with Airlift Device, Asiatic-Pacific Service Medal, American Service Medal, WWII Victory Medal, Longevity Service Ribbon with four oak leaf clusters, Air Force Reserve Medal with Hourglass Device, Armed Forces Service Medal.

He and his wife, LuAnn, have two daughters, Leslie and Kathleen, and three grandchildren. Today he is president of the Berlin Airlift Veterans Association.

WAYNE R. HESKEW, Colonel, Commander, 619th Airlift Support Group Hickam AFB, HI (current assignment as of July 1992).

Previous assignments: OTS 1968; UPT Vance AFB, 1969; C-141s 58th MAS, Robins AFB, 1970-71; C-130s 21st TAS, CCK AB, Taiwan, 1971-73; C-5 flight examiner 75th MAS, Travis AFB, 1973-78; C-5 flight examiner/assistant operations officer 56th MAS, Altus AFB, 1978-81; student/faculty Army Command and General Staff College, Ft. Leavenworth, KS, 1981-83; exchange officer Royal Australian Air Force, Canberra, Australia, 1983-86; chief wing inspections 443rd MAW, Altus AFB, 1986-87; squadron commander 56th MAS, Altus AFB, 1987-89; deputy commander/commander 443rd Support Group, Altus AFB, 1989-92.

Educational achievements: BA History; MA Public Administration; SOS; Army CGSC; AWC.

Military Achievements: Command pilot with 5,000 flying hours. Decorations include Distinguished Flying Cross, Meritorious Service Medal with four oak leaf clusters and Air Medal with one oak leaf cluster.

Personal date: Married to the former Jolinda Lower of Midwest City, OK. They have two children, Shannon and Jason.

JOHN M. HIGH III, Colonel, born Jan. 22, 1935, in Riverdale, NY. He graduated from John Burroughs High School in Burbank, CA, and went from there to the University of California at Berkeley, where he obtained a bachelor of science degree in physiological psychology. A distinguished graduate of Squadron Officers School in 1963, High also attended the Air War College in 1975, and obtained his MBA degree from Auburn University during that same year.

Following pilot training, High began his active duty career with the 49th Tactical Fighter Wing, Spangdahlem, Germany. He subsequently served tours in Air Rescue Service and HQ MAC, Directorate of Aircrew Standardization. During the Vietnam War in 1965-66, he was standardization pilot for the Jolly Green Giants (Air Rescue helicopters). He subsequently was assigned as Commander, Det. 14, 41 ARRWg with rescue responsibility for U.S. Forces in the Ryukyu Islands. Following a tour in the Pentagon, High was assigned as Commander, 22nd MAS. This unit operated C-5a aircraft on global airlift missions from Travis AFB, CA. High's final assignment before retirement was as Director, Programs and Policy, HQ MAC.

Among his awards and decorations are the Legion of Merit, Bronze Star, Air Medals and Meritorious Service Medal.

High and his wife, Dorette, own and operate a resort and marina in Sandpoint, ID. They have two sons, John IV and Steven.

GEORGE W HILLYER, "BILL", Colonel, USAF, (Ret.), born July 18, 1934, in Kilgore, TX. He graduated from Tyler High School in Tyler, TX, in 1952, and Texas A&M in 1956.

He retired from the U.S. Air Force after 30 years in July 1986. During his last assignment he was base commander of Norton AFB, CA. Other assignments included Commander, 774th TAS, DCS Operations 834th ALDiv., Director of Operations 314th TAW, and flying assignments all over the C130 world.

Highlights of his career include the great association with Airlifters worldwide, Combat flying in Vietnam 1964-68, Apollo Recovery operations 1968-1972, and of course his commands of the 774th TAS and 63rd ABGp.

Among his many awards and decorations are the Legion of Merit, Meritorious Service Medal with four oak leaf clusters, Air Medal with eight oak leaf clusters, and both the Air Force Commendation Medal and the Navy Commendation.

Bill and his wife, Marilyn, reside in Tyler, TX. They have two daughters Susan and Frances.

DAVID S. HINTON, Colonel, USAF, (Ret.), of Clarksville, TN, was commissioned through the ROTC at the University of Tennessee in 1960. In 1961, following pilot training at Craig AFB, AL, he was assigned to Travis AFB, CA, flying C-124s.

In 1963, he transferred to Tachikawa AB, Japan, and in 1964, moved with the 6th MASq to Hickam AFB, HI. After transition to the C-141 in 1966, he returned to Travis and in 1968 moved to Altus as the chief of C-141 training. In 1973, he was a FAC with the 23rd TASS Nakhom Phanom, Thailand, flying OV-10s.

Returning to airlift in 1974, Dave worked in DOO at MAC HQ. In 1976, he became Chief, Studies and Analysis Branch, DCS Air Transportation. Graduating from Air War College in 1978, he went to HQ USAF at the Pentagon. He worked in the Airlift Forces Division, DCS Plans and Operation and later in the Transportation Plans, DCS Logistics. He ended his Pentagon tour with a year on the JCS.

In 1982, he became DO of the 438th MAWg, McGuire AFB, NJ. A year later he became the 21 AF DO. His next tour was as Vice Commander, 322nd Airlift Division, Ramstein AB, Germany. Back to Scott in 1987, he became the first Chief of Staff for USTRANSCOM. He retired in 1989.

Dave was a Command pilot with over 9,000 hours of flying. His decorations include the Legion of Merit, Defense Meritorious Service Medal, Meritorious Service Medal, and Air Medal. Dave and his wife Gail now reside in Fairfax, VA, where he works for Paramax Systems Corp.

ROBERT E. HITE JR., graduated from Hopkinsville High School in Hopkinsville, KY, in 1949, and entered the Air Force in 1950. Upon his discharge in 1954, he entered the University of Tennessee, and was a Distinguished Military Graduate in 1958, with a degree in industrial engineering. He received a master's degree from Golden Gate University in 1974. He completed Squadron Officer's School, Air Command and Staff, and Air War College.

Hite spent his operational carrier in Tactical Airlift and Special Operations, and the last nine years with the Defense Logistics Agency. He retired from the USAF in March 1989. His last assignment, he served as Commander Defense Contract, Management Area, Atlanta.

Among his awards and decorations are: Defense Superior Service Medal, Legion of Merit, Distinguished Flying Cross, Air Medal with six oak leaf clusters, Defense Meritorious Service Medal, Meritorious Service Medal with oak leaf cluster, Air Force Commendation Medal with oak leaf cluster and the Army Commendation Medal.

Hite and his wife, Mary Lou, reside in Cartersville, GA.

CLIFF HODGE, born June 12, 1937, in Denver, CO. He graduated from Denver Manual High School in 1955. Hodge entered pilot training class 62F at Reese AFB, and graduated in class 67F at Craig AFB. During those years, he also served at Keesler AFB, MI, Incirlik AB, Turkey, and Whitman AFB, MO. Hodge transitioned to the C-130 and was in Southeast Asia during 1967 and 1968, then on the flight line all over the world for the next 20 years, with some wonderful people. Col. Hodge's last assignment and only staff job was AFSOC/CS, Hurlburt Field, FL. Many memorable experiences are greatly appreciated from all who provided them!

Hodge retired April 1, 1991, with 30 years of service, 6,700 flying hours, a Legion of Merit, Distinguished Flying Cross, three Meritorious Service Medals, four Air Medals, Humanitarian Service Medals, 17 overseas rotational deployments, five overseas tours, and two parachutist badge. He and wife Mary reside in Ft. Walton Beach, FL, children: Julie, Laura, Amy and Stephen have scattered. All who are friendly please call or stop by.

GERALD JOSEPH HOLLAND, Chief Master Sergeant, USAF, born Feb. 21, 1942, in Oklahoma City, OK. He joined the service Oct. 9, 1960, serving with the U.S. Air Force as a loadmaster, AC-47, C-124, C-130E, C-141A/B. He served in Techikawa, Japan; Charleston, SC; McGuire AFB, NJ; Pleiku, Vietnam; Pope AFB, NC; Altus AFB, OK.

While assigned to McGuire AFB, he was sent to Ft. Benning, GA, to participate in a Army Airborne exercise. While there he was diverted to participate in the filming of the movie *The Green Beret.* Meeting the star John Wayne was a memorable experience.

He was discharged May 1, 1987, with the rank of chief master sergeant.

Awards/Medals: Distinguished Flying Cross, Expeditionary Service Medal, National Defense Medal, Vietnam Cross of Gallantry with Palm, Vietnam Campaign Medal, Longevity Achievement Commendation, Meritorious Service Medal.

He is married and has one son. Graduated from Capitol Hill Sr., OKC, bachelor of science, Southern Illinois University. Today he is chief loadmaster instructor, Hughes Technical Services Company C-141 Aircrew Training System, Altus AFB, OK.

GENE R. HOLLRAH, born July 28, 1937, in Enid, OK. He is a 1954 graduate of Pioneer High School, Waukomis, OK, and received a bachelor's degree in agronomy from Oklahoma State University in 1959.

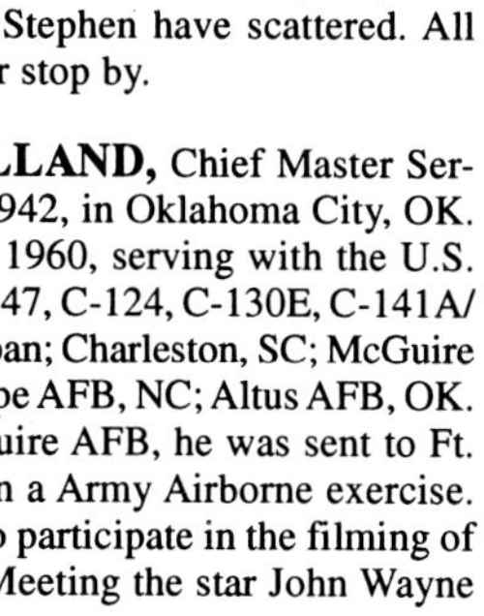

The majority of Hollrah's Air Force career was associated with tactical airlift with assignments at Sewart AFB, TN; Dyess AFB, TX; Udorn RTAB, Thailand; and Rhein-Main AB, Germany. While commander of the 772 TAS at Dyess AFB in 1979, his squadron was selected as Military Airlift Command's Tactical Airlift Squadron of the Year. Additionally, he served a tour at Robins AFB, GA, in the Directorate of Materiel Management as the C-130 System Manager.

Hollrah retired from the U.S. Air Force in 1988. In his final assignment, he was the base commander at Altus AFB, OK. He was honored as Altus Man of the Year in 1987. His numerous awards and decorations include the Legion of Merit, Meritorious Service Medal with two oak leaf clusters, and Air Medal with eight oak leaf clusters.

Hollrah and his wife, Christine, reside in Abilene, TX. They have three children: Donna, Patrick and Rhonda.

JAMES J. HOOGERWERF, "JIM", born Dec. 25, 1943, Detroit, MI. He joined the U.S. Air Force on Nov. 30, 1966. He served in UPT Class 68-D Webb AFB, TX, 1966-67; 773 TAS Clark AFB, Philippines, 1968-70; and 41 TAS Pope AFB, NC, 1970-71.

Memorable experiences include flying the C-130 in Vietnam never being boring. Flew anytime, anywhere under any conditions. Assault landed and air dropped food, fuel, ammo, troops, etc. He is proud to have served as a trash hauler and salutes all airlifters, past, present and future.

As a pilot flying C-130 transports in Vietnam, he flew over 1,000 sorties. One he remembered as being especially satisfying.

On April 20, 1970, we were diverted to Pleiku (in the Central Highlands) to fly an emergency airevac mission to Qui Nhon (on the coast). Since the evacuee had a serious head wound, his medics asked if we could limit cabin pressure changes as much as possible.

Normally we would have climbed steeply after takeoff to get above enemy small-arms fire and then flown high over our own free-fire artillery zones. That night though, to minimize pressurization adjustments, we stayed low after takeoff and flew directly through the friendly artillery areas. We didn't know if they were actually firing or not, but we felt the risk of being hit by them or the enemy was worth taking to help this soldier. They got him to Qui Nhon OK

Awards/Medals: Distinguished Flying Cross, Air Medal with five oak leaf cluster, Presidential Unit Citation, Air Force Outstanding Unit Award with V and one oak leaf cluster, Combat Readiness Medal with one oak leaf cluster, National Defense Service Medal, Armed Forces Expeditionary Medal, Vietnam Service Medal with five Bronze Service Stars, Air Force Longevity Service Award, RVN Gallantry Cross with Palm Unit Citation, and RVN Campaign Medal.

He was discharged Dec. 16, 1971, with the rank of captain. Today he is MD-88 Captain with Delta Air Lines.

He and his wife, Arlene, have children Brian and Karen.

MALCOLM P. HOOKER, Brigadier General, USAF, (Ret.), born in Memphis, TN, Jan. 25, 1920. He received his pilots license in 1939 and was given a direct commission as a second lieutenant in 1942. He was assigned to the Ferrying Division of the Air Transport Command and flew all types of aircraft all over the world. He was chief pilot of the Central African Division of the Air Transport Command; flew on the Berlin Airlift; and later commander a C-19 Group.

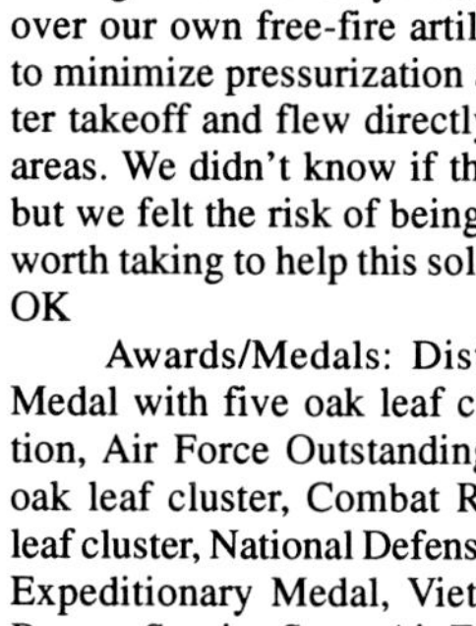

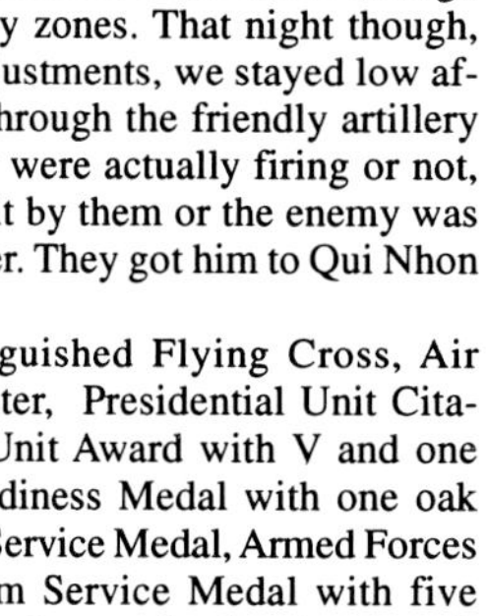

He served at staff level at all Echelons of Command including four years in the Pentagon. He attended Air Command and Staff School and the Air War College. He was stationed three years in each Germany and Japan and 16 months in Vietnam. He commanded a C-130 wing and was promoted to brigadier general. He retired in 1975, from duty as vice commander of the Military Traffic Management Command. His 23 decorations include the Distinguished Service Medal and the Legion of Merit with three clusters. Since retirement he served as chairman of the Airlift Association and later as president, chairman, and chairman emeritus of the National Defense Transportation Association. He is married to the former Margaret "Bunny" Wulbern. They have two daughters, Jean Winters and Cathy Moser, a son Malcolm Jr., and seven grandchildren.

THOMAS A. HOOSS, Lieutenant Colonel, born April 9, 1927, in Dallas, TX. He received his BA from Syracuse University.

He joined the Navy, April 16, 1945. Served Navy, 1945-46, NTC, Millington, TN; NTC, Norman, OK, Great Lakes NTC, NAS, St. Louis, MO. Served with unit NTC.

Served USMCR, 1946-48, NAS, St. Louis, MO. Served with unit VMF221.

Served U.S. Air Force, 1948-74, Goodfellow AFB, TX; Williams AFB, AZ, 1948-49; Turner AFB, GA, 1949; Sewart AFB, TN, 1949-50; Ashiya, Japan, 1950-52; Miami International Airport, FL, 1952; Charleston AFB, SC, 1953-55; Kodiak NAS, AK, 1955-56; Donaldson AFB, SC, 1956; Shaw AFB, SC, 1957-63; CINCPAC, HI, 1963-66; Pentagon, Air Staff 1966/JCS 1966-69, Washington, DC; Tan Son Nhut, Vietnam, 1969-70; Hanscom AFB, MA, 1970-74. Served units: 31TFG, 308 FTRSq; 334 TCSq; 2601 ASLT Sq.; 61st TCSq.; 62nd TCSq.; 314 TCGp.; 78 TCSq.; 746th TCSq.; 456 TCGp.; 18th AF; 9th AF; CINCPAC, Air Staff; Joint Chief of Staff; 7th AF; AFCRL (Hanscom).

His memorable experiences include two years in Korean War (Inchon, Sunchon-Mansanni, chosen reservoir drops); Dragnet Project; year in Vietnam; flying in Alaska, Korea and Vietnam.

He was discharged Sept. 30, 1974, with the rank of lieutenant colonel.

He and his wife, Dorothy L. Meetze Hooss, have three sons, and four grandchildren. As a civilian he was employed with South Carolina Electric and Gas Company, Columbia, SC, as assistant manager. He retired Jan. 18, 1988.

HUBERT V. HOPKINS JR., Lieutenant Colonel, born Jan. 12, 1923, in Ft. Monroe, VA. He enlisted in the U.S. Army Reserve in September 1942, at the University of Arizona, Tucson, AZ. He was commissioned in the Army Air Corps in April 1944. As an intelligence officer during WWII, he subsequently served as a general's aide in postwar Washington, DC, where he obtained his Regular Air Force commission. Upon graduation from Pilot Training School in 1949, He began his initial tour in Airlift as a pilot in the 62nd Troop Carrier Group at McChord flying C-82s. He saw action in Korea and flew the Berlin Airlift. Next in his career, he experienced a four year Caribbean and South American tour of duty in Airway and Air Communication Service (AACS) flying C-54s; this was followed by a four year tour of duty flying KC-97 tankers for SAC. Heeding the call for Air Force Civil Engineers in 1960, Lt. Col. Hopkins volunteered for and served the next five years as staff officer/Sq. commander of Civil Engineering in the CONUS and 322 Air Division in France. He returned to the CONUS in late 1964, with the 317th TAW reassignment to Lockbourne AFB, OH, serving as the director of Training and C-130 RTU commander. Upon completion of his sea tour at HQ MACV in 1969, he began his final Air Force assignment as Chief, Airlift Systems Division, HQ TAC. Lt. Col. Hopkins has been awarded the U.S. Army Bronze Star, the Air Medal, the Joint Service Commendation Medal, and 35 other decorations. His retirement on July 1, 1972, culminates 30 years of distinguished service to the nation.

He married Orene Johnson of Virginia Beach, VA, on May 31, 1947, at the U.S. Navy Chapel, Norfolk, VA. They have four children: Pamela, Vincent, James and Cynthia.

WILLIAM D. HUBBARD, graduated from John Burroughs High School, Burbank, CA, June 1961. He entered the Air Force, August 1961, trained in aircraft maintenance. Served in SAC and MAC maintaining bomber, tanker and airlift aircraft. He completed his service obligation with 21 years in 1982.

Hubbard earned an As from the CCAF in May 1981 and received a bachelor's degree with university honors in June 1981 from Southern Illinois University. His decorations include the Meritorious Service Medal, Vietnamese Service Medal with four stars, Outstanding Unit Award with five clusters and Valor Device.

Hubbard worked as a quality assurance specialist on the Bradley Fighting Vehicle and the Hubble Space Telescope before joining the 2nd Satellite Tracking Group, Onizuka AFB, in 1987. He is currently the Deputy Chief, Space Systems Control Branch.

He was selected Civilian of the Quarter and the year for 21st Space Operations Squadron and Civilian of the Quarter for Onizuka AFB in 1991. He participated in the Industrial Development Education in Acquisition (IDEA) program at McDonnell Douglas in Quality Engineering for the C-17 Airlifter program.

He and his wife Elsie have five children.

JOHN M. HUDSON, Colonel, graduated from Oklahoma State University in 1968, with a bachelor of science degree in business administration. Upon graduation he was commissioned a second lieutenant through the Reserve Officers' Training Corps and entered active duty on July 1, 1968. In 1975, he received a master of public administration from the University of Oklahoma.

Col. Hudson served in Southeast Asia, various assignments in MAC, TAC, SAC, ATC, PACAF, USAFE and the Aerospace Rescue and Recovery Service as well as an advisory tour in Saudi Arabia. During Desert Shield/Storm he served with the 60th Airlift Wing at Travis AFB, CA. He is currently assigned to the 81st Fighter Wing.

Awards and decorations include the Bronze Star, Meritorious Service Medal with seven oak leaf clusters, and the Air Force Commendation Medal with two oak leaf clusters.

He is married to the former Lynda Wixom of Durant, OK. They have two children.

ROBERT E. HUYSER, General, Chairman Emeritus, The Airlift-Tanker Association, born June 14, 1924, in Paonia, CO. He was drafted into the Army in 1943; graduated from the Aviation Cadet Program in 1944 and was commissioned a second lieutenant.

During WWII, Gen. Huyser flew B-29s out of North Field, Guam. Following the war he served tours of duty at several bases; including a tour at Castle AFB, where he became the wing chief of training in 1950.

During the Korean War, Gen. Huyser served as the Far East Air Forces chief of Bomber Combat Operations. He also flew combat missions in the B-29 with the 98th Bomb Wing.

Following the Korean War, General Huyser served in many positions including wing commander before returning to Headquarters SAC, where he served in many high level positions including Director of Operations Plans; and Chief of Single Integrated Operations Plans Division, Joint Strategic Target Planning Staff. During the Vietnam War, he administered the contingency war plans for Headquarters SAC. He was responsible for planning B-52 strikes throughout the theater. He also managed tanker support for the Southeast Asia area. In the performance of these duties, he flew combat missions over Vietnam in the B-52 and tanker support missions out of Thailand.

Following his duties at Headquarters SAC, Gen. Huyser was chosen to serve as the Deputy Chief of Staff, Plans and Operations, at USAF Headquarters. His work there directly contributed to the designation of the Military Airlift Command MAC as the Department of Defense's third specified Command.

Gen. Huyser served as the Deputy Commander in Chief of the United States European Command, Germany, where he became very involved in the global use of MAC airlift capabilities. In 1979, he assumed Command of MAC and served as its commander in chief until his retirement in July 1981.

Gen. Huyser is a Command pilot with over 10,000 flying hours: over 5,000 in SAC bombers, approximately 2,000 in SAC tankers, over 1,400 in single engine jet aircraft and over 1,500 in various light military aircraft including B-25s, C-54s, T-39S and such.

General and Mrs. Huyser were married on Nov. 15, 1942; they celebrated their Golden Anniversary in 1992. They have two daughters and live in San Jose, CA.

DONALD B. HYDE, born at home on Jan. 29, 1924, in Pawtucket, RI. He graduated from West (now Shea) High School in May 1942. He enlisted for weather in the ERC (Enlisted Reserve Corps) on Oct. 26, 1942, while attending Brown University. On Feb. 16, 1943, he reported to Ft. Devens, MA and six days later to Amherst College. He was a weather observer during the war in England at St. Mawgan, an ATC base and also in France and Germany. He was discharged on Jan. 1, 1946, and graduated from Brown in 1949. On Oct. 6, 1949, he reentered the service as a Pfc., and after a few months as a weather observer in Texas went to Chanute AFB to forecaster school graduating in January 1951, as a sergeant. He had, including WWII, 13 years in England, one year in France and Germany and three years in Spain.

The remaining years in service were spent in Massachusetts, North Carolina, Illinois, Texas, Colorado, California and New Hampshire. His wife died in 1968, leaving him four children. He married again in 1970. Decorations include the Air Force Commendation Medal, the Army and Air Force Good Conduct Medals, the ETO Ribbon with two battle stars and the WWII Victory Medal among others. Retired from the Air Force as a SMSgt. on Dec. 1, 1974, at Pease AFB. He resides in Colorado Springs with his wife.

PHILIP A. IANNUZZI JR., Captain, born in Philadelphia, PA, on July 6, 1961. He received a bachelor of science degree from Pennsylvania State University on May 19, 1984. He received a master of science degree from the University of Southern California on Dec. 21, 1989.

Capt. Iannuzzi entered active duty on Sept. 29, 1984. He completed pilot training at Reese AFB, TX, on Sept. 20, 1985. He was then assigned to Plattsburgh AFB, NY, where he held positions as KC-135 standardization/evaluation (stan/eval) copilot, aircraft commander, instructor aircraft commander, instructor flight examiner and chief, stan/eval operations. He was then assigned to Castle AFB, CA, where he taught students in the KC-135 combat crew training school (CCTS) and combat flight instructor course (CFIC). He flew missions in Operations Desert Shield, Desert Storm and Proud Return. He flew JCS-directed special operations low-altitude, night, communications-out aerial refueling missions.

Capt. Iannuzzi's awards include: ROTC Vice Commandant Outstanding Achievement honors, Air Force Association Outstanding Second Lieutenant, 380th Bomb Wing Crew of the Quarter, Distinguished Graduate of KC-135 aircraft commander class, 380th Air Refueling Squadron (ARS) Top Gun airmanship award, 93 ARS Instructor of the Month, 398th Operations Group (OG) Instructor of the Month, 97th Training Squadron Company Grade Officer of the Quarter, 97 OG Instructor Pilot of the Quarter, and Airlift/Tanker Association 1994 Young Leadership award.

Capt. Iannuzzi is married to the former Maria Palmer from Ventnor, NJ. They have one son, Philip.

PETER H. JACKSON, born April 29, 1939, in Bridgewater, MA. He graduated from Bridgewater High School in 1957. Pete received a bachelor's degree from Boston University in 1961. He attended the U.S. Army War College at Carlisle Barracks, PA, in 1980.

During his Air Force career, Pete served various flying tours as a pilot, in TAC, PACAF, SAC and MAC in the C-130, AC-130 and B-52 aircraft, logging over 1,600 hours combat time.

He also served staff tours at the U.S. Readiness Command, MacDill AFB, FL, and Air Force Legislative Liaison, Pentagon, Washington, DC.

Among his many awards and decorations are: Legion of Merit, Distinguished Flying Cross with two oak leaf clusters, Meritorious Service Medal with one oak leaf cluster, Air Medal with 16 oak leaf clusters, and the Joint Service Commendation Medal.

He retired from the USAF in August 1985, with the rank of colonel. Pete is employed by CAE-Link, C-130 Aircrew Training System, Little Rock AFB, AR. He resides in Sherwood, AR.

HANSFORD T. JOHNSON, born Jan. 3, 1936, in Aiken, SC. He was the outstanding graduate in thermodynamics and aerodynamics from the U.S. Air Force Academy in 1959, received a master's degree in aeronautics from Stanford University in 1967 and a master's degree in business from the University of Colorado in 1970. The general completed Army Command and General Staff College in 1972, National War College in 1976

Johnson retired from the United States Air Force in August 1992. During his last assignment, he served as Commander in Chief of the U.S. Transportation Command and the Air Mobility Command. He led these commands in their vital roles during Just Cause in Panama (1989), Desert Shield in the Middle East (1990-91), and every humanitarian effort around the world while maintaining high levels of peacetime service. He was instrumental in launching Quality journeys in both commands, refining DOD Transportation policy, greatly expanding the role of USTRANSCOM through performance and charter expansion, and shaping the direction of Sealift Programs and integration.

Johnson and his wife, Linda, live in San Antonio, TX. They have three children: Richard, Mrs. Elizabeth Trojan and David.

STEPHEN K. JOHNSON, born Feb. 5, 1941, in Detroit, MI, where he graduated from Highland Park High School in 1959. He earned an associate degree in transportation/traffic management from the Community College of the Air Force in 1980, and completed his bachelor's degree in occupational education from Southern Illinois University in 1984. He is a 1976 graduate of the USAF Senior NCO Academy.

He retired from the Air Force in November 1992. During his last assignment he served as the Chief Loadmaster 435th Airlift Wing at Rhein-Main AB, Germany. During this assignment he developed cargo compartment configurations and load plans for the deployment of the Wing to Desert Shield, and during Desert Storm he supervised the loading of over 875 XVIII Airborne Corps sorties supporting their surge into western Iraq.

His decorations include the Distinguished Flying Cross

with two oak leaf clusters, the Meritorious Service Medal with three oak leaf clusters, and the Air Medal with 11 oak leaf clusters.

He and his wife Anita now resides in Suisun City, CA.

KELLY M. JONES, Chief Master Sergeant, AFSC: M11300, Flight Examiner Flight Engineer Superintendent, hails from Arlington, TX. Assigned at Yokota AB, Tokyo, Japan since August 1989 as Wing Chief Flight Engineer, Standardization/Evaluation Division, Deputy Commander for Operations. Arrived at Yokota AB from his former assignment at Scott AFB, IL, where he was Command Chief, Flight Engineer Training, Directorate of Aircrew Training and Resource Management, Deputy Chief of Staff, Operations.

Chief Jones enlisted in the United States Air Force on June 4, 1965, and has had previous assignments at Chanute AFB, IL, Amarillo AFB, TX, England AFB, LA, Bien Hoa AB, Republic of Vietnam, McGuire AB, NJ, Ching Chuan Kang AB, Taiwan, Republic of China, Kadena AB, Japan, a previous assignment to Yokota AB, JA, and Dyess AFB, TX. He also served briefly in the Persian Gulf during Operations Desert Storm.

He has over 6,400 hours of flying time in the C-130 and C-141 aircraft including 660 hours in combat. Chief Jones' military decorations include the Distinguished Flying Cross with oak leaf cluster, the Meritorious Service Medal with oak leaf cluster, the Air Medal with four oak leaf clusters, the Air Force Commendation Medal with two oak leaf clusters, Presidential Unit Citation, the Air Force Outstanding Unit Award with V device and two oak leaf clusters, the Air Force Organizational Excellence Award with oak leaf cluster, the Combat Readiness Medal with seven oak leaf clusters, Air Force Good Conduct Medal with eight oak leaf clusters, National Defense Service Medal with oak leaf cluster, the Armed Forces Expeditionary Medal with two devices, the Vietnam Service Medal with six devices, the Humanitarian Service Medal with two devices, the Overseas Short Tour Ribbon with two oak leaf clusters, Overseas Long Tour Ribbon with oak leaf cluster, Air Force Longevity Ribbon with six oak leaf clusters, Air Force Professional Military Education Ribbon with oak leaf cluster, Small Arms Expert Marksmanship Ribbon with device, the Air Force Training Ribbon, the Philippine Presidential Unit Citation, Republic of Vietnam Gallantry Cross with Palm, and the Republic of Vietnam Campaign Medal.

Chief Jones is a graduate of the Senior NCO Academy, Command NCO Academy, and MAC Airlift Operations School. Chief Jones was the 463rd TAW NCO of the Year in 1980, and 463rd TAW Senior of the Year in 1981 and again in 1983. He is a member of the National Board of Advisors of the Airlift Association, life member of the Air Force Association, and president of the Shogun Chapter of the Airlift Association. He is also a girl scout leader. He is married to the former Miss Sharon Doudna of Gulfport, MS. The couple has no children.

Most memorable experiences: serving as a C-130 flight engineer during Operation Eagle Pull, evacuating non-combatants from Phenom Phen, Cambodia; serving as a C-130 flight engineer during Operation Frequent Wind, the evacuation of Saigon, Vietnam; instrumental in effecting the regulation changes allowing female flight engineers to be allowed to fly unilateral airdrop missions on C-141s; and serving as a flight engineer on the first C-130 mission to Vladivostok, Russia, supplying the newly opened American Embassy.

LOWELL W. JONES, born Jan. 11, 1937, in Redfield, SD. Joined the service June 7, 1961, serving in the U.S. Air Force as pilot C-130s, TAC and MAC. Military locations include Vance AFB, OK; Pope AFB, NC; Langley AFB, VA; Pentagon; Little Rock AFB, AR; Ramstein and Rhein-Main AB, GE; Scott AFB, IL; Vietnam two tours (Saigon, April - September 1963; Philippines, Japan, Taiwan September 1965-66); Sq. Commander of 1st PCS C-130 Squadron in Europe, 37 TAS, Rhein Main AB, Germany.

Awards/Medals: Legion of Merit, Meritorious Service Medal, Air Medal (12), Air Force Commendation Medal, Air Force Expeditionary Medal, Vietnam Service Medal, Air Force Longevity Service Award (four), RVN Campaign Medal, Air Force Outstanding Unit Award, Combat Readiness Medal, Small Arms Expert Marksmanship Ribbon, Humanitarian Service Medal, National Defense Service Medal. He retired Aug. 1, 1985, with the rank of colonel.

He and his wife Sandra have sons, Lance and Kevin (both married). Today he is senior analyst with ANSER, Inc. Arlington, VA, working on C-17.

SHARON T. JONES, Staff Sergeant, AFSC: 73270, Personnel Specialist. Hails from Gulfport, MS, the daughter of Retired Technical Sergeant Robert F. and Edith M. Doudna. Assigned at Yokota AB, Tokyo, Japan since August 1989, as Assistant Non-Commissioned Officer In Charge, Outbound Assignments, Consolidated Base Personnel Office, and Non-Commissioned Officer In Charge, Personnel Actions for the 374th Morale, Welfare, Recreation and Services Squadron. She arrived at Yokota AB from her previous assignment at the Air Force Military Personnel Center, Randolph AFB, TX, where she was the Non-Commissioned Officer In Charge, MAC Aircrew Training.

She enlisted in the USAF on Oct. 9, 1980. She has had previous assignments at Columbus AFB, MS, and San Vito de Normanni, Italy.

Military decorations include the Air Force Commendation Medal w/one oak leaf cluster, the Air Force Achievement Medal, the Air Force Outstanding Unit Award, Organizational Excellence Award, the Good Conduct Medal w/ three oak leaf clusters, the National Defense Service Medal, the Air Force Overseas Long Tour Ribbon w/one oak leaf cluster, the Air Force Longevity Award w/two oak leaf clusters, the Air Force PME Ribbon and the Air Force Training Ribbon.

She is a life member of the Air Force Association, vice-president of the Mt. Fuji Chapter, Services Society, Public Affairs Representative for both the Shogun Chapter, Airlift Association and the Tokyo Chapter, Air Force Association. She is a girl scout leader and is married to CMSgt Kelly M. Jones from Arlington, TX.They have no children.

NORMAN E. JUKES, attended schools in San Bruno, CA, and received an As degree from College of San Mateo in aeronautics.

He enlisted in USAF Reserves, 349th FBWg, Hamilton AFB, in 1954 and remained in the Reserves until retirement as a master sergeant in 1978. He initially served as crew chief on F-80s. The Air Force then decided to "Modernize" the Reserve and 349th converted to C-46s, providing an opportunity to fly as a flight mechanic on C-46s then on to C-119Gs. In 1965, the 349th converted to C-124s and MAC. In 1968, he was recalled to active duty as a flight engineer performing airlift missions to SEA and Europe. In 1969, to C-141s with 710th MAS at Travis until retirement. He had a parallel career with United Airlines for 36 year and is currently a staff engineering representative.

He and his wife, Helen have two children and reside in Burlingame, CA.

WILLIAM A. KEHLER, born in Cincinnati, OH, on Aug. 4, 1937. He graduated from Lockland High School in 1955. Earned a bachelor's degree in personnel management and Air Force commission from AFROTC at Ohio State University in 1960. An AFIT-sponsored MBA was received at Ohio State University in 1969.

After receiving his pilot wings in 1961, Bill spent his next 25 years divided evenly between flying, airlift (C-123, C-130 and C-141) and special operations (AC-130H); and personnel duties at AFMPC and the Pentagon. Combat tours were served at Da Nang AB, RVN, and Ubon RTAFB, Thailand. From 1987 to retirement from active duty in 1990, Bill was the professor of air science at the University of Virginia. Most memorable assignments were the AC-130H flying tour with SPECRTE and as Wing Commander, 314th T AWg.

Bill's awards include the Legion of Merit with oak leaf cluster, Distinguished Flying Cross, Air Medal with 11 oak leaf clusters, and the Meritorious Service Medal with two oak leaf clusters. He's a life member of the Airlift Association, Air Force Association and Daedalians.

Since his 1990 retirement, Bill has been the aerospace science instructor at the Cabot High School AFJROTC in Cabot, AR. Bill and his wife Barbara, live in Jacksonville, AR. They had two sons, Tim (USAF, deceased) and Kris, a travel agency manager in Annandale, VA.

JAMES D. KELLIM, Major General, Deputy of Staff for Operations, Headquarters Military Airlift Command, Scott Air Force Base, IL. He is responsible for all worldwide operations for the command, which include deployment and employment of U.S. combat forces and their support equipment; logistical resupply of these forces; aeromedical evacuation; presidential airlift; aerial search and rescue of downed flyers; U.S. Air Force special operations forces; and other missions assigned by the Department of Defense.

Gen. Kellim was born Oct. 3, 1936, in Pocahontas, AR, and graduated from Delaplaine High School, Arkansas, in 1954. He received a bachelor's degree in accounting, management and marketing from Memphis State University in 1958, and a master's degree in management from Auburn University in 1971. He completed Squadron Officer School in 1965, Air Command and Staff College in 1971 and the Army War College in 1976.

Commissioned through the Reserve Officer Training Corps program and entered the U.S. Air Force in October 1958 with an assignment to Lackland AFB, TX. Gen. Kellim attended undergraduate navigator training at Harlingen AFB, TX, from November 1958 to July 1959, and electronic warfare officer training at Keesler AFB, MS, until April 1960. The general remained at Keesler as an instructor in the Electronic Warfare Officer School until attending pilot training at Webb AFB, TX, in June 1963. He graduated from pilot training in June 1964 and was assigned to the 41st Military Airlift Squadron at Charleston AFB, SC, flying C-130E's and C-141A's. In May 1968, he was selected as a member of the initial cadre for the C-9A aeromedical evacuation aircraft and served at Scott AFB from June 1968 until July 1970.

After graduating from Air Command and Staff College in June 1971, the general was assigned to the Republic of Vietnam, serving at Cam Ranh Bay and Tan Son Nhut air bases, flying C-7A's. During this tour of duty, he also served as a squadron operations officer and commander of the 310th Tactical Airlift Squadron. He returned to the United States in September 1972 and was assigned to Military Airlift Command Headquarters, Scott AFB, in the office of the deputy chief of staff for Operations, where he served in the Current Operations Division, successively as director of special assignment airlift missions, director of airlift operations, chief of the Capabilities Branch, and chief of the Resource Management Branch. During this tour of duty he concurrently maintained aircraft commander status in T-39s.

After graduating from the Army War College in July 1976, Gen. Kellim was assigned to the 62nd Military Airlift Wing, McChord AFB, WA, as assistant deputy commander for operations until August 1977, when he became deputy commander for operations. In June 1978 he was assigned as wing vice commander and in February 1979 he assumed command of the wing.

From June 1980, until August 1981 Gen. Kellim represented the Military Airlift Command as senior representative for C-X (now designated C-17) source selection. Subsequently was assigned to Headquarters U.S. Air Force, Washington, DC, in the office of the inspector general as Chief, Office of Review and Oversight. In March 1982, the general was named vice commander of the Lowry Technical Training Center at Lowry AFB, CO. He served as commander of the Air Reserve Personnel Center, Denver, from May 1983 to July 1985, when he returned to Washington, DC, as vice commander of the Military Traffic Management Command. In August 1986, he became inspector general for Military Airlift Command, Scott AFB. Present duties assumed in June 1987.

A command pilot with more than 6,700 flying hours, including 328 combat hours in the Republic of Vietnam his decorations and awards include: Defense Superior Service Medal, Legion of Merit w/one oak leaf cluster, Bronze Star Medal, Meritorious Service Medal w/two oak leaf clusters, Air Medal w/one oak leaf cluster, Air Force Commendation Medal w/one oak leaf cluster, Air Force Outstanding Unit Award, Vietnam Service Medal w/three service stars and Republic of Vietnam Gallantry Cross with Palm Promoted to major general April 1, 1987, with same date of rank.

General Kellim has two sons, James Jr. and Mike.

THOMAS R. KETTLER, was born on April 18, 1957, in Blytheville, AR. Graduating from the Pennsylvania State University, he was commissioned through Officers Training School in 1983. He attended pilot training at Columbus AFB, and was assigned to the 18th MAS, McGuire AFB, where he flew C-141 aircraft. In 1989, he was assigned to the 57th ALS, Altus AFB, the command's only strategic airlift schoolhouse. As an instructor/examiner and wing exec at Altus, he represented the top two-percent of the command's C-141 pilots. A senior pilot with more than 3,500 flying hours, military decorations include the Meritorious Service Medal, the Air Force Aerial Achievement Medal, the Air Force Commendation Medal, the Combat Readiness Medal with two oak leaf clusters, the Armed Forces Expeditionary Medal, the National Defense Service Medal, and the Southwest Asia Service Medal with two service stars. He received the Airlift Association's Young Leadership Award for 1992 as well as the Air Force Association's Officer Leadership Award for 1992.

CLARK J. KHOLOS, Colonel, USAF, born Dec. 23, 1936, in Pittsburgh, PA. He entered the Air Force in 1955, and was an airborne radio operator at Ashiya AB, Japan. He was commissioned in 1960 and assigned to McGuire AFB, NJ. He earned acclaim in aerial delivery techniques at Norton AFB, CA, and led his crew to a clean sweep victory of the Military Airlift Command's 1971 Volant Rodeo. He coordinated and directed AC-130 Spectre gunship missions in Southeast Asia while at Ubon RTAFB, Thailand. A master navigator, he accumulated 8,600 flying hours, including 560 combat hours on 269 missions, on seven different aircraft. The colonel retired from the Air Force in July 1989, where he was the comptroller at Travis AFB, CA.

Decorations include the Legion of Merit, Distinguished Flying Cross, Meritorious Service Medal with two oak leaf clusters, Air Medal with eight oak leaf clusters, and the Air Force Commendation Medal. He is married to the former Alberta Grossman of San Antonio, TX. They have two children, Cheryl and Alan.

EARL L. KIELGASS, born Aug. 26, 1921, in Sedgwick, CO. Received his MA in language and literature from the Arizona State University. His hometown USA is Miami, AZ. Pilot training was with Class 43-K (Santa Ana, CA; Dos Palos, CA; Lemoore, CA; Williams Field, AZ). P-47 Thunderbolt RTU at Dover, DE. He went to the United Kingdom in May 1944 and after a short period at Atcham joined 368th Fighter Group at A-3 near Isigny, France on July 1, 1944. He was assigned to 396th fighter Sq. and stayed with it for remainder of war against the Reich (other European WWII bases were Chartre, Laon, Chievres, Rheims, Metz, Frankfurt, and Nuremburg).

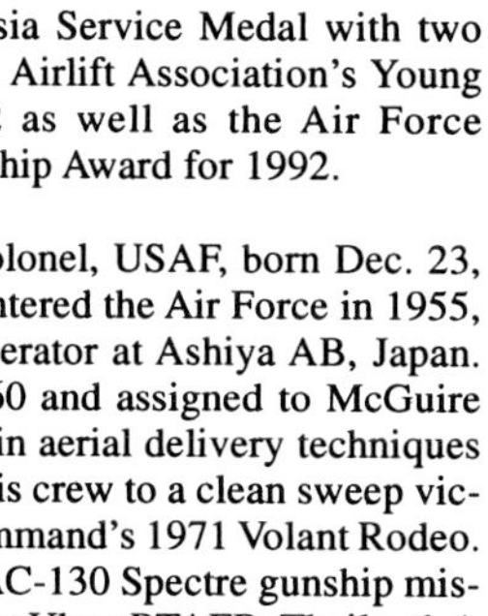

Separated from service in 1946, after serving as adjutant of pilot training at Luke and Williams bases. Joined Arizona National Guard in early 1947 and flew P-51 Mustangs at Luke until recalled to active Regular Air Force in the fall 1947. He was assigned to 1st Ftr. Wing March Field until wrangled a Berlin Air Lift assignment in early 1949. Went thought C-54 pilot transition at Great Falls, MT, then joined 53rd Troop Carrier Sq. at Rhein Main and hauled supplies into Berlin for next several months. Spent several years in USAF Training Command (Enid, OK; Big Spring, TX) with duty as Pilot Tng. Gp. and Wg. Adjutant and thousands of hours as an instructor pilot in AT-6, C-47, B-25, AT-7, T-28 and T-33 aircraft. He was commander of Pilot Training Sections I and III for about two years at Webb AFB. He was selected by USAF HQ for a SHAPE HQ assignment and spent 39 months at Villacoublay (outside Paris) with Air Training Advisory Group which had the NATO responsibility for all pilot training within the alliance. In 1959, he joined AACS after attending the Air Traffic Control Staff Officer Course at Keesler AFB. In AACS he was squadron commander (Whiteman AFB), communications region inspector general (Chanute AFB), and also spent two years in Southeast Asia as director of air traffic control services for the SEA areas. In 1965, he was detailed by USAF to the FAA in Washington, DC, and served three years with Research and Development Division as program manager for ATC Systems. In 1968, he was assigned to UK Comm Region (3rd Air Force South Ruislip) as director of USAF ATC Services. Stayed in England for four years then came back to Williams AFB in May 1972 and retired from active duty, with the rank of lieutenant colonel. He and his wife, Loraine, have two children and five grandchildren.

JOHN OMAR KISER, Colonel, Headquarters Special Operations Command, Logistics Directorate, Plans, Operations and Exercises Division Chief.

Col. Kiser was born on Aug. 22, 1938, and began his military career in September 1955, by enlisting in the United States Air Force. He rose to the rank of staff sergeant and attended officer candidate school, Lackland AFB, TX, earning his commission as a second lieutenant on Dec. 22, 1961. Immediately following commissioning, he attended undergraduate navigator training at James Connally AFB, Waco, TX, earning his wings on Oct. 7, 1962. From 1963-1973, he served in various capacities in the airlift operations field, navigator (instructor and flight examiner), command post duty officer, wing training officer, wing plans officer and numbered Air Force combat tactics and techniques officer. In September 1973, Col. Kiser was awarded a scholarship to attend the University of Tennessee to complete a master of business administration degree. He was awarded the degree on March 1975, and was assigned to McGuire AFB, NJ, as the passenger service officer for the 438th Aerial Port Squadron (APS). His next assignment as the air terminal manager came in November 1976, followed by the announcement of his promotion to lieutenant colonel in January 1977 with a date of rank of October 1977. He was subsequently given command of the 438th Transportation Squadron. While serving in that capacity, Col. Kiser was selected as aide-de-camp to Commander-in-Chief, Military Airlift Command, Scott AFB, IL. The following assignment was to the Air Transportation Deputate where he filled the roles of training division chief, service quality division chief and deputy director, service quality, training and standardization. Subsequently, he was assigned to the Operations Deputate where he performed duties as an airlift director. In June 1981 he was transferred to West Germany and given command of the busiest aerial port in Military Airlift Command at Rhein Main AB. After serving two years as the 435th Aerial Port Squadron Commander, Col. Kiser was assigned to the 322nd Airlift Division and given the duties of the director for Air Transportation at Ramstein AB, Germany. Upon returning to the U.S., he was assigned as the director for Logistics with Headquarters Joint Special Operations Command.

Col. Kiser has over 5,000 flying hours accumulated during 11 years of flying assignments in C-124 and C-141 aircraft.

Medals/Awards: Meritorious Service Medal, Air Force Commendation Medal, Combat Readiness Medal, Army Good Conduct Medal, National Defense Service Medal, Air Force Expeditionary Medal, VN SUC and VN Campaign Medal.

Col. Kiser is married to the former Mariette M. Schmeisser of Metz, France. They have a son, Patrick Lee Kiser, and a daughter, Michelle Louise Kiser. His hometown is Hampton, VA.

JAMES E. KITTS, born April 18, 1939, in Greentown, IN. He was commissioned through Navigation Cadets Class 62-21 at Harlingen AFB. After completion of Bomb Nav Training at Mather AFB, assigned to 345TCS, Dyess AFB as C-130E navigator. In 1964, transferred to 4486 Test Squadron, Tactical Air Warfare Center, Eglin AFB. Reassigned to the 345 TCS in 1965 when it was sent to CCK AB, Taiwan. Left active duty to return to college and flew C-119G's with 71 TCS, Air Force Reserve, Bakalar AFB, then assigned to instructor duties on AC-119G with 4413 CCTS Cockbourne AFB, when recalled to a active duty in 1968. After return from active duty joined 122 CSS, Indiana ANG, as transportation officer. Transferred to Ohio ANG, 164 TAS / 179 TAG as C-130B flight examiner, intelligence technician, group plans officer and chief of operations training. Retired in 1990, with the rank of lieutenant colonel.

Awards/Medals: Meritorious Service Medal, Air Medal with three oak leaf clusters, Air Force Commendation Medal with one oak leaf cluster, Air Force Outstanding Medal, Air Force Outstanding Unit Award, Presidential Unit Citation, Combat Readiness Medal with two oak leaf clusters, National Defense Service Medal with star, Armed Force Expeditionary Medal, Vietnam Service Medal, Humanitarian Service Medal, Armed Forces Reserve Medal, RVN Gallantry Cross with Palm, Republic of India Campaign Medal.

MICHAEL J. KOBOLD, born on Jan. 4, 1946, in Omaha, NE. He graduated from Omaha Central High School, received a BA in broadcast communications from Ohio State University, and earned an MBA from Golden Gate University. He completed Air Command and Staff College in 1978 and Air War College in 1986.

During his 21 year career in the United States Air Force he flew the C-130, the CT-39, and the C-141. He also served with the Aerospace Audiovisual Service as a producer/director, operations officer and MAC's audiovisual manager. In addition to serving in Vietnam, he participated in many important MAC airlift operations including Nickel Grass, Baby Lift, New Life and Urgent Fury. Deeply involved in airdrop and aerial refueling operations, he was chosen to lead the airdrop of the 82nd Airborne Division into Panama during Operation Just Cause. His late assignment was as operations officer of the 14th Military Airlift Squadron.

Mike and his wife, Dellita, reside in Redlands, CA. They have two children: Troy and Megan. After retiring in 1990, he continued his aviation career as a pilot for American Airlines.

VERNON J. KONDRA, Lieutenant General, Commander of the Military Airlift Command's 21st Air Force, with headquarters at McGuire Air Force Base, NJ. The 21st Air Force is responsible for all Military Airlift Command airlift operations from the Mississippi River eastward to the eastern borders of Iran and the Arabian peninsula, including Greenland, Iceland, NATO Europe, Africa, the Middle East, the Caribbean, and South America. Within this region, more than 30,000 Military Airlift Command people and 280 airlift aircraft operate from more than 50 locations.

Gen. Kondra was born June 1, 1938, in Kelso, WA. He graduated from Kelso High School in 1956 and earned a bachelor of science degree in biology from Washington State University in 1961. The general completed Squadron Officer School as a distinguished graduate in 1964, Industrial College of the Armed Forces in 1974, Armed Forces Staff College in 1975, Air War College in 1981 and the Graduate School of Industrial Administration at Carnegie-Mellon University in 1985.

He was commissioned as a second lieutenant through the Reserve Officer Training Corps program and entered the Air Force in July 1961. After graduating from pilot training at Williams AFB, AZ, in August 1962, he was assigned to the 8th Troop Carrier Squadron, McChord AFB, WA, flying C-124s. Upon completion of C-123 qualification training at Pope AFB, NC, Gen. Kondra was assigned on temporary duty to Da Nang AB, South Vietnam. From September 1963 to November 1969, he flew C-124s and C-141s at McChord. During this tour of duty, he served as a co-pilot, aircraft commander, instructor pilot and wing standardization flight examiner. He then returned to Da Nang as a command post duty officer and, later, executive officer for the 617th Military Airlift Support Squadron.

In July 1970, the general was assigned to Headquarters Military Airlift Command, Scott AFB, IL, as a member of the command briefing team. From July 1972 to January 1975, he served as aide and executive officer to the MAC vice commander. He completed Armed Forces Staff College in July 1975 and returned to McChord AFB as a C-141 flight commander, 4th Military Airlift Squadron. He then served as operations officer with the 8th Military Airlift Squadron. In February 1977, while at McChord, Gen. Kondra became chief of the command post, 62nd Military Airlift Wing. He assumed command of the 8th Military Airlift Squadron in April 1978. His next assignments were as assistant deputy commander and, then, deputy commander of operations, 62nd Military Airlift Wing, until August 1980.

He completed Air War College in May 1981 and was

assigned as assistant director for forces and then as chief of the Mobility Forces Division, Directorate of Programs and Evaluations, Office of the Deputy Chief of Staff for Programs and Resources, Headquarters U.S. Air Force, Washington, DC. From June 1983 to March 1984, he served as vice commander, 63rd Military Airlift Wing, Norton AFB, CA. He then became commander of the 62nd Military Airlift Wing, McChord AFB. In June 1986, he was assigned as deputy director for operations, National Military Command Center, Organization of the Joint Chiefs of Staff, Washington, DC. In August 1987, he assumed command of the 834th Airlift Division, Hickam AFB, HI. He also served as special assistant for airlift to the command in chief, Pacific Air Forces, and played an airlift liaison role to the commander in chief, U.S. Pacific Command. He was assigned as deputy chief of staff for Plans and Programs, Military Airlift Command Headquarters, in July 1989, and in August 1990, he became deputy chief of staff for Operations, responsible for directing the airlift operations for Operations Desert Shield and Desert Storm. He assumed his present command in July 1991.

The general is a command pilot with more than 7,000 flying hours. His military awards and decorations include the Distinguished Service Medal, Legion of Merit with oak leaf cluster, Bronze Star Medal, Defense Meritorious Service Medal, Meritorious Service Medal with three oak leaf clusters, Air Medal with two oak leaf clusters, Air Force Commendation Medal with oak leaf cluster, National Defense Service Medal, Armed Forces Expeditionary Medal, Vietnam Service Medal with five service stars, Air Force Longevity Service Award Ribbon with five oak leaf clusters, Small Arms Expert Marksmanship Ribbon, Republic of Vietnam Gallantry Cross with Palm, and Republic of Vietnam Campaign Medal. He also was awarded the MAC Order of the Sword. He was promoted to lieutenant general July 17, 1991, with same date of rank.

Gen. Kondra is married to the former Sharon Linton of Norfolk, VA. They have five children: Sheryl, William, Michael, Mark and Scott.

ROBERT M. KRAFT, "BOB", born on Aug. 3, 1954, at Luke AFB, AZ. Enlisted in the Air Force delayed enlistment program on Dec. 12, 1975, and entered basic training at Lackland AFB, TX, on March 2, 1976. After basic training, he began the Basic Loadmaster Training Course at Sheppard AFB, TX, in April 1976. Then attended physiological training, basic survival training courses and basic water survival course at Fairchild AFB, WA, in July 1976, en route to permanent duty station Dyess AFB, TX (773rd Tactical Airlift Squadron).

Participated in numerous joint, combined and allied operations. These include Gallant Crew, Volunt Shield, Solid Shield 77, the NATO directed Reforger 77, and participated in airlifting supplies and personnel throughout hurricane devastated Dominican Republic, September 1979. Participated in two rotational deployments to Europe based at Rhein-Main AB and Mildenhall Royal Air Station.

Selected Airman of the Quarter for 773rd TAS April-June 1977 and October-December 1977. Also selected Airman of the Month for 463rd Tactical Airlift Wing, June 1977. Was awarded Air Force Marksmanship Ribbon for both M-16 Rifle and 38 Caliber Pistol. Received Humanitarian Award for participating in airlifting supplies into blizzard devastated Midwestern Cities (USA), February 1978. Awarded Senior Airman Below the Zone, December 1978. Aided 773rd squadron in achieving Best C-130 Squadron of the Year Throughout Military Airlift Command for Year of 1978, January 1979. Aided 773rd Squadron in achieving Best C-130 Squadron of the Year Throughout USAF for Year of 1978, February 1979. Awarded Air Force Good Conduct Medal, March 1979. Had over 1,500 flying hours. Was honorably discharged as a sergeant in March 1980.

Now resides in Pasadena, TX, with wife, Donna, of 17 years and two sons where he is a process technician for Fina Bayport. Also is a certified EMT and volunteer firefighter for the city of Pasadena.

GEORGE R. KUZMA, born Dec. 11, 1918, in Portland, OR. Joined the USAF on Jan. 7, 1941. Military locations and stations at which he served: Kelly AFB, TX; 5th Ferry Wing, Dallas, TX; Tezpur, India (Hump Pilot KC-97); KC-135 Pilot, Castle AFB, CA.

Memorable experiences include fling in a five element KC-97 tanker, flight from Castle AFB, CA, when lead tanker threw #2 prop completely off causing engine to become disengaged from its housing and fall into #1 causing aircraft to crash killing all 18 aboard. He had to take offer the lead on into Castle AFB. Flew the plane that went down on a previous flight several days prior to this and wrote up the same engine as having excessive vibration. He had to shut down the engine, feather the prop, and return the aircraft to home base. It was ground checked and vibration was thought to be a defective alternator. It was changed and the aircraft was released for flight. Had he not shut down the engine and returned to base, he could have been killed. That gave him time to pause and reflect "Why him and not me." Life takes some strange turns!

He retired in 1964, with 23 years of active duty, with the rank of major.

He and his wife, Dorothy (now deceased), were married for 49 years. They had three children and seven grandchildren. After retirement from the service he bought a farm and raises black angus cattle.

ALLEN W. LANDAICHE, "PAT", was born in Gonzales, LA, on Feb. 11, 1941. He graduated from St. Gabriel High School in St. Gabriel, LA, in 1959. In 1964, he graduated from Louisiana State University with a bachelor of industrial technology and from Websters University in 1978 with a master's degree in management/public administration.

He completed pilot training at Reese AFB, TX, in 1965, with an assignment to C-133Bs at Travis AFB, CA, flying primarily Pacific routes to and from Vietnam. He served in Vietnam as a Forward Air Controller in 1969-70, flying D-1Es. With the phase out of the C-133B in 1971, he was assigned to the C-5 at Travis AFB, CA, with following assignments to Altus AFB, OK, as a formal school instructor and Headquarters Military Airlift Command at Scott AFB, IL, in the Directorate of Training. Landaiche retired from the Air Force in November 1984.

His awards include the Distinguished Flying Cross, Meritorious Service Medal, Air Force Commendation Medal, Air Medal with nine oak leaf clusters, and the Vietnamese Cross of Gallantry with Bronze Star.

Landaiche and his wife, Carol, live in Valrico, FL.

DERWENT LANGLEY JR., born Sept. 11, 1930, Augusta, GA. Served in the USAF Dec. 23, 1953 to Jan. 1, 1984, achieved the rank of colonel. Education: Academy of Richmond County, Augusta, GA, Diploma; University of Georgia, Athens, GA, BS in education; ROTC Commission 1953 Middle Tennessee State University, Murfeesboro, TN, MA in education, 1964.

Military location, stations and assignments: January 1954 - December 1954, pilot training, Primary - Bainbridge AF, GA, Basic - Webb AFB, TX. January 1955 - July 1957, Charleston AFB, SC, transport pilot with MATS, Air Police officer, 1608th Air Police Sq. (16mo). July 1957 - July 1961, Tachikawa AB, Japan, Flying Air Evac Missions, Aide-de-Camp B/G T.G. Kershaw. July 1961 - January 1967, Sewart AFB, TN. January 1967 - December 1967, Jedda, Saudi Arabia. January 1968 - March 1970, Sewart AFB, TN. March 1970 - October 1970, Little Rock AFB, AK. October 1970 - July 1972, Langley AFB, VA. July 1972 - July 1974, Pope AFB, NC. July 1974 - July 1976, Headquarters, USAF (Pentagon) Washington, DC. July 1976 - February 1981, Dobbins AFB, GA. February 1981 - December 1983, Dobbins AFB, GA. January 1984, Retirement.

Awards/Medals: Legion of Merit, Meritorious Service Medal with two oak leaf clusters, Air Force Commendation Medal with two oak leaf clusters.

Memorable experiences: Large formation troop/heavy equipment drops at Ft. Campbell, KY; Ft. Bragg, NC; Central Germany. Commander 39th TAS that flew the first C-130 nine ship formation personnel drop on Sicily Drop Zone (DZ) in instrument flight conditions using all Weather Aerial Delivery System (AWADS) equipped aircraft at Pope AFB, NC. All the great men and women I've met in those 30 years.

Married to former Sally E. Tucker of Tifton, GA, three children: Derwent Kent, Elizabeth R., and Judson R.; and two grandchildren, Samuel Kent and Brittney Lauren. Today he is enjoying life playing golf, fishing and hunting with his rich friends.

CHARLES B. LATHAM, born Dec. 27, 1950, in Washington, NC. He graduated from Washington High School. He received a bachelor of science degree from the United States Air Force Academy in 1974, and a master of science in systems management from the University of Southern California in 1983. He completed Air Command and Staff College in 1983 and Air War College in 1987.

Latham is currently the HQ Air Force Materiel Command Program Manager for Arms Control at Wright Patterson AFB, OH. Previously he was the deputy director of Contracting, Propulsion System Program Office and a contracts manager in the F-16 SPO, Aeronautical Systems Division, WPAFB. He is a master navigator with over 3,000 hours in the C-130 with tour of duties with the 374TAW, Clark AB, Philippines, 61TAS, Little Rock, AR, and HQ 22nd AF, Travis AFB as ALCE Branch Chief.

Among his awards and decorations are the Air Force Meritorious Service Medal with two oak leaf clusters, Air Force Commendation Medal with one oak leaf cluster, Air Force Achievement Medal and the Air Force Outstanding Unit Award with valor devise.

Latham and his wife, Kathy, reside in Beavercreek, OH. They have four children, David, Dan, Matt and Amanda.

MELVIN N. LEDBETTER, USAF, (Ret.) of Hendersonville, TN, entered the United States Air Force as a distinguished ROTC graduate from Baylor University in 1955. His assignments during 28 - 1/2 years of military service include tours of duty at Squadron, Group, Wing, HQS USAF (The Pentagon), Joint and Unified Command levels of which over 12 years were spent overseas.

Col. Ledbetter is a graduate of the Inter-American Defense College in Washington and a distinguished graduate of the Air Command and Staff College. He is the recipient of over 30 military awards and decorations include: Distinguished Flying Cross with oak leaf cluster, Air Medal with 17 oak leaf clusters, Defense Superior Service Medal, Defense Meritorious Service Medal, and Vietnamese Cross of Gallantry, Legion of Merit with oak leaf cluster, Joint Service Commendation Medal, Armed Forces Expeditionary Forces Medal with four oak leaf clusters, Vietnamese Cross of Gallantry with Palm and Gold Star, Brazilian Order of Santos Dumont, Chilean Armed Forces Military Star, Spanish Cross of Honor, First Class.

He flew the C-119G in the 40th TAS at Neubiberg, Germany 1956-57, and the C-130 in the 7406th Support Squadron at Rhein-Main AB, Germany 1958-59, in the 50th TCS at Sewart AFB in 1959-63, while at 314th TCW Hqs. in 1963-65 and later with the 50th TCS at CCK in 1966.

He is a command pilot who has flown over 20 types of transport and fighter aircraft including tactical airlift in Vietnam in 1965-66, and 320 attack fighter combat missions with Special Operations Forces in Vietnam in 1971-72.

He is married to his childhood sweetheart, Willie; and is now serving as President of Armed Services Mutual Benefit Association, located in Nashville, TN.

JAMES R LEE, "RANDY", Director of Communication-Computer Systems at the Oklahoma City Air Logistics Center, Tinker AFB, OK. He also serves as the Commander, 1985th Communications-Computer Systems Group.

Col. Lee was born Nov. 27, 1945, in Pine Bluff, AR. He graduated from White Hall High School in Pine Bluff and received his bachelor's degree in education from the University of Arkansas in 1968. In 1977, Col. Lee earned a master's degree in computer science from the University of Southwestern Louisiana through the Air Force Institute of Technology. His military education includes resident attendance at Squadron Officer School, 1975, where he was a Distinguished Graduate, and seminar programs for the Air Command and Staff College, 1979, and the Air War College, 1987. Col. Lee has also attended the Airlift Operations School and the Communications and Computer Systems Executive Forum.

He received his commission as a distinguished graduate of the Reserve Officer's Training Corps at the University of Arkansas in 1968 and earned his pilot wings in June 1969 at Williams AFB, AZ.

During his career, Col. Lee has flown a variety of aircraft including the C-130, T-38, T-39 and C-21. In 1969, he was stationed at Ching Chaun Kang (CCK) AB, Republic of Taiwan, and flew 150 combat missions in the Republic of South Vietnam in the C-130 aircraft. He is a previous Chief of T-38 Academics, instructor pilot and assistant flight commander in the T-38 aircraft at Laughlin AFB, TX; chief pilot and assistant operations officer of the 37th Tactical Airlift Squadron, Rhein-Main AB, Germany; and the operation officer of the 58th Military Airlift Squadron, Ramstein AB, Germany. He is a command pilot with over 3,500 flying hours.

Besides these duties at CCK, Laughlin, Rhein-Main, and Ramstein, Col. Lee has seen duty in the Air Force Data Services Center at the Pentagon, where he served as a systems analyst, division chief, and executive officer to the commander; as the Chief of Wing Inspections and the Chief of Wing Operations Plans, 435th Tactical Airlift Wing, Rhein-Main AB, Germany; and as the Chief of Operations Plans, 322nd Airlift Division, Ramstein AB, Germany. In July 1987, he was assigned to Scott AFB, IL, where he assumed duties as initial cadre of the newly formed unified United States Transportation Command serving successfully as the Chief, Program Management Branch; Chief, Plans and Policy Branch; Chief, Operations Division: then prior to his current assignment as Chief, Global Transportation Network (GTN) Development Division - all within the Command, Control, Communications and Computer (C4) Systems Directorate.

Col. Lee's decorations include the Defense Superior Service Medal, the Distinguished Flying Cross with one oak leaf cluster, the Meritorious Service Medal with two oak leaf clusters, the Air Medal with three oak leaf clusters, the Air Force Commendation Medal, the Army Commendation Medal, the Joint Service Achievement Medal and the Air Force Achievement Medal with one oak leaf cluster.

Col. Lee is married to the former Elizabeth Lewis, a home economist and social services worker from Fayetteville, AR. Colonel and Mrs. Lee have three children: Mary Jane, James and Beth Ann.

BILL LEEDS, Major, military career began in March 1967, when he enlisted in the United State Army. Following basic training at Ft. Polk, LA, he was accepted into the Warrant Officer Candidate Flight Training Program culminating in his appointment as a warrant officer and the award of his Army helicopter pilot rating. He served with distinction in Vietnam for 13 months as a UH-1 Huey helicopter pilot, and later in Germany as a CH-34 Choctaw pilot before returning to the United States in 1970 to complete his college education. Major Leeds was commissioned in the Air Force through Officer Training School at Lackland AFB in 1974, and attended the Multi-National pilot training course at Sheppard AFB, TX, where he was the distinguished military graduate. From 1975 to 1980, he was a C-130 Hercules pilot for the 314th Tactical Airlift Wing, Little Rock AFB, AR, and then was one of a very small group of pilots selected to cross-train into the C-5 Galazy. He spent the next seven year flying the C-5 worldwide and broadening his career as a senior command post duty officer, operations executive officer and C-5 instructor pilot for the 60th Military Airlift Wing Travis AFB, CA.

Major Leeds' final active duty assignment was operations officer, 2873rd Flight Test Squadron, Kelly AFB, TX, where he was instrumental in leading that unique unit to the highest unit effectiveness evaluation score ever attained by a flying squadron in Air Force Logistics Command.

Major Leeds is from Daytona Beach, FL, a bachelor, and has pursued a second aviation career as an L1011 Tristar commercial airline pilot since retiring from the Air Force in 1990.

Awards/Medals: Distinguished Flying Cross, Meritorious Service Medal, Bronze Star with Valor Device, Air Medal (five awards), Air Force Outstanding Unit Award, Air Force Commendation Medal, Army Good Conduct Medal, Vietnamese Cross of Gallantry, Republic of Vietnam Campaign Medal, Combat Readiness Medal, Humanitarian Service Medal, Air Force Longevity Service Award, National Defense Service Medal, Air Force Overseas Short Tour Ribbon, Air Force Overseas Long Tour Ribbon, Air Force Training Ribbon, Vietnamese Service Medal.

Bill is a life member of the Airlift/Tanker Association, Daedalians and Quiet Birdmen. He is a 1973 graduate of Embry-Riddle Aeronautical University with majors in aeronautical science and aviation management, a master's degree from Webster College-St. Louis rounds out his education accomplishments.

MICHAEL G. LEONARD, born in St. Louis, MO, on Dec. 14, 1941, one week after Pearl Harbor. Too young to enlist, he elected to wait for the Vietnam "conflict" (sure looked like a war to me). He attended Christian Brothers College High School (CBC) in St. Louis, graduating in 1959. He attended St. Benedict's College in Atchison, KS, graduating in 1963, with a degree in political science. Mike attended the Air Force Officers Training School (OTS) and was commissioned a second lieutenant in September 1963. He began his career as a weapons controller assigned to the 770th Radar Sq., Ft. Meade, MD. In the Spring of 1965, he was assigned to the 964th AEW & C Sq. (ADC) at McClellan AFB, riding backwards in the EC-121 (an Airborne Weapons Controller). While in the 964th, Mike pulled several TDYs, including Vietnam, Taiwan, Panama and Johnston Island, just to name a few. His weapons controller (W.C.) duties lasted from September 1963 to March 1967.

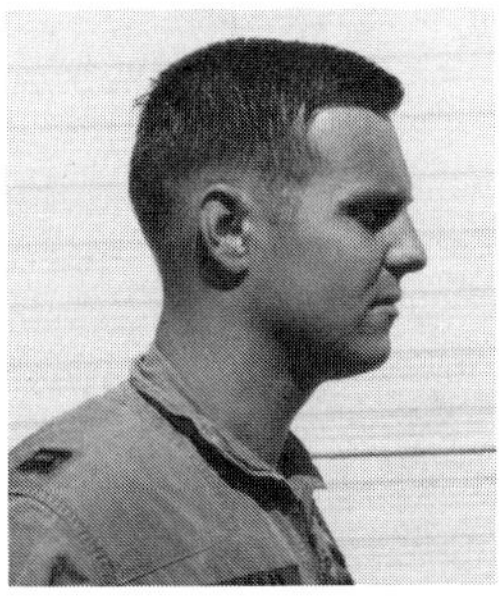

It took Mike almost four years to realize that there was a "glitch" in the promotion system, and that he could use his nose to count the number of general officers wearing that W.C. badge (and the toes on his left foot to count the Eagles). The light went on and Mike was off to pilot training at Vance AFB, OK, from April 1967 to May 1968. Following Flight training he was assigned to the 963rd AEW & C Sq. flying the beloved EC-121, this time out of Otis AFB, MA. (It serves him right for closing the bar on too many exam nights; and listening to the ex-Navigators in Class 68-G tell war stories). June of 1969 found Mike beginning the trek to SEA, this time in an 0-1 (Birddog) as a forward air controller (FAC).

Mike answered to the call signs; "Walt 70" (Bao Loc), "Walt 22" (Gia Nghia), both in the Central highlands, and finally to "Dart 09" (Chief, 0-1 Stan-Eval, 504th TASG). Nearly 275 sorties later he hopped the "freedom bird" and an assignment the newly formed C-5A Sq. at Charleston AFB, SC. Lt. Col. Leonard earned an MBA from South Illinois University during his three years at Charleston.

Mike served as chief of the Student Activities Division at AFOTS from April of 1973 to June 1975. From 1975 to June of 1979, he flew "Jetstars" for the 89th MAW, Andrews AFB, MD. This group provided airlift support for U.S. Government and foreign dignitaries. The curtain fell on his Air Force career after a final four year stint with the 75th Military Airlift Sq. (C-5A's), Travis AFB, CA (1979-83).

Lt. Col. Leonard's career took exactly "20 years and 20 minutes." He walked away with three Distinguished Flying Crosses and nine Air Medals (to name a few), and many fond memories and good friends. He retired Oct. 1, 1983. Mike is employed as marketing director by Space Systems Loral, Palo Alto, CA. He and his wife Lindy reside in San Ramon, CA.

VAUN N. LINIGER, born Aug. 1, 1921, in Ft. Wayne, IN. He received his BA at Chapman College. He joined the USAF on March 17, 1942, serving in missions 322 BG/ 449 BS 189 trips. He served in WWII ETO B-26 Marauder 68.

Memorable experiences include Berlin Airlift Douglas, C-54, Basic Instructor Randolph AFB, B-29 Boeing Training for Korea, 10,000 hours in B-47s Boeing, Southeast Asia Task Force Alpha, 13 different aircraft, 189 combat hours.

He was discharged May 30, 1970, with the rank of lieutenant colonel.

He has four children and two grandchildren.

GEORGE M. LIVERS, Colonel, USAF, (Ret.), Enlisted Army Air Corps in October 1941, Maltiusburg, WV. Assigned Keesler Field, Biloxi, MS. Completed pilot training, Moody Field, GA. December 1942. Assigned 4th Ferrying Group Memphis, TN, ATC. Subsequently assigned 3rd Combat Cargo Group, 12th Squadron, CBI, flying C-47s supporting Gen. Stilwell's efforts in Burma/China. Other: two years with the Navy; four years 1254th ATC (Special Air Missions) Washington. Awarded USAF Flying Safety Award by Gen. Joe Smith, Commander MATS.

Commander, 61st C-130 TAC Airlift Squadron, Smyrna, TN. Survived Airborne Jump School age 45, Directed assignment HQ Air Force, Deputy Chief Airlift Division. Retired October 1971, Base Commander, Pope AFB, NC. Completed Air TAC School, Air Command and Staff School, Air War College.

Awards: Distinguished Flying Cross with two oak leaf clusters, Air Medal with five oak leaf clusters, Commendation Medal with one oak leaf cluster, Meritorious Service Medal. Loving wife and successful realtor Ann, six children, 12 grandchildren.

JERRY D. LIVINGSTON, Colonel, (Ret.), born July 10, 1932, in Las Vegas, NM. He graduated from Santa Fe High School in Santa Fe, NM in 1951. In 1955, he received a bachelor's degree in business administration from the University of New Mexico and was commissioned a second lieutenant through the USAF ROTC program.

Livingston retired from the Air Force in July 1986, with the rank of colonel. His last assignment was the Military Airlift Command Liaison Officer to Red Flag at Nellis AFB, NV. During this assignment he was given the responsibly of establishing and conducting the Military Airlift Command's "Combat Aircrew Training School" (CATS).

Livingston's assignments include the 314TCW Sewart AFB TN, 64/516 TCW Dyess AFB TX, 316TCW Langley AFB VA, 504TASG Bien Hoa AB Vietnam, Headquarters Tactical Air Command Langley AFB VA, Alaskan Command Headquarters Elmendorf AFB AK, 314MAW little Rock AFB AR, Headquarters Military Airlift Command Scott AFB IL, USAF Airlift Center Pope AFB NC and 4440TFTG (Red Flag) Nellis AFB, NV.

He was in the initial cadre (61TCS/17TCS) that operated C-130D aircraft from Sondrestrom AB, Greenland to the Greenland Icecap during the period March 1959 to February 1963. Jerry and his wife Jimmye Lou have five children and currently live in Las Vegas, NV.

DARRYL V. LUNDGREN, born Dec. 21, 1948, in Duluth, MN. A 1970 U.S. Air Force Academy graduate with a bachelor's degree in Asian History, he also graduated from Air Command and Staff College in 1983, the Naval War College in 1984, and the Air War College in 1989. In 1990-91, he was a National Defense Fellow at the University of Pittsburgh.

Following UPT, Darryl flew C-130s with the 348 TAS773 TAS from Dyess AFB, TX, with several rotations to Europe and Southeast Asia. After ACSC, he flew with the 345 TAS from Yokota, Japan, where highlights included being the mission commander for the unit's support for President Reagan's trip to China in 1984, and for the 40th Anniversary of the Battle of Iwo Jima Reunion of Honor.

The Intelligence Area Specialist Program sent him to the University of Southern California for a master's degree in Asian History in 1978. Several current intelligence assignments with the Defense Intelligence Agency in Washington, DC, followed where he currently is duty director for Intelligence in the National Military Command Center.

His decorations include the Defense Meritorious Service Medal with two oak leaf clusters, the Air Force Meritorious Service Medal, and the Combat Readiness Medal with one oak leaf cluster. He and his wife, Kathy, have two children: Carrie and Mark.

GARY J. MACHOVINA, Colonel, Vice Commander, 317th Airlift Wing, Pope AFB, NC.

Col. Machovina was born Nov. 30, 1945, in Berea, OH. Graduated, in 1968, as a distinguished AFROTC graduate from Bowling Green State University with a BS degree in education. He received his master's degree from Webster University in 1983. His PME includes SOS, ACSC, Naval War College and National Security Management.

He attended pilot training at Laughlin AFB, TX, and then was assigned as a C-130B pilot in 773 TAS, Clark AB, RP. He served as a flight examiner in the 316 TAW at Langley AFB and the 36 TAS at McChord AFB. The colonel's next assignment was to 22nd Air Force at Travis AFB as a flight standards officer.

Over the next 10 years, he served at Headquarters MAC as Chief of Tactics, in the Philippines as operations officer and commander 21 TAS and assistant director, 374 TAW Clark AB. He returned to HQ MAC as the director of Aircrew Stan Evaluation. He moved to Pope AFB and became deputy commander for Special Capabilities, 317 AW and then assumed current position.

A command pilot with over 4,800 hours including 615 combat hours. Qualified in the C-130A/B/E/H/LC and has flown the C-9, C-12, C-21, C-29, C-140 and C-141.

His decorations include the Distinguished Flying Cross, the Meritorious Service Medal with three oak leaf clusters, the Air Medal with three oak leaf clusters, and the Air Force Commendation Medal with one oak leaf cluster.

Col. Machovina is married to the former Joan M. Habluetzel. They have two daughters, Krista and Hilary.

BRUCE S. MAHAFFEY, Major, USAF, born May 20, 1956, in Bernkastel, Germany. A 1974 graduate of Wagner High School, Clark AB, Philippines, he received a bachelors degree in behavioral sciences from USAFA in 1978. He holds a master's degree in human resources development from Webster University and completed the Army Command and General Staff College in 1991.

Currently chief of Combat Tactics and Technique, 437 AW, Charleston AFB, his earlier positions at Charleston were chief pilot and assistant operations officer, 17 ALS. Past assignments include chief of Strategic Inspection Plans, Headquarters MAC IG, Scott AFB; director of operations, Korean Airlift Control Center, Osan AB; Special Operations Low Level II (SOLL II) pilot, flight examiner, wing tactics officer and command post duty officer, Charleston AFB. He received the 1986, Tunner Award for extracting Haiti's then-president Jean Claude Duvalier.

His decorations include the Air Force Meritorious Service Medal with two oak leaf clusters, the Air Medal, the Air Force Commendation Medal with two oak leaf clusters, and the Air Force Achievement Medal.

He and his wife, Karen, have two daughter, Jennifer and Caitlin.

BEN MAINA, born July 12, 1930, Canandaigua, NY, graduated from Canandaigua Academy 1948, one semester at Rochester Institute of Technology. Joined the U.S. Air Force in January 1951. Military stations at which he served were Sheppard AFB, TX, A&E School; Donaldson AFB, SC, 343rd TCS; Sewart AFB, TN; Ashiya AFB, Japan, 61st TCS (Green Hornets); Hancock Field, Syracuse, NY, 32nd AD; Tan Son Nhut Airfield, Saigon, (33rd Camron); Griffiss AFB, NY, (RADC). He retired in August 1965, with the rank of Tsgt.

Memorable experiences include his time serving with a great bunch of guys at Sewart AFB, TN, and as a flight engineer during the Korean Airlift out of Ashiya, Japan.

At the conclusion of the Korean Conflict being assigned as an advisor to the French Air Force in Haiphong prior to the fall of Dien Bien Phu in 1954.

Married in January 1956, to the former Elinor Sawers, and had three children and two grandchildren. His civilian employment for 28 years was with Niagara Mohawk Power Company, Syracuse, NY, as a flight attendant and mechanic. He retired in August 1993.

CHARLES E. MALCOM, Colonel, (Ret.), born Dec. 27, 1939, in Sac City, IA. He received a bachelor of science degree from Iowa State University in 1961, and a master of science degree in systems management from the University of Southern California in 1972. He was commissioned through Officers Training School in June 1962. Attended Industrial College of the Armed Forces 1980-81.

Upon graduation from navigator training, he was assigned to Tactical Air Command in the 778th Troop Carrier Squadron (TCS), Pope AFB, NC, flying new C-130E's. He flew in operations in support of the 1964 Congo and 1965 Dominican Republic crises, several composite air strike force deployments, and rotations to Evereux, France; Moron, Spain; Kadena, Okinawa; and Mactan, Philippines. May 1966, he moved to the 815 TCS, Tachikawa, Japan. He flew over 1,200 C-130 combat sorties in the RVN highlighted by several CDS and LAPES missions during the Battle of Khe Sanh. In May 1969, he was assigned to the newly formed Directorate of Airlift in DCS/Operations, Headquarters PACAF, Hickam AFB, HI. During this tour he was honored to escort two POWs to the states on their repatriation in 1973.

Arriving at Andrews AFB, MD, in 1974, he flew VIP special missions in VC-135 and VC-137 aircraft. He then commanded the 89th Organizational Maintenance Squadron for two years. He spent two years in the Pentagon. In 1983, he returned to his alma mater in Ames, IA, and was commander and professor of aerospace studies, AFROTC, Det. 250. He ended his career at Scott AFB, IL, where he was deputy commander for maintenance, 375 AAW and later was assigned to the Inspector General, HQ MAC where he later became the director of Readiness.

He is a master navigator with over 5,900 hours and was awarded the Senior Aircraft Maintenance Badge. Married to the former Karen Dethlefsen of Royal, IA, have two sons, Greg and Doug.

Units: 778 Troop Carrier Squadron, Pope, AFB, NC; 815 Troop Carrier Squadron, Tachikawa AB, Japan; HQ PACAF/Directorate of Airlift, Hickam AFB, HI; 89 Military Airlift Wing, Andrews AFB, MD; 375 Aeromedical Airlift Wing, Scott AFB, IL; HQ MAC/Inspector General and Quality Support and Readiness, Scott AFB, IL.

Memorable Experiences: Several CDS and LAPES missions during the Battle of Khe Sanh. Escort officer for two POWs from Clark AB to the states, 1973. Flew in operaitons in support of the 1964 Congo and 1965 Dominican Republic Crises.

EVERETT L. MARSHALL, born Feb. 14, 1919, in Crown Point, IN. He received his education at Montana State University. He joined the U.S. Air Force on June 25, 1940, serving with the 1042nd Twin-Engine Training SQ, 68th Bomb Wg.; 97th AREFS. Military locations at which he served were Brooksfield, TX; Lake Charles, LA; and Great Falls, MT.

Memorable experiences: three years as flight line chief, pilot training, flight engineer, special air mission in Puerto Rico Hqs, flight engineer for 20 years with 14,000 hours.

Discharged June 24, 1970, with the rank of master sergeant.

He married the former Elinor Lois Barnett, and has three children and two grandchildren. As a civilian he was employed with Lockheed Aircraft Company as an electrician. He retired Jan. 1, 1968.

JAMES A. MASSERANO, Lieutenant Colonel, USAF, (Ret.), retired on June 1, 1992, and is currently working as a senior analyst for Science Applications International Corporation in the Washington, DC, area.

Lt. Col. Masserano was born April 21, 1947, in New York City. He spent his early life in New York and graduated from St. Raymond's High School in the Bronx, NY. The colonel earned a BA from Manhattan College, NY, in 1970, majoring in accounting. He also obtained a MA degree in management and public administration from Webster University in 1977. Lt. Col. Masserano is a graduate of Air Command and Staff College, National Security Management College, and Air War College.

Following his commission through Officer Training School, Lt. Col. Masserano completed undergraduate navigator training at Mather AFB, CA, where he was a distinguished graduate. He then completed C-141 aircraft upgrade training at Altus AFB, OK, where again he was a distinguished graduate. During the next three years, he was a C-141 navigator and instructor assigned to the 6th Military Airlift Squadron at McGuire AFB, NJ. He participated in all aspects of the Wing's mission and was initial cadre in the primary nuclear airlift force (PNAF). He flew airlift combat support missions into SEA and participated in Operation Nickel Grass, the resupply of Israel during the 1973 Arab-Israeli War.

In June 1975, Lt. Col. Masserano entered Squadron Officers School and graduated in September 1975. He was then assigned to the 443rd Military Airlift Wing at Altus AFB, OK. He served there as an instructor navigator, academic classroom instructor, and flight examiner navigator at both squadron and wing level. He was initial cadre in C-141A station keeping equipment (SKE) airdrop, the C-141A improved navigation system (INS), and the C-141B stretch aircraft. He was the first aerial refueling rendezvous qualified C-141 flight examiner navigator in the Military Airlift Command (MAC).

Lt. Col. Masserano was reassigned to the USAF Airlift Center in May 1980 and served as Chief, C-141 Navigation Branch, Test and Evaluation Division. He was responsible for directing MAC operational tests and evaluations involving simplified fuel savings systems, initial chemical warfare defense equipment, C-141 color radar enhancements, C-141 instrument condition airdrops and austere airfield approach improvements, and certain classified special operations projects.

From April 1984 to May 1985, he served in Korea at Osan AB as the Chief of the Command Advisory Function for the 314th Air Division (now 7th Air Force). He led an around-the-clock, all source information center responsible for obtaining, evaluating, and disseminating special intelligence information in support of the Air Division Commander and Air Component Command battle staff. He was also responsible for monitoring and protecting Peacetime Aerial Reconnaissance Program (PARPO) assets operating within the Korean theater and had contingency responsibility for all PARPRO assets in the Far East.

In May 1985, he returned to the USAF Airlift Center and in December 1985, assumed the duties of Chief, Readiness and Applications Division leading 15 project directors in the conduct of operational test and evaluation (OT&E) in the diverse areas of airlift tactics, transportation, communication, weather, and airdrop systems, special operations, electronic warfare, intelligence and aeromedical evacuation. He was responsible for the management and conduct of over 50 operational tests of equipment, concepts, doctrine, and tactics for the employment of airlift forces worldwide.

In May 1989, he was assigned to the Pentagon as the Chief of the Airlift and Special Operations Forces Branch, Operational Test and Evaluation Division, Directorate of Operations, HQ U.S. Air Force. He became deputy division chief in September 1990. In September 1991, he was assigned as the deputy chief of the Policy and Programs Division, in the newly established Directorate of Test and Evaluation, HQ USAF. He retired from this position in June 1992.

Lt. Col. Masserano is a master navigator with 4,500 flying hours, predominantly in the C-141. His military decorations and awards include the Meritorious Service Medal with two oak leaf clusters, Air Force Commendation Medal with one oak leaf cluster, Air Force Achievement Medal, Air Force Outstanding Unit Award with two oak leaf cluster, Air Force Organizational Excellence Award with one oak leaf cluster, Combat Readiness Medal with three oak leaf clusters, National Defense Service Medal with one Bronze Star, Armed Forces Expeditionary Medal with one Bronze Star, Vietnam Service Medal with one Bronze Star, and Humanitarian Service Medal with two Bronze Stars.

Lt. Col. Masserano is married to Dace Krastkalns of Chicago, IL, and they have two children, Amanda and John. His hometown is the Bronx, NY.

CHARLES M. MASTERSON, Chief Master Sergeant, born in Pittsburg, CA, and graduated in 1952, from Mt. Diablo Union High School, Concord, CA. He entered the USAF the same year and was Distinguished Graduate of both Aircraft and Engine School at Sheppard AFB, TX, and basic Flight Engineer School at Chanute AFB, IL.

Chief Masterson retired after 30 years of service from Scott AFB, in 1982. His last assignment was at HQ Military Airlift Command; Directorate of Aircrew Standardization. He was the Chief Flight Engineer C-141. During his career he was assigned to the 62nd and 63rd Troop carrier/Military Airlift Wings and was an instructor/flight examiner engineer at Altus AFB, OK (HTTU).

An Airlifter his entire 30 year military career, Chief Masterson amassed 17,000 hours in C-124 and C-141 aircraft. Among the many missions he participated in over the years were Dewline, the distant early warning radar sites; Deepfreeze, Antarctic airdrop missions for the international geophysical year; The Lebanon and Congo crises of the 1960s and over 60 combat support missions in Vietnam.

Among his decorations are: Defense Meritorious Service Medal, Meritorious Service Medal, Air Medal with one oak leaf cluster, Air Force Commendation Medal with one oak leaf cluster, Air Force Expeditionary Medal with three oak leaf clusters, Antarctic Service Medal, Vietnam Service Medal with three oak leaf clusters.

Masterson and his wife Mary currently reside in Charleston, SC, where he is a part of the McDonnell-Douglas team setting up the initial C-17 Squadron. They have three children.

LAURENCE MAXTON, Airlifters exist to move troops and cargo. In the days of C-82s and C-119 flying boxcars, "troop carrier" provided a number of one way rides to Laurence Maxton.

Born Dec. 9, 1929, in rural Bottineau County, ND, educated at Lansford grade and high school, he enlisted in the U.S. Army on July 28, 1947. During basic training at Ft. Knox, KY, he volunteered for parachute duty and qualified as a gliderman and parachutist at Ft. Benning, GA.

He was assigned to 82nd Airborne HQ CO. Motor Pool at Ft. Bragg, NC, and served the rest of his three years enlistment there.

Employed 30 years as a truck driver in the automobile transport industry, he retired in February 1990.

He is an amateur military historian with a primary interest in airborne and troop carrier units.

Father to six and grandfather to six he lives in Belleville, IL.

DONALD J. MCCARTHY, was born on Jan. 28, 1923, in Brockton, MA. He received a BA and MBA from Golden Gate University.

He joined the USAF in April 1943. Military locations and stations include: San Marcos, TX; Bassingborn, England; Sewart AFB; Langley AFB, VA; Mather and AFB, CA; McCarthy served with the following units: 91st BG, 62TCSq Sewart AFB/Ashiya, Japan, HQ TAC, AFAT 1 Tan Son Nhut AB, 3541 FTSQ.

McCarthy's memorable experiences include the 8th AF bombing raid on Berlin, para-drop north of Seoul in the fall of 1950, resupply landings at Hungnam Reservoir, Korea in the winter of 1951, and night flare drops around outposts in Vietnam.

He achieved the rank of lieutenant colonel and was discharged in November 1972. He is married to Lorraine, they have two children and two grandchildren. He was a Tac Collection Supervisor for the State of California. He retired in September 1983.

CARL L. MCPHERSON, Colonel, USAF, born June 2, 1942, in Lisbon, OH. He joined the USAF Aug. 8, 1960, in Youngstown, OH. Geodetic Surveyor 1960-62; USAF Academy Prop School, 1962-63; USAF Academy Cadet, 1963-68; Pilot Training, Vance AFB, OK; 776 TAS, CCK AB Taiwan 1970-71; 773 TAS Dyess AFB, 1972-73; Wpns Ctl, Mongil Son, Korea, 1973-74; AFMPC, Randolph AFB, 1974-79; Armed Forces Staff College, 1979-80; 39 TAS Pope AFB, 1980-82; 772 TAS, Dyess AFB, 1982-84; Naval War College, 1984-85; USCENTCOM/JS, MacDill AFB, FL, 1985-86; C-130 System Program Manager, Robins AFB, GA, 1986-88; 463 TAW, Vice Commander, Dyess AFB, TX, 1988-90; HQ AFSOC, DCS Personnel, Hurlburt Field, FL, 1990-92; HQ AFSOC Chief of Staff, Hurlburt Field FL, 1992 to present.

Memorable experiences include flying through the ridge saddle to land at Kham Duc, RVN, uploading heavy equipment and hanging on the props to depart through the ridge saddle.

Awards/Medals: Distinguished Flying Cross, two; Defense Meritorious Service Medal, one; Meritorious Service Medal, three; Air Medal, four; Air Force Commendation Medal, three.

He and his wife have a son Greg, born January 1970; daughter Alison, born May 1973; and a son Chad, born January 1992.

Today he is HQ Air Force Special Operations Command (APSOC), Chief of Staff, Hurlburt Field, FL.

DONALD L. MCSWAIN, born June 13, 1947, in Tampa, FL. He joined the U.S. Air Force June 28, 1965, serving as a command pilot at Vandenberg AFB, McChord AFB, Laredo AFB, Moody AFB, Scott AFB, Rhein Main AB, Little Rock AFB, Pentagon, Maxwell AFB.

His memorable experiences include Commander, 50 TAS; Chief, Air Force Long Range Plans Div.; Chief, Air Force Doctrine Div.; MAC Chief, Air War College.

Awards/Medals: Meritorious Service Medal, Air Force Commendation Medal, Humanitarian Award.

He was discharged Dec. 31, 1990, with the rank of colonel. Today he is a pilot for UPS/CRAF.

He and his wife, Gerrie, have two daughters, Amy and Jennifer.

PAUL J. MERCADANTE, entered the Air Force in 1975, as a law enforcement specialist. He cross-trained in 1979 to become a loadmaster on the C-141A/B and C-130E/H aircraft. His assignments were the 729 MAS and 53 MAS at Norton AFB, CA and is currently assigned to the 345 ALS at Yokota AB, Japan as their chief loadmaster.

He is an honor graduate from all flying courses he attended from Sheppard AFB, TX, Altus AFB, OK, and Little Rock AFB, AR. He holds or held qualification as special operations low-level I, joint airdrop inspector, instructor and flight examiner.

He was selected the Base NCO of the Year at Norton AFB, earned an associate degree from the community college of the Air Force, and bachelor's degree from Embry-Riddle Aeronautical University. His decorations include the Air Force Commendation Medal with one oak leaf cluster, Air Forces Expeditionary Medal with one star, Southwest Asian Service Medal with two stars, Humanitarian Service Medal, and the Kuwait Liberation Medal.

FELIX MEYER III, Colonel, USAF, Deputy Command Surgeon, U.S. Atlantic Command, Norfolk, VA.

Col. Meyer was born May 22, 1943, in Phoenix, AZ, and graduated from Westside High School, Omaha, NE, in 1961. He received a bachelor's degree in business administration from Southern Methodist University in Dallas, TX, in 1965, and as an ROTC cadet was commissioned into the Air Force. He attained a bachelor of science degree in aviation from Louisiana Tech University in 1971 and an AFIT sponsored master's degree in health care administration from the Medical College of Virginia in 1974.

Col. Meyer entered active duty at Gunter AFB, AL, in May 1965. He was initially assigned to the USAF School of Aerospace Medicine, Brooks AFB, TX, from 1966 to 1967. Later assignments included 12th USAF Hospital, Cam Rahn Bay, Republic of Vietnam, from 1967 to 1968; USAF Hospital Barksdale, Barksdale AFB, LA, from 1968 to 1971; Headquarters Air Defense Command from 1971 to 1972, prior to his AFIT selection; Aerospace Medical Division as Aide-de-camp to the Commander at Brooks AFB, TX, from 1974 to 1976; and USAF Occupational and Environmental Health Laboratory as Chief, Support Services Division at Brooks AFB, TX, from 1976 to 1978.

From 1978 to 1981, Col. Meyer was the administrator of the USAF Clinic Aviano, Aviano AB, Italy. He was then assigned to Pope AFB, NC, as the commander of the 1st Aeromedical Evacuation Squadron. During this period, he participated in over 25 JCS exercises worldwide, and served as the senior Air Force medical officer during Operation Urgent Fury in Grenada.

Col. Meyer was then transferred to the Medical Wartime Hospital Integration Office, Office of the Surgeon General, Headquarters, USAF, Ft. Detrick MD, where he served from 1985, until reassignment as commander of the 24th Medical Group, Howard, AFB, Panama, in August 1987. While assigned at Howard, he was the director for all aeromedical evacuation operations in Central and South America for the U.S. Southern Command. He remained in this capacity until 1991, during which period he functioned as the senior Air Force medical officer for Operation Just Cause. From 1991 until 1993, Col. Meyer served as Director of Medical Plans and Operations, Office of the Command Surgeon, Air Combat Command, Langley AFB, VA.

Col. Meyer is a Fellow of the American College of Healthcare Executives and of the Aerospace Medical Association. His military decorations include the Bronze Star Medal, Meritorious Service Medal, Air Medal, Air Force Commendation Medal, National Defense Service Medal, Armed Forces Expeditionary Medal, Vietnam Service Medal, Humanitarian Service Medal, Combat Readiness Medal, Republic of Vietnam Campaign Medal, and Republic of Vietnam Gallantry Cross.

Col. Meyer has accumulated over 800 hours of operational support flying in the Air Force as an aeromedical evacuation operations officer. He holds current FAA Civil Aviation certificates, including commercial and instructor pilot with instrument and multi-engine ratings.

Col. Meyer was promoted to colonel on June 1, 1986. He is married to the former Linda Neumann of Alexandria, MN. They have two sons, Jeffrey Allan and Gregory Thomas.

ROBERT H. MEYERS, 314th Operations Group Commander, Little Rock AFB, AR.

Col. Meyers was born Jan. 8, 1951, at Scott AFB, IL. He graduated from Morris High School, Morris, MN, in 1969, and then attended the U.S. Air Force Academy. He graduated with a bachelor of science degree in civil engineering and was commissioned on June 6, 1973. He was awarded a master of science degree in business administration from Boston University in 1980. He has completed Squadron Officer School, Air Command and Staff College and National War College.

Following commissioning, he attended pilot training at Sheppard AFB, TX. His first operational assignment was to the 41st Tactical Airlift Squadron, Pope AFB, NC. He served as a C-130E pilot and aide-de-camp to the 317th Tactical Airlift Wing Commander. In 1977, he was reassigned to Rhein-Main AB, Germany with the permanent deployment of the 37th Tactical Airlift Squadron and C-130E Adverse Weather Aerial Delivery System aircraft to Europe. He served this tour as an aircraft commander and later executive officer to the 435th Tactical Airlift Wing Commander.

In 1981, Col. Meyers was assigned to Headquarters Military Airlift Command, where he served as the assistant executive agent to CINCMAC for the National Defense Transportation Association Airlift Committee and later as the assistant executive officer to the MAC Deputy Chief of Staff for Plans. In 1982, he was reassigned as the assistant for Senior Officer Management, MAC Personnel Directorate, where he was responsible for MAC colonels assignments.

In 1985, he was assigned to the 773rd Tactical Airlift Squadron, Dyess AFB, TX, as instructor aircraft commander and assistant operations officer. In May 1987, Col. Meyers took command of the 772nd Tactical Airlift Squadron, and in 1989, became the assistant deputy commander for Operations, 463rd Tactical Airlift Wing.

After graduating from National War College in June 1990, Col. Meyers was assigned as special assistant for General and Flag Officer (G/FO) Matters, Joint Staff and responsible to the director and the chairman of the Joint Chiefs of Staff for managing all joint/outside service general and flag officer positions and related matters.

Col. Meyers' decorations include the Defense Superior Service Medal, the Meritorious Service Medal with two oak leaf clusters and the Air Force Commendation Medal. He is a command pilot with more than 2,700 hours in the C-130E/H. He was promoted to colonel April 1, 1991.

Col. Meyers is married to the former Rosalind Anna Bryant of Colorado Springs, CO. They have one daughter, Laura.

THOMAS R. MIKOLAJCIK, Brigadier General, is the Director of Transportation, Headquarters U.S. Air Force, Washington, DC. Gen. Mikolajcik was born in Norwich, CT, on Aug. 17, 1946. He is a 1969 graduate of the U.S. Air Force Academy. He completed pilot training in 1970. He has served as squadron commander, wing vice commander and commander of two wings (Rhein Main and Charleston). He was the U.S. Air Force component commander in Somalia from December 1992 to March 1993, during Operation Restore Hope. As the first C-17 Wing Commander at Charleston, he directed preparations for training and fielding the C-17 and its first year of operation. He is a command pilot with more than 4,000 flying hours and has flown in the C-9, CT-39, C-130, C-141 and C-17 aircraft. His assignments have taken him from numerous locations in the States to Germany and Italy. He is married to the former Mary Carmen Heft of Denver and has two daughters and one son.

KENNETH G. MILLER, Brigadier General, USAF, Director of the C-17 System Program Office, Aeronautical Systems Division, Air Force Systems Command, Wright-Patterson AFB, OH.

Gen. Miller was born July 29, 1944, in Bryan, TX. He earned a bachelor of science degree in aeronautical engineering from Purdue University in 1966, and a master of science degree in systems management from the University of Southern California in 1970. The general completed Armed Forces Staff College in 1976, National War College in 1986 and the Executive Development Program at the University of Virginia in 1988.

He was commissioned through the Air Force Reserve Officer Training Corps program as distinguished graduate in August 1966. His first assignment was with the F-4 Program Office, Aeronautical Systems Division, where he was the project engineer on the F-4 Wild Weasel Electronic Warfare Program.

In January 1970, Gen. Miller was assigned to the Electronic Systems Division, Hanscom AFB, MA, as project engineer on the C-130 Hurricane Hunter Program. During this assignment, he flew on numerous hurricane penetration missions with the 53rd Weather Reconnaissance Squadron, Ramey AFB, Puerto Rico.

He reported to Carswell AFB, TX, as a project manager for Strategic Air Command's Aircraft Engineering Division in June 1972, and worked on a variety of reliability and maintainability projects on B-52, KC-135, FB-111 and SR-71 aircraft. In June 1974, he transferred to Robins AFB, GA, as a branch chief in the Service Engineering Division and, later as aide-de-camp to the Warner Robins Air Logistics Center Commander.

Upon graduation from the Armed Forces Staff College in August 1976, Gen. Miller reported to Headquarters U.S. Air Force, Washington, DC, as a foreign military sales action officer. In June 1977, he became executive officer to the director of security assistance and sales, and in January 1980 was assigned to the office of the Secretary of Defense to work on fighter aircraft sales to foreign customers.

Miller returned to Wright-Patterson AFB in August 1981 as chief of the 4950th Test Wing's Engineering Division and, later, deputy director of aircraft modification. In April 1983, he moved to the F-16 System Program Office as director of field operation sand headed Falcon 80, an effort that raised the fleet's mission capable rate from 72 percent to 90 percent.

In 1986, the general reported to McClellan AFB, CA, as chief of the Engineering Division, Sacramento Air Logistics Center. In May 1987, he was assigned as director of materiel management, Ogden Air Logistics Center, Utah, with primary support responsibilities for the F-4 and F-16 fleets, intercontinental ballistic missiles, non-nuclear munitions and trainer-simulator systems.

Gen. Miller assumed duties as vice commander of the Acquisition Logistics Division, Wright-Patterson AFB, in June 1989. He commanded the Air Force Systems Command's Contract Management Division, Kirtland AFB, NM, from March to June 1990, closing down that organization and transferring its components to the new Defense Contract Management Command. In July 1990, he assumed command of Defense Contract Management District West, heading a new joint-service activity with 5,000 people in eight states, managing more than 100,000 contracts valued at more than $235 billion. He spent the early part of 1991 on the team that planned the integration of Air Force Systems Command and Air Force Logistics Command into the Air Force Materiel Command, with specific responsibility for the integrated weapon system management concept. He assumed his present position in September 1991.

The general's military awards and decorations include the Defense Superior Service Medal, Legion of Merit, Defense Meritorious Service Medal and Meritorious Service Medal with four oak leaf clusters. He has also received the Freedom Foundation's Award of Merit.

He was promoted to brigadier general Nov. 1, 1990, with same date of rank.

Gen. Miller is married to the former Ann Perpich. They have three children: Keith , Deborah and Craig.

ERNEST M. MILTON, born Nov. 1, 1960, in Orlando, FL. He was commissioned through Officer's Training School in 1982, and entered navigator training. His first assignment was to Dyess AFB, TX, flying C-130s, where he upgraded to instructor and worked in Wing Current Operations. His next assignment was to Little Rock AFB and where he instructed and evaluated students in the 16th Tactical Aircrew Training Squadron and the 34th Combat Aircrew Training Squadron. His next assignment was to the Tanker Airlift Control Center at Scott AFB, IL. After 16 months in the TACC Capt. Milton moved to Operations and was Chief, Plans and Programs Branch. He is moving back to the 62nd Tactical Airlift Training Squadron, Little Rock AFB, AR. He is married to the former Gail Langfeldt of Prescott, AZ. They have a daughter Amy, and two sons, Tyler and Jacob.

ROBERT P. MIMS, Colonel, Commandant, USAF Air Mobility School, received a bachelor of science degree from Mississippi State University and master of science degree in operations management from the University of Arkansas. His professional military education includes Squadron Officers School, Armed Forces Staff College, and Air War College.

Col. Mims entered the Air Force Officer Training School and was commissioned on March 11, 1970. He attended undergraduate Pilot Training at Williams AFB, AZ, and was assigned to the 62nd Tactical Airlift Squadron, Little Rock AFB, AR. He flew as an instructor and flight examiner in the AC-130H in the 16th Special Operations Squadron, Ubon, Thailand and the 16th Tactical Airlift Training Squadron, Little Rock AFB, AR. He moved to the 314th Tactical Airlift Wing Current Operations before being assigned to HQ MAC, Scott AFB, IL, as a tactical airlift director and later as an operations inspector for the MAC Inspector General. Following graduation from the Armed Forces Staff College, Col. Mims was assigned to joint staff duty at the Headquarters United States Pacific Command, Camp H.M. Smith, HI, as the political military officer for Australia and New Zealand. He returned to MAC as chief pilot, 773rd Tactical Airlift Squadron, Dyess AFB, TX, and then operations officer, 772nd Tactical Airlift Squadron. He the commanded the 772nd Tactical Airlift Squadron. Col. Mims was the Commandant of the Airlift Operation School when he was promoted to colonel on March 1, 1992.

Decorations include the Defense Meritorious Service Medal with one oak leaf cluster, the Meritorious Service Medal with two oak leaf cluster, the Air Force Commendation Medal and the Combat Readiness Medal with one oak leaf cluster. He is a command pilot with over 3,500 flying hours. Col. Mims and his wife, Diane, have one son, David.

TROY E. MINTON, born April 13, 1945, to Edith and Charlie Minton of Geneva, TX. After completing college, he joined the Air Force in 1968, and attended OTS at Medina AFB, TX, then UPT at Laughlin AFB, TX. Assigned to the 11 AREFSq at Altus AFB, OK, flying the KC-135/A, he served five Young Tiger tours in SEA, logging over 190 combat missions. Other units of assignment include the 305 ARW and the 931st ARG (Reserves) Grissom AFB, IN, and the 2nd BMW, Barksdale AFB, LA. Later in his career, while serving with the 9th SRW at Beale AFB, CA, he flew missions in support of Operation Just Cause and Desert Shield. His awards include the Distinguished Flying Cross, Meritorious Service Medal with oak leaf cluster, Air Medal with four oak leaf cluster, Air Force Commendation, Air Force Achievement, Air Force Outstanding Unit with Valor and SOLC, Combat Readiness Medal with oak leaf cluster, National Defense Service Medal with Bronze Star, Southwest Asia Service Medal, Armed Forces Expeditionary Medal with Bronze Star, Humanitarian Service Medal, Armed Forces Reserve Medal, Small Arms Expert with Bronze Star, Republic of Vietnam Gallantry Cross with Device, Republic of Vietnam Campaign Medal with two Bronze Stars, and others. He was married to the late Mary Ann "Sunny" Ryder for 14 years. Maj. Minton retired in September 1992, after almost 24 years of service, and lives in Albuquerque, NM.

GEORGE J. MITCHELL, born July 21, 1929, in Milwaukee, WI. He graduated Syracuse University, with a BS degree, NY. He joined the USAF on Jan. 7, 1951. Military history includes 1954-56, 816th TCS, Troop Carrier Sq., Ashiya, AB, Japan; 1956-61, 50th TCS, Sewart AFB, TN; 1961-63, 18th Troop Carrier Sq., Dyess AFB, TX; 1963-68, 7406 Support Sq., Rhein-Main AB, Germany; 1968-70, 774th TAC Airlift Sq., Clark AB, Philippines (Vietnam); 1970-71, 348 TAC Airlift Sq., Dyess AFB, TX. Command Pilot USAF, 8,000 plus, Civil Air Transport Pilot at Prating T-33; KB50; C-119; C-130A,B,D,E.

Memorable experiences include five years of recon. flying in Europe, 18 months, flying in Vietnam.

Awards/Medals: Distinguished Flying Cross and the Air Medal with nine oak leaf clusters, for service in Vietnam and Europe. He was discharged Feb. 1, 1971, with the rank of major.

He and his wife, Gwendolyn Joy, have three children and six grandchildren. As a civilian he was employed as a supervisor for L.A. Co. Sanitation Districts, 18 years. Retiring Feb. 1, 1971.

MICHAEL A. MOFFITT, born Sept. 13, 1946, in Oakland, CA. He graduated from Oakland Technical High School in 1964, and the U.S. Air Force Academy in 1968. He earned a master's degree in management from Central Michigan University in 1976. He completed the Armed Forces Staff College in 1981, and the Air War College in 1987.

Moffitt has accumulated over 3,700 flying hours, mostly in the C-130, with tours at Dyess AFB, Korat RT AFB, and Little Rock, AFB. He has also served five years in aircraft maintenance, working with fighters at Seymour-Johnson AFB and airlifters at RAF Mildenhall.

He commanded the 436th Airlift Wing at Dover AFB during Operations Desert Shield and Storm and piloted the first aircraft into Moscow during Operation Provide Hope.

Among his many awards and decorations are: Legion of Merit, Distinguished Flying Cross, Meritorious Service Medal with three oak leaf clusters, Air Medal with one oak leaf cluster, and the Air Force Commendation Medal.

Moffitt and his wife, Theo, have two daughters: Jennifer and Sarah.

WILLIAM G. MOORE JR., received his pilot rating and second lieutenant commission in the Army Air Corps in May 1941. He completed Air War College in 1957, and National War College in 1962.

Gen. Moore retired from the U.S. Air Force in 1979, as Commander-in-Chief, Military Airlift Command. He commanded a B-24 Squadron in Italy during WWII, a B-26 Combat Group in Korea during the Korean War, and the 834th Airlift Air Division in Vietnam. Other airlift assignments included command of the 314th Tactical Airlift Wing, the 839th Tactical Airlift Air Division, and 22nd Air Fore (MAC). While commanding the 314th Wing, he also directed "Close Look" a project designed to modernize and improve the capability of Tactical Airlift Forces.

He participated in the initial organization of the Airlift Association.

Decorations include the Defense Distinguished Service Medal, Air Force Distinguished Service Medal with one cluster, Legion of Merit with five clusters, Silver Star, Distinguished Flying Cross with two clusters, and Air Medal with nine clusters.

Currently, he is president of the Metropolitan Nashville Airport Authority.

Gen. Moore and Mrs. Moore (Marjorie) reside in Franklin, TN.

HAROLD C. MORGAN, General, born March 9, 1930, in Pittsburgh, PA. He attended the University of Pittsburgh for both undergraduate and graduate studies, obtaining his medical degree in 1955. He interned in McKeesport Hospital, McKeesport, PA and obtained his obstetrical/gynecological residency at Wilford Hall, Lackland AFB, TX. His assignment included Etain-Rouvres AB, France, Bitburg, Germany, Wilford Hall U.S. Air Force Hospital, March AFB, CA, and Evreux AFB, France.

He entered private practice in obstetrics and gynecology in 1966, and continued with the New Jersey National Guard, where he was hospital commander of the 108th attack fighter unit, state air surgeon, his last assignment being the National Guard Assistant to the MAC surgeon. As a civilian, Dr. Morgan continues in the private practice of obstetrics and gynecology in Bricktown, NJ, along with his son David Morgan. Two of his other sons are also physicians in the field. Gen. Morgan is married to the former Diane Halaburda of Mineville, NY. He is former president of the Ocean County Medical Society.

WILLIAM G. MORLEY, Army Air Corps, 1943-47, USAF, 1947-73, retired. He was born Stockton, CA, 1923. Graduated Georgetown University, Washington, DC, 1956. He served in the following aircraft: B-17, B-25, C-47, C-54, SA-16, C-119, C-130, U3-A, T-33.

After WWII various assignments in Air Transport Command., MATS, MAC, Air Rescue Service AFIT, USAFE (Greece), AFROTC, MAAG (Morocco), 322nd Air Div. (Normandy, FR), HQ, USAF, and 834th Air Div. (Vietnam).

Used low-altitude parachute ejection in pioneer effort with Moroccan C-119s for flood relief at Taroudant, 1962. Later a standard procedure LAPES.

Executive Administrator, Arnold Air Society and Angel Flight, 1973-1991. Retired to Emerald Bay CC. Currently, member of the board of trustees Arnold Air Society.

He married to Marian Elise (Franck) 1945. Four children and eight grandchildren. Mrs. Morley served as executive secretary to Arnold Air Society, 1976-91. Recognitions include: Air Force Commendation Medal, two; Meritorious Service Medal, two; Outstanding Unit Award (Leopoldsville, Belgian Congo, hostage evacuation 1967), Bronze Star Medal, Exceptional Service Award, 1991. U.S. Army, 8th Infantry Division Outstanding Performance Citation, 1966, Southern Arrow. Greece, 1st Airborne Battalion Medallion, Deep Furrow, 1965. Air Force ROTC, Outstanding Service Award, 1987, also a similar recognition to Marian Elise Morley. Air Force Association Presidents Citation, Airlift Association Distinguished Service Recognition as National Secretary 1978-82.

CHARLES W. MORRELL JR., born April 10, 1929, in Knoxville, TN. He has a BAAS. He joined the USAF Sept. 17, 1947, serving with the 305 and 306 Bm Wg.; 23rd Camron; 314th FMS; 64th FMS; 401st TFW; 354th TFW. He was stationed at Amarillo AFB, TX; MacDill AFB, FL; Bunker Hill AFB, IN; Stewart, TN.

Memorable experiences include winning SAC Bombing Competition, 1958, flying Mach 2, in a B-58, making master crew chief five times; flying missions over Laos and North Vietnam; Serving almost four years in Spain.

He married Margaret Ann Hatley of Braintree, Essex, England. They have two children and eight grandchildren. As a civilian he was employed at Northrop Acft., as instructor, Hallmark Institute of Tech.-Supv. of Aviation Training. He retired Oct. 1, 1973.

ERNEST L. MUELLER, born Feb. 2, 1935, in Dover, OH. He graduated from Sugarcreek-Shanesville, High School in Sugarcreek, OH, in 1953. He received a bachelors degree in dairy plant management from the Ohio State University in 1957. He received his master of arts degree in management from Webster University, St. Louis, MO, in 1977. He completed Air Command and Staff College in March 1972, and the Industrial College of the Armed Forces in July 1975.

Mueller retired from the U.S. Air Force in November 1981. During his last assignment, he served as deputy commander for Operations of the 89th Military Airlift Wing, Headquartered at Andrews AFB. During his tenure at Andrews, he organized and was elected the first president of the Capitol Chapter of the Airlifter Association. He is currently a life member of the Airlift Association.

After retirement, Col. Mueller was employed with Midway Airlines in Chicago and accumulated over 19,000 pilot hours of flying, 9,000 military and 10,000 commercial aircraft flight hours.

Among his many awards and decorations are Legion of Merit, Bronze Star, Meritorious Service Medal, Air Medal, Joint Service Commendation Medal, Vietnam Service Medal, Vietnam Armed Forces Staff Service and Vietnam Air Service Medal.

Mueller and his wife, Peggy, reside in Hinsdale, IL. They have two children: Ernest James and Elise Louise.

ROBERT M. MURDOCK, born Sept. 27, 1947, in Montclair, NJ. He received a bachelor of arts from Davis and Elkins College in 1969, and master of arts degree from Central Michigan University. His first operational assignment was to Ching Chuan Kang (CCK) Air Base, Taiwan, as a C-130 pilot. In December 1972, he reported to the 436th MAW, Dover AFB, as a C-5 pilot. In September 1977, Col. Murdock was assigned to the Air Force Military Personnel Center on an Air Staff Training Assignment. Following this tour, he was assigned to HQ MAC as C-17 Program Manager and executive officer to the C-17 Source Selection Team. Selected as the aide to the chief of staff of the Air Force, he subsequently attended the Armed Forces Staff College. In June 1985, he became commander of the 22 MAS, and then attended the U.S. Army War College. He was next assigned to U.S. European Command as the deputy inspector general. He then served as vice commander of the 436 AW, Dover AFB. Col. Murdock is currently assigned as the Senior Defense Fellow to the Atlantic Council, Washington, DC. He and his wife, Ann, have a daughter.

DENNIS L. MURPHY, born Feb. 2, 1947, in Washington, IA. He received a bachelor's degree in graphic arts from the School of Journalism, Drake University in Des Moines, IA, in 1969. He entered the U.S. Air Force at Reese AFB, TX, in August 1969. He received a master's degree in government and international politics in 1972, from Webster University in St. Louis, MO.

Promoted to the rank of colonel in August 1991, he has served as Air Force Director, Airlift Concepts and Requirements Agency at Scott AFB. At the Department of State as Military Director for Foreign Assistance Programming and Policy. At Headquarters USAF as program director for USAF Security Assistance programs for Australia, New Zealand, Dominican Republic, and other U.S. interests in the South Pacific and Caribbean and special projects officer for security assistance joint policy and legislative programs. His other experience includes assignments to Headquarters MAC on the MAC Command Briefing Team and as a JCS Exercise Airlift Scheduler and numerous wing level jobs in operations and maintenance. He performed flying duties as pilot and instructor pilot in the C-141, C-9, and CT-39 aircraft. In addition to assignments at Washington, DC, and Scott AFB, he served at Goose Bay, Newfoundland, Canada; Charleston AFB, SC, and Dover AFB, DE. Among his many awards are the Airman's Medal, Defense Meritorious Service Medal with one oak leaf cluster, Meritorious Service Medal with two oak leaf clusters, State Department Award for Heroism, Vietnam Service Medal, and the Vietnam Campaign Medal.

He is married to Marylyn Kimlin of Poughkeepsie, NY, and has two daughters Melissa and Kelly.

He holds Life Membership in the Daedalians; Air Force Association; and the Airlift Association, now Airlift Tanker Association; and has served the Airlift Association as national convention planner/manager more than 10 years.

BILLIE EUGENE MYERS, born Nov. 3, 1930, in Maynardville, TN. He served the U.S. Army, 378th Ord HMCO, 1951-53, Camp Irwin, CA; 1953-56, Air National Guard, 119th AC&W Sq., McGee Tyson AFB, Knoxville, TN; 1956-59, 1999th AACS SQ., Sewart AFB, TN; 1959-61, 6912th Security Service Sq., Tempelof AFB, Berlin, Germany; 1961-65, 401st Field Maintenance Sq., England AFB, LA; 1965-66, 612th TAC Fighter Sq., Misarva AFB, Japan; 1966-68, 314th Supply Sq., Sewart AFB, TN; 1968-70, 6109th Special Services Sq., Cegli AFB, Turkey; 1970-74, 437th Supply Sq., Charleston AFB, SC.

His memorable experiences include serving with the USAF Security Service in Berlin, Germany.

He was discharged Nov. 1, 1974, with the rank of E-6.

His wife died Aug. 31, 1993, they had a child, now deceased also and three grandchildren.

He was employed as a neuclar missile mechanic Polaris Missile Facility, Charleston, SC. He resigned in July 1983.

PHYLLIS J. NAGLE, born November 6, in Middletown, PA. Joined the USAF in 1961, serving as a flight nurse. Has 2,500 hours flying time with Air Evac; flew DC-9, C-141, C-130. Her memorable experiences include five missions returning POWs from Vietnam and many missions with injured. Retired in May 1981, with the rank of lieutenant colonel.

CHARLES W. NEELEY, Lieutenant Colonel, USAFR, born Dec. 15, 1949, in Bossier City, LA. He graduated from Mt. Tahoma High School in Tahoma, WA, in 1968, received a bachelor's degree in business administration from Washington State University in 1972, and an MBA in finance in 1985, from Golden Gate University. He completed Air Command and Staff College in 1982, and National Security Management in 1988.

Col. Neeley's career has included assignments to Dyess AFB, TX, and Kulis ANG Base, AK, where he flew C-130s;

and Travis AFB, CA, where he flew C-141s and C5s. His diverse airlift background has taken him to every continent and has included certification as chief of standardization/evaluation. Since 1983, he has served as an Individual Mobilization Augmentee (IMA) at McClellan AFB, CA, in the comptroller and logistics career fields. In 1991, he was nominated as IMA Field Grade Officer of the Year.

His decorations include the Meritorious Service Medal, the Joint Service Commendation Medal, and the Air Force Commendation Medal.

He and his wife, Candy, reside in Sacramento, CA. They have three sons: Mark, Phillip and Adam.

NELSON L. NEIL, born March 20, 1925, in Derry, PA. He graduated from Derry Township High School and attended the University of Pittsburgh for three years. He earned a bachelor of science degree from Southeastern Oklahoma College in 1956.

He entered the Army Aviation cadet program in August 1943, and graduated pilot in December 1944. He transferred to the U.S. Air Force in 1947, with a career focus of commanding flying units.

From February 1970 to September 1975, he served in five flying command assignments.

His most memorable command was his last combat assignment to AC 130 gunships in Southeast Asia from December 1972 to December 1973. His management of this sophisticated weapons system produced an unparalleled record of 1327 successive completed combat missions without a combat loss. For this effort he was awarded the Legion of Merit.

A command pilot with 8,000 flying hours he flew a combat tour in Korea and two tours in Southeast Asia. He retired in September 1975. Some additional awards and decorations include: Meritorious Service Medal with cluster, Air Medal, Air force Commendation Medal with cluster, Air Force Outstanding Unit Award with cluster. He and his wife, Jean, have four children: Scott, Michael, Patti Jo and Kirk.

WARNER E. NEWBY, Major General, born June 28, 1919, in Reedley, CA, commissioned in U.S. Army Air Corp. March 1942, following graduation from Aviation Cadet Program. Served in Middle East and North African campaigns with 12th Bomb Group (B-25S) and 26th Air Depot Group. In October 1943, flew a captured German JU 88 bomber from Cairo, Egypt to Wright Field, OH. Served rest of WWII with Flight Test Division at Wright Field as Chief of Maintenance, Chief of Bomber Flight Test Branch and Chief of Accelerated Service Test Branch.

Assigned Far East Air Forces 1951, and in 1954, to HQ U.S. Air Force as Chief Industrial Planning Division of the Directorate of Procurement and Production. Gained approval of the Aerospace Machine Tool Modernization Program which led to design and production of high powered automated machine tools needed to produce high performance aircraft. Was assigned Air Force Plant Representative, Boeing, Wichita, June 1958. Responsible for production of B-52Fs and development, test, production, and support of all B-52G's and B-52H's.

September 1962 to September 1969, served at HQ AFSC, DSA, and HQ USAF in Procurement, Production, and Programming staff assignments. Last assignment was with Joint Chiefs of Staff as Deputy J-4 (Logistics). September 1969 assigned as Command, Air Force Contract Management Division.

In June 1970, became C-5A Program Director. Renegotiated the contract for C-5A 81 aircraft program and abandoned the Disengagement Management Concept and instituted Engaged Program Office Management and a teamwork relationship with the contractor. Successfully completed design, test, production, and support of program with effective cost control. Managed the development of the Wing H Modification Program to achieve 30,000 hour plus service life by retrofit of the C-5A fleet and in the C-5B production program. In June 1973, following delivery of last C-5A production aircraft, he was transferred to Military Airlift Command as deputy chief of staff for Logistics.

In June 1975, assigned as Commander of Space and Missile Test Center at Vandenburg AFB, CA. Retired in February 1977, having completed 35 1/2 years of military service.

Gen. Newby was a command pilot with 4,500 flight hours. He attended school through junior college level at Reedley, CA, University of California, Santa Barbara; Harvard Advanced Management Program (48th AMP), and Air War College.

Military decorations and awards include: The Distinguished Service Medal, Legion of Merit with three oak leaf clusters, Meritorious Service Medal, Air Medal, Joint Service Commendation Medal, Air Force Commendation Medal, Army Commendation Medal, and Distinguish Citation.

Gen. Newby married Virginia Jeanne Rowe of Maders, CA, in June 1942, and they have four children: Howard, Candace, Stephen and Laura.

Current Status: Retired, Consulting, and Community Service.

THOMAS E. NEWTON JR., graduated from Polytechnic High School, Ft. Worth, TX, in May 1941. He attended evening classes at Texas Christian University until being transferred to West Texas with Western Electric Corp. He registered for the draft in Big Springs, TX, in February 1942, as he reached his 18th birthday. He tested for aviation cadet training and was sworn into the Army Air Corps in Dallas, TX, in August 1942. He reported to San Antonio Aviation Cadet Center in February 1943, progressed through pilot training with Class 43K and graduated as a second lieutenant pilot in December 1943. He trained in high altitude bombing in the B-24 Liberator and went to England in May 1944. He and his crew were selected from the aircrew replacement pool to go to Harrington, England and train in Black B-24s flying at night in support of the office of Strategic Services supplying the many underground organizations. This became his first involvement in the Airlift Mission. Flying over the continent dropping supplies and human organizers from an operational altitude of 400 feet. One tour was completed from England and the aircrew voted to go to Italy in December 1944 and completed a second tour in April 1945. He has subsequently served in the Korean Conflict, Vietnam and other contingencies.

He has flown several aircraft in his airlift career. The C-46, C-54, C-124, T-39 and the C-130.

He has served in Tactical Airlift at Squadron, Wing, Division, Numbered Air Force, Major Command and in Air Force Studies and Analysis at the Pentagon. Col. Newton and retired in September 1975 from duties as the airlift officer, J-4, CINCPAC, Camp Smith, Hawaii.

His 19 decorations include the Legion of Merit, Bronze Star with oak leaf cluster, Air Medal with five oak leaf clusters, and the Air Force Commendation Medal with two oak leaf clusters. He and his wife, Hoa, reside in Central Florida.

JOHN S. NOLEN, born May 29, 1932, in Houston County, TN. He has a master of art degree. Joined the U.S. Air Force in January 1954, serving with Wheelus, Tripoli, Libya 5th APO loaned to 1615 MATS. Military locations at which he served include Lackland AFB, Sewart, TN, Donaldson, SC.

He was separated in January 1956, with the rank of A/3C.

He and his wife, Joan McIntosh Nolen, have two sons and five grandchildren. As a civilian he was employed as manager quality control with Jack Daniel Distillery, retiring in June 1993.

ROBERT S. NORRIS, Lieutenant Colonel, USAF, (Ret.), born Jan. 18, 1949, in Jonesboro, AR. He graduated from high school there in 1967. Attending the University of Arkansas in Fayetteville, he completed a bachelor's degree in business administration in 1972. He received a master's degree in management from Webster University in 1978.

Lt. Col. Norris retired from the Air Force in July 1992. He received navigator wings at Mather AFB in August 1973, being named the recipient of the ATC Commander's Trophy for finishing first in his class. His entire operational career was in airlift. Initially at Dover AFB, he was selected as the first lieutenant to become a C-5 instructor at MAC's airlift training center at Altus AFB, progressing to wing stan/eval navigator. He became the command's Chief C-5 Navigator at Scott AFB in 1980. Overseas for five years from 1984, he was Chief, European ALCC at Ramstein, GE and later MAC detachment commander at Rota NAS, Spain. Completing his career at McGuire AFB, NJ, he was transportation squadron commander for three years, including Desert Storm, and later the wing executive officer. Decorations include the Meritorious Service Medal with four oak leaf clusters and the Air Force Outstanding Unit Award with six oak leaf clusters.

Norris and his wife, Ginny, reside in Moorestown, NJ with two children: Robert Jr. and John.

CORNELIUS NUGTEREN, Major General, USAF, Commander, Warner Robins Air Logistics Center, Robins AFB, GA. The center is one of the U.S. Air Force's five air logistics centers and is the state of Georgia's largest industrial complex. He is responsible for worldwide support of most transport aircraft, F-15 air superiority fighters, helicopters, air-to-air missiles, surface motor vehicles and high-technology airborne electronics.

Gen. Nugteren was born Feb. 7, 1928, in Colton, SC, and completed high school in Marion, SD. He received a bachelor of arts degree in education from Central College, Pella, IA, in 1951.

He enlisted in the Air Force in January 1952, and received his commission as a second lieutenant in March 1953, through the aviation cadet program. The general was awarded his pilot wings after completing advanced flying training at Williams AFB, AZ. In September 1953, he was assigned to the 36th Tactical Fighter Wing, Bitburg AB, West Germany, where he flew F-86s and F-100Cs. Gen. Nugteren returned to the United States in December 1956, and served with the 4th Tactical Fighter Wing at Seymour Johnson AFB, NC, as wing quality control officer. While there he flew F-86H's and F-100s. In June 1958, he transferred to the Air Force Academy as a flight test and maintenance officer.

In August 1960, the general was assigned to the 49th Tactical Fighter Wing at Spangdahlem AB, West Germany, as organization maintenance officer and as chief of field maintenance. In August 1964, he returned to the United State and was assigned to HQ 12th Air Force at Waco, TX, as a special project officer with the F-105 program. In May 1966, he transferred to McChord AFB, WA, as a C-141 pilot and flight test officer with the 62nd Military Airlift Wing.

From March 1968 to October 1970, Gen. Nugteren was assigned to Headquarters Military Airlift Command at Scott AFB, IL, where he served as an aircraft maintenance inspector and then as chief of the Materiel Branch, Inspection Division, Office of the Inspector General. He then joined the U.S. Air Force Advisory Group in the Republic of Vietnam as senior aircraft maintenance and materiel adviser.

The general became deputy commander of maintenance for the 50th Tactical Fighter Wing at Hahn AB, West Germany, in November 1971. In February 1974, he became wing commander of the 322nd Tactical Airlift Wing, Rhein-Main AB, Germany. In August 1975, Gen. Nugteren transferred to Ramstein AB, West Germany, as commander of the 86th Tactical Fighter Wing and served as the Air Force's first Kaiserslautern Area Community Commander.

In July 1977, he became vice commander of the Ogden Air Logistics Center, Hill AFB, UT. During this assignment he was awarded the missile badge and also received the Air Force award for commander action in the Equal Employment Opportunity Program.

Returning to Scott AFB, in September 1979, Gen. Nugteren took command of the Aerospace Rescue and Recovery Service. In this capacity he was responsible for worldwide search and rescue missions, the National Rescue Coordination Center, worldwide weather reconnaissance and Strategic Air Command missile site support.

In September 1981, the general was assigned as chief of the Joint U.S. Military Advisory Group in Greece. While there he served as the representative of the secretary of defense and the commander in chief, United States European Command, in the discharge of his assigned security assistance responsibilities. He also served as contract officer in Greece for the commander in chief, United States European Command, as the senior U.S. Military representative in Greece and as a member of the U.S. Country Team. He assumed his present command in September 1982.

The general is a command pilot with more than 4,000 flying hours. His military decorations and awards include the Distinguished Service Medal, Defense Superior Service Medal, Legion of Merit with one oak leaf cluster, Bronze Star Medal, Meritorious Service Medal, Air Force Commendation Medal, Air Force Outstanding Unit Award with V device and one oak leaf cluster, Air Force Organizational Excellence Award, Combat Readiness Medal, Good Conduct Medal, Army of Occupation Medal, National Defense Service Medal with one service star, Vietnam Service Medal with seven service stars, Air Force Overseas Ribbon-Short, Air Force Overseas Ribbon-Long with three oak leaf clusters, Air Force Longevity Service Award Ribbon with seven oak leaf clusters, Republic of Vietnam Air Service Medal Honor class, Republic of Vietnam Armed Forces Honor Medal 1st Class, Republic of Vietnam Gallantry Cross with Palm and Republic of Vietnam Campaign Medal.

Gen. Nugteren was inducted into the Air Force Logistics Command Order of the Sword by the enlisted men and women of the Warner Robins Air Logistics Center in May 1985. In September 1986, he was presented the Air Force Association Ira Eaker Fellow Award.

He was promoted to major general March 1, 1980, with date of rank July 1, 1976.

Gen. Nugteren is married to the former Liane Albrecht of Schwelm, Germany. They have two daughters, Cecile and Aneli.

DAVID J. ORTH, Enlisted Air Corps Oct. 14, 1942. Crew Chief Stockton Field's production line; maintenance BT-13-15-AT6-9-11-17, UC78 Avro Anson; requested CBI duty; maintenance, pre-flight Martin B-26 at ORD; Gen. A.E. Anderson 32 days to Bombay; train to Calcutta; train guard to Ledo (PX supplies and 500 lb. bombs); hitched 10 minute Combat Cargo C-46 flight to assignment. 1st Troop Carrier Sqdn. at Sookerting (ate water buffalo that tasted like swamp water smell!); 25 Hr. inspections every-other night; C-47-A flashlight maintenance dictated by Zeros, 30 minutes away that shot down three C-47s Christmas Day, 1944, over same drop target we had the day before.

Re-equipped with C-46Ds; drove a 6x6 over Ledo Road to Warazup (rained 14" in two hours one afternoon); jungle and beautiful timber swallowed men and planes alike. Shingbwiyang undershot washing out COs and my Queen of the Tiger Fleet; another #13 crashed at Calcutta, Navy Weathermen aboard.

Back to Dinjan. Transported gasoline and ammunition for B-29s and Chennault's Flying Tigers. An engine changed at Chanyi to Tokyo Rose's music.

Surrender - the thrill of waving back the V for Victory sign to cheering crowds of thousands showing their gratitude (with 14 and 24 course dinners) in Shanghai; Chihkiang and Hankow's unforgettable troop movements; flying over the Great Wall; kissing Chinese soil after the last flight and again the dock at Seattle. Discharged Dec. 30, 1945.

"I've never regretted enlisting Oct. 14, 1942, in the Air Force. Stationed at Stockton Field. Then TDY Yuma and requested duty in CBI. Six APO numbers still remind me of the odds of coming back clean, with no crews lost. Nothing was as rough as thinking about all those still over there. After trading the tools of war for agri-business, food production, taking up railroads, etc. Eloise gave me three sons before her death. Now there are eight grandchildren and one great-granddaughter. I'm now married to Darla and her German Shepherds and love them all. In 1982, with former ATC pilot Harry Howton, I watched the DC-8s wings dip for those still below in the middle of the hump. Another single, thrilling easy wing dip by a westbound 707 over this former C/C, aerial engineer as I looked up and saluted a former Troop Carrier pilot making the past seem almost like yesterday.

JAMES A. OSTREM, born in Lanesboro, MN. Enlisted Dec. 1, 1953, serving with the USAF SAC B-47 for three years and TAC C-130, C-123 for 15 years airlift. Military locations at which he served include MacDill, Pope, Langley, Vietnam, JTF LEO.

Memorable experiences include joint task force LEO.

Discharged April 1, 1974, with the rank of lieutenant colonel.

He and his wife, Ruth, have four children and four grandchildren. Today he is farming, raising cattle, corn and hay.

LYNN R. OVERDEER, Lieutenant Colonel, born July 11, 1945, Orange, NJ. Rank: lieutenant colonel; DOR: July 1, 1984; service: USAF; date/SRC COMSN: March 28, 1968/OTS; SEC CLRNC: TS; SPEC QUAL: JOPS; Aero rating: command pilot; total flying time: 3,300 hours.

Awards/Medals: Distinguished Flying Cross, Defense Meritorious Service Medal with two oak leaf clusters, Meritorious Service Medal, Air Medal with nine oak leaf clusters, Joint Service Commendation Medal, Air Force Commendation Medal, Joint Service Achievement Medal.

Civilian education: Georgetown University, national security studies, MA degree 1987; University of Arkansas, operations management, MS degree 1981; University of Colorado, psychology/sociology, BA degree 1967.

Military education: AIS Management for Intermediate Execs., Residence, 1988; Foreign Affairs Interdept Sem, Residence, 1986; National Security Management, Seminar, 1985; Air War College, Seminar, 1983; Air Command and Staff College, Correspondence, 1979; Squadron Officers School, Residence, 1974.

Assignments: December 1988 - August 1991, Joint Training Officer, USCENTCOM, MacDill AFB, FL. July 1988 - December 1988, Chief, JOPES Branch, USCENTCOM, MacDill AFB. September 1987 - June 1988, Operations Plans Officer, USTRANSCOM, MacDill AFB, FL. June 1984 - September 1987, SWA Planner, OJCS, Pentagon, Washington, DC. July 1981 - June 1984, War and Contingency Plans Officer, HQ MAC, Scott AFB, IL. May 1977 - July 1981, Ops Plans Officer, C-130E Pilot, Little Rock AFB, AR. October 1972 - May 1977, T-37B Asst. Sec. Cmdr., Flt. Cmdr., RSU Cont., SOF, Instructor Pilot, Reese AFB, TX, Laredo AFB, TX. August 1970 - October 1972, C-130E (AWADS) Instructor Pilot, Pope AFB, NC. August 1969 - August 1970, O-2A Forward Air Controller, DaNang AB, RVN.

His off-duty areas of interest include car racing, scuba diving, computer, skiing and reading.

ALBERT L. PAGANI, born Feb. 19, 1936, in Fairview, NJ. He entered the Naval Academy and transferred to the Air Force upon graduation in 1957. Navigator and bombardier training was followed by a tour in SAC B-47s. In 1962, pilot training preceded a tour in C-118s at McGuire AFB. That was followed as the CINC Alaska pilot. After a tour in the Tan San Nuit ALCC, he returned to Scott AFB on the Commander Communications Command crew. Al transferred to MAC Headquarters DOO staff. He served as Commander, 627th Airlift Support Squadron and the new 435th Tactical Airlift Group at RAF Mildenhall. He went to Rhein Main AB, Germany as the deputy base commander. He returned to Headquarters, MAC as chief of Airlift Management, followed by assignments as director of Technical Plans and Concepts and as director of Command and Control. He attended Squadron Officers School, Air War College and Armed Forces Industrial College. Among his Military Awards are: Legion of Merit, Bronze Star, Defense Meritorious Service Medal, Air Medal, Meritorious Service Medal with four oak leaf clusters, Air Force Commendation Medal with two oak leaf clusters.

Al retired in May 1985, and joined Lockheed in Sunnyvale, CA, where he held several jobs including program manager, manager Advanced Command and Control Systems and Advanced Program Development. Al is now director of Data Applications for PAR Government Systems Corporation.

Al is married to the former Beverly Good of Medford, Oregon. They have three children, Penni, Deborah and Michael. They reside in Fremont, CA.

JAMES J. PARGOE, born March 30, 1907, in Norfolk, VA. He has a BS in electrical engineering. He joined the USAF Feb. 25, 1942, serving with the ATC, Ferry Command, 1345 AFB Unit Westover Air Evac., Special Air Hassity Sq. Military locations at which he served include Long Beach, CA; Kermitola, India; Westover AFB; Washington, DC.

Memorable experiences include flying the South Atlantic, Fly Hump in C-109 and G54s, SAM flying generals, admirals, senators and congressman all over the world.

He was discharged Sept. 15, 1951, with the rank of major.

He and his wife, Martha, have three children and six grandchildren. As a civilian he was employed as police officer and electrical engineer, retiring Oct. 31, 1969.

RONALD PARSONS, Captain, USAF, started his military career as an airman in 1966. He served with the 60th Aerial Port SQ at Travis AFB in 1966 and 1967 as an air freight specialist. He was assigned to the 15th Aerial Port SQ based at Da Nang AFB, Republic of Vietnam from February 1968 to September 1969. He was part of the Air Mobility Team that supported the 3rd Marine Division at the Khe Sanh combat base during the siege. Many hot offloads of C-130s, 123s and CH-34 helicopters were done under fire. The fastest offload from touchdown to takeoff (including a three point turn at the end of the runway) was a minute and a half.

In 1981, he enlisted in the U.S. Army Reserves. He attended the Army Medic School and made SP5. Concurrently he attended Massasoit Community College and received an associates degree in nursing. In 1983 he was commissioned a second lieutenant in the Army Nurse Corps, U.S. Army Reserves and in 1987 received a BSN from the University of Massachusetts/Boston.

His decorations include the Air Force and Marine Presidential Unit Citations, Air Force Outstanding Unit Citation, Army Commendation Medal, Army Achievement Medal, Air Force Good Conduct Medal, Vietnam Service Medal, and the National Defense Service Medal.

Currently he is a coordinator of the Day Treatment Program at the Brockton VA Medical Center, Brockton, MA.

Capt. Parsons and his wife Darlene have two daughters, Heather and Ashley. They now reside in Middleboro, MA. He is a member of the Reserve Officer's Association and the Association of Military Surgeons of the U.S.

ROBERT B. PATTERSON, born Oct. 18, 1933, in Mebane, NC, and graduated from Chapel Hill High School in 1952, and the University of North Carolina in 1956. He earned a master's degree in business administration from Webster College and is a graduate of Columbia University's executive program in business administration. He completed Armed Forces Staff College in 1971 and the Air War College in 1976.

Patterson retired in October 1989. His last assignment was Commander, Military Airlift Command's 23rd Air Force and Commander, Air Force Special Operations Command, the Air Force component of United States Special Operations Command. Gen. Patterson enjoyed a long history of airlift command positions including: Commander, 50th Tactical Airlift Squadron, Commander 463rd Tactical Airlift Wing, Commander, 317th Tactical Airlift Wing, Commander, 21st Air Force, and commander, 322nd Airlift Division. He was the senior Air Force officer on Grenada during the student rescue operations, Urgent Fury.

He retired in the grade of major general, a command pilot with over 9,000 flying hours in 52 types and models of aircraft. Among his many awards and decorations are: Distinguished Service Medal, Defense Superior Service Medal, Legion of Merit, Distinguished Flying Cross, Bronze Star Medal, Defense Meritorious Service Medal, Air Medals, and Republic of Vietnam Gallantry Cross with Palm.

Patterson and his wife, Sandy, reside in Peachtree City, GA. They have three children, Robert, Michael and Kristin.

BERYL D. PECK, born Feb. 21, 1924, in Britt, IA. Raised on a farm and after grade school employment was with area farmers. When WWII began, most farm labor was deferred. Since draft boards wouldn't draft critical farm labor his three brothers and he were deferred until after the European War ended.

Being drafted on Sept. 10, 1945, he entered infantry training at Ft. McClellan, Aniston, AL. After 17 weeks of training, six more weeks of advanced training near Atlanta, GA, a voluntary transition to Army Air Force was accepted, with assignment Europe.

A seven day travel to Le Harne, France, by a converted troopship German luxury liner Wensback, then a four day train ride to Dachau, Germany. (the train ride from Le Harne, France to Dachau, Germany was in a 40 man or eight horse box car, there is one on static display between Waverly and Cedar Falls, IA.) Vehicle transportation and maintenance were duties at Ansbach AB, Erding Air Deport and Oberweasenfelt Air Depot Southern Germany.

Bases during the Berlin Airlift were Oberweasenfelt, Rhein-Main and Oberphenhoffen.

Arriving back into U.S. midsummer 1949, he was transferred to Japan, December 1949, as a search radar supervisor, with two years on islands of Japan, Korea and China. He returned for one year duties at Highlands, NJ. Being discharge from active service in 1953, he finished his military career with Air Reserves and Guard. Other state side base were at Greenboro and Goldboro, Carolina, McGuire AFB and Salina Air Force Base, KS. Major commands were 3rd Army, 8th Air Force, 5th Air Force, 12th and 15th Air Force.

Memorable experiences include tracking a B26 Bomber from Korea, belly land and rescue pilot in bay 1/8 mile from our radar installation on island of Fukishima located between Japan and China. A staff sergeant sank with the plane.

After 49 years he doesn't recall everything. He retired from Military at 133rd ACWS Ft. Dodge, IA, in 1970.

He is a widower with one child. As a civilian he was employed as a John Deere Farm Implement technician-supervisor. Retired 1953-1985, now semi retired, home children daycare.

ROBERT C. PECK, born July 27, 1940, in Miles City, MT. He graduated from high school in 1958. He received a bachelor's degree in business administration from Montana State University in 1958, and a master's degree in counseling from Troy State University in 1976. He completed Air Command and Staff College in 1974, and the Army War College in 1984.

Peck retired from the United States Air Force in January 1988. During his career he completed two tours in Vietnam, one in C-130s and one as a FAC. His background includes being rated as an Air Traffic Controller, Communications Squadron Commander, C-130 Squadron Commander, Military Assistance Group Commander, and C-130 Training Group Director of Operations. He is presently the training manager for CAE Link at the C-130 School in Little Rock.

Among his awards and decorations are the Silver Star, Legion of Merit, Defense Distinguished Service Medal, Distinguished Flying Cross, and the Air Medal with 23 oak leaf clusters.

Peck and his wife, Mary reside in Jacksonville, AR. They have two children, Mary Beth and Carol.

DAVID M. PELLETIER, born Oct. 15, 1937, in Detroit, MI, and attended schools in St. Louis and graduated from Detroit's Redford High School. He enlisted in the U.S. Navy in 1954, and after his sea duty enlisted in the USAF. He attended the USAF Senior NCO Academy and Southern Illinois University and retired as a chief master sergeant from the USAF Airlift Center in March 1978. His decorations include the Air Medal with oak leaf cluster, Meritorious Service Medal with oak leaf cluster, and the USAF and Army Commendation Medals.

Upon retirement, he joined Brooks and Perkins and remains with them in their Military Marketing Group. In 1978, he was appointed to an Ad Hoc committee to reorganize the Airlift Association and in 1979 was elected vice president and from 1983 to 1985 served as the president of the Association.

Chief Pelletier and his wife, Evelyn reside in Livonia, MI, in close proximity to their three sons, Mike, John and Bob while enjoying their three grandchildren Michael Jr., Danielle and Lindsey.

LEONARD WILLIAM PENTON, "BILL", born Aug. 31, 1938, in Baton Rouge, LA. Education includes BSME 1965;, MSIE 1972, Oklahoma State University; MBA 1976, University of Dayton. He joined the USAF on Sept. 21, 1955, serving with USAFSS, 1956-62; 776 TAS, 1966-68; 62 TAS 1968-71; AFSC 1973-77. Military locations include Japan/Germany/Turkey; CCKAB, Taiwan; Sewart, Little Rock AFB; WPAFB.

Memorable experiences include basic training 1955, Officer Training School 1965, Undergraduate Navigator Training 1966; Squadron Officer School 1971; too many good times to list.

He was discharged March 1, 1977, with the rank of major.

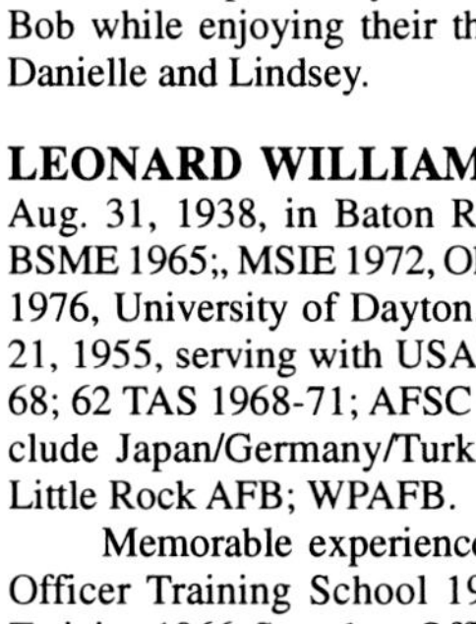

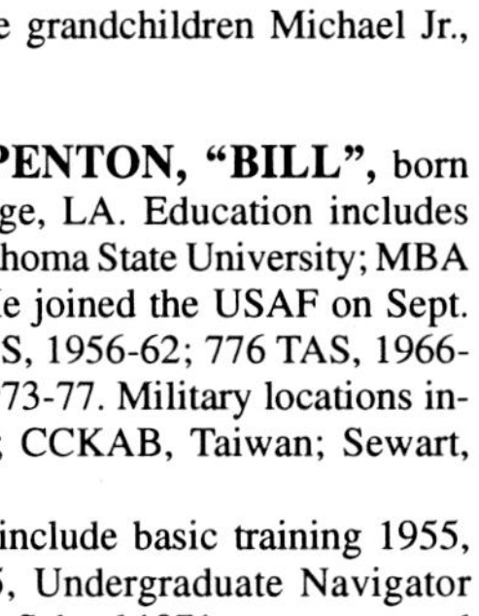

He and his wife, Christel Doerfler, have four children and eight grandchildren. As an civilian he is employed with General Dynamics; LTV Aerospace/Loral Vought Systems as senior staff engineer.

VINCENT R. PERRIN JR., born July 4, 1924, in New Orleans, LA. He received his BS from Tulane University. He joined the USAF on Dec. 10, 1942, serving with the 11th Combat Cargo Sqdn., 3rd Combat Car. Gp., 332 Troop Carr. Sqdn, etc. Military locations at which he served include Dindan, India; Yunanyi, Lulliang, Shanghai - China; and the U.S.

Memorable experiences include flying C-47 over Hump (Burma to China) load Chinese Troop, on instruments still climbing to altitude, lost engine gave crew bailout choice, made emergency landing, Paoshan, China. Plane (1,300 hours on Eng.) junked.

Left active duty Jan. 27, 1946. He retired from the USAFR April 15, 1972, with the rank of lieutenant colonel.

He and his wife, Joyce Menant Perrin, are separated and have six children and 11 grandchildren. Civilian employment includes consulting geologist, prior to consulting Texaco & Freeport Sulphur.

HARVEY E. PETERSON, born Nov. 6, 1943, in Grand Forks, ND. He graduated from Northwest School of Agriculture in Crookston, MN, in 1961. He received a bachelor of science in education degree from the University of North Dakota in 1966. He completed Squadron Officer School in 1975, Air Command and Staff College in 1982, and received a master of aviation management degree from Embry-Riddle Aeronautical University in 1986.

He went to Undergraduate Pilot Training at Williams AFB, AZ, where he received his pilot wings in June 1967. He served as an instructor pilot at Vance AFB, OK, and C-130 Aircraft Commander at CCK AB, Taiwan. He served as an instructor pilot and field training detachment commander at Randolph AFB, TX, and C-130 instructor pilot at Pope AFB, NC. He retired in July 1987. After retirement, he served as a C-130 flight simulator instructor pilot until January 1991, when he went to work for Chrysler Technologies Airborne Systems as a C-27 Test and Instructor Pilot.

His decorations include the Air Medal with two oak leaf clusters, Meritorious Service Medal with one oak leaf cluster, and the Air Force Commendation Medal with one oak leaf cluster.

He and his wife, Georgia, reside in Woodway, TX.

ROBERT D. PHILLIPS, Colonel, USAF, born on Jan. 18, 1946, in Junction City, KS. He received his bachelor of science degree and was commissioned a second lieutenant from the U.S. Air Force Academy in 1969. He graduated from flight school at Laughlin AFB, TX, in 1970, and transitioned into the C-141 at Altus AFB, OK. His first operational assignment was with the 14 MAS at Norton AFB, CA, where he flew missions in and out of Vietnam and Southwest Asia for three years. Col. Phillips was selected for a career broadening assignment in 1974, to Titan II missiles at McDonnell AFB, KS. During this assignment Col. Philips completed a master's degree in political science at Wichita State University. From October 1977 until May 1981, Col. Phillips served as an aircraft commander and C-5 instructor pilot with the 9 MAS and wing plans officer with the 436 Military Airlift Wing at Dover. Col. Phillips then had a remote tour at Osan AB, Korea where he served as an Airlift Coordination Center Duty Controller until May 1982. From June 1982 until June 1985, Col. Phillips served as a planning and readiness analyst, Operations Plans, HQ MAC,

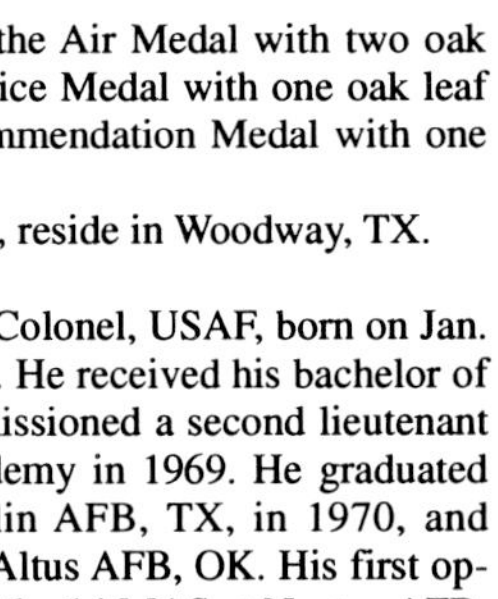

Scott AFB, IL. Col. Phillips returned to C-5 flying duties at Dover AFB, DE, in June 1985, as the assistant operation officer of the 3 MAS. After one year in this position, he was named commander of the 436 Field Maintenance Squadron at Dover responsible for the welfare and morale of that 1,000 member unit as well as the production of specialist maintenance on 36 assigned C-5s. After his commander tour, Col. Phillips was competitively selected in 1988 to be an Air Force Research Associate at the American Enterprise Institute (AEI), in Washington, DC, where he served as the Air Force Fellow at the institution. Col. Phillips was assigned to the Joint Staff in J-5 as a political military planner in the Persian Gulf branch in June 1989. He was the desk officer for Saudi Arabia prior to and during Desert Shield and Desert Storm. He served as part of the SECDEF directed U.S.-Saudi Arabian Joint Security Review in Saudi Arabia and upon his return, Col. Phillips became Chief, Middle East Branch, Deputy Director for Political-Military Affairs, J-5, Joint Staff in June 1991. In July 1992, Col. Phillips was reassigned to the Defense Intelligence Agency and became special assistant for Joint Matters for the director of Systems Management.

In June 1993, Col. Phillips assumed assignment as the vice commander of the 60 AW at Travis AFB, CA.

His professional military education includes: Squadron Officers School, 1974; Air Command and Staff College, 1977; National Security Management, 1982; Air War College, 1985; and Air Force Research Associate, 1989.

Col. Phillips is a command pilot with over 3,400 flying hours primarily in the C-141 and C-5.

His personal decorations include: the Defense Meritorious Service Medal with one oak leaf cluster, the Meritorious Service Medal with one oak leaf cluster, the Air Medal with one oak leaf cluster, the Joint Service Commendation Medal with one oak leaf cluster, the Air Force Commendation Medal with two oak leaf clusters, and the Combat readiness Medal.

Col. Phillips and his wife, the former Lynn Whitaker of Kansas City, have four children. Daughters Cosette and Meredith and sons, David and Brett.

WALTER W. PINE, graduated from Chino High School, Chino, CA, in 1941. He received a bachelor's degree from San Francisco State University in 1958, graduated from the, U.S. Army Command and General Staff College in 1963, and Industrial College of the Armed Forces in 1970.

He retired from the U.S. Air Force January 1979, with the rank of colonel. His last 20 years of a 36 year military career he was involved with the C-130 aircraft. He retired from the Screen Actors Guild in 1990, as a motion picture stunt pilot and continues to fly the B-25 camera plane. His most recent films include Iron Eagle 3, Aces and Cliff Hanger.

Among his many awards and decorations are: Silver Star, Legion of Merit, Distinguished Flying Cross, Bronze Star, Meritorious Service Medal, Air Medal with 11 oak leaf clusters and the Air Force Commendation Medal.

Pine and his wife, Marilyn, live in Keller, TX. They have three daughters, six grandsons and three granddaughters.

LEE R. PITZER, graduated from the University of Missouri in 1955, and entered navigator training. His operational assignments as a navigator were on Okinawa and at Charleston AFB, SC. In 1962, he completed the Army parachute course and became OIC of a Combat Control Team (CCT) at McChord AFB, WA. He was CCT and Aerial Delivery (AD) Staff Officer with Headquarters MAC and 2nd Aerial Port Gp.; RVN, before attending the Army's Command and Staff Course at Ft. Leavenworth, KS. Upon graduation he was assigned as OIC CCT/AD with the 69th MAW Travis AFB, CA. In 1974, he became commander of the 316th Aerial Port Squadron, Langley AFB, VA, with a follow on posting as Commander 3rd Mobile Aerial Port Squadron, Little Rock AFB, AR. In 1978, he was Commander, 374th Aerial Port Sq., Clark AFB and in late 1979, was reassigned as Deputy Chief of Staff, Air Transportation, HQ MAC, Scott AFB, IL.

Col. Pitzer retired from the Air Force in April 1982. For the past eight years, he has been with Merrill Lynch in St. Louis, MO, and was recently promoted to the position of assistant vice president. Col. Pitzer and his wife, Peggy, have three grown children and raise Thoroughbred race horses near O'Fallon, IL.

RICHARD F. POE, born Oct. 1, 1927, Burlington, NC. Drafted into Army, Ft. Bragg, NC, in 1946, as M.P. then went to Camp Killmer, NJ. In 1947, Yokohoma Japan during WWII occupation. In 1949, he enlisted in USAF, Pope Air Force, NC, as Air Police, Ottis AFB, MA. In 1951, Floyd Bennett NAS, Brooklyn, NY, Sandi AFB, NM; Langley AFB, VA; Schulthrope AFB, England. In 1954, Sewart AFB, Smyrna, TN; 1962, Bitburg, Germany; 1964, North Bend Air Station, OR; 1965, Lackland AFB, Lackland, TX; 1971, Pope USAF, NC. In 1974, October 1, retired from Pope AFB. In 1987, October 1, retired from Metropolitan Nashville Airport, Nashville, TN. In 1994, he lives three miles from Old Smyrna Sewart AFB, off Weakley Lane, which runs behind base.

Memorable experiences include met many important people, new friends seen lots of beautiful country, met his wife while at Sewart AFB.

He and his wife, Florella Si Poe, have three children and eight grandchildren. As a civilian he was employed as a airport policeman, retiring as Oct. 1, 1992.

RONALD G. PORTER, Colonel, born Nov. 5, 1945, in St. Joseph, MO. Raised near Viola, KS, his education includes a bachelor of science degree in business administration from Kansas State College of Pittsburg, KS, in 1968, and a master's degree in human relations from Webster College in St. Louis, MO, in 1982. He completed Squadron Officer School in 1974, Air Command and Staff College in 1982, Air War College in 1985, and the Industrial College of the Armed Forces in 1987.

Commissioned from OTS, Col. Porter completed pilot training at Williams AFB, AZ, in 1970 and was assigned to McChord AFB, WA, flying C-141 and T-39 aircraft. In July 1974, the Col. Porter moved to Anderson AFB, Guam for two years as Group Headquarters Squadron Section Commander and flew the C-118 aircraft. Follow-on tour was to the 9 MAS, Dover DE, where he became a C-5 Wing Standardization Flight Examiner. In September 1979, he was assigned to HQ MAC, Scott AFB, IL, as special assistant to CINCMAC on aircrew matters and later as Chief, Special Projects Branch in DCS Operations. After attending ACSC, he was assigned to the 3 MAS, Dover AFB, DE, as assistant operations officer and in 1984 became the field maintenance squadron commander. In July 1986, he attended ICAF was then assigned to the Air Staff where service in DCS Plans and Operations, included Division Chief in the Operations Resources Division. Reassigned to the 443 MAW, Altus AFB, OK, Col. Porter was deputy commander of Operations from September 1989, until assignment to McGuire AFB, NJ. Col. Porter is the Chief, Operations Division, at Headquarters 21st Air Force, McGuire AFB.

His date of rank is July 1, 1988. He is a command pilot with more than 3,600 flying hours in C-141, T-39, C-118 and C-5 aircraft. Decorations and awards include the Defense Meritorious Service Medal, w/ oak leaf clusters, Air Medal, and Air Force Commendation Medal. Col. Porter and his wife, Norma, have two children, Heather and Jason.

EDWARD R. PRINCE, born on Aug. 15, 1922, in Sherwood, TN. He joined the USAF on Sept. 24, 1940. Military locations at which he served included: Maxwell Field, AL; Orlando AB, FL; Chanute Field, IL; Edin Field, FL; Rio Hatio AB, Panama; Albrook Field, Canal Zone; Rhein Main AFB, Germany; Bergstrom Field, TX; Sewart AFB, TN; Ashyia AFB, Japan; Sandustrom AFB, Greenland; McMurdo Sound Navel Air Station, Antarctic; Edwards AFB, CA.

Units he served with include: 23rd Composite Group and the Air Proving Ground Command at Eglin Field, FL, during WWII and with the 97th Bomb Sq. and 20th Troop Carrier Sq. in Panama and Germany from 1946 though 1948. Served with the 314th Troop in Japan during the Korean War. Served briefly at Edwards AFB, CA, after Korean War, and about two year at Eglin AFB, FL, and transferred to Sewart AFB, TN, from which he retired in 1960.

Memorable experiences would be being on and flying during the Berlin Airlift. Being with the 61st Troop Carrier Sq. in Greenland during 1959, and 1960, when they were hauling building material and supplies to the ice cap to build two early warning radar stations. We also went to the Antarctica to have equipment to the South Pole and Marie Byrd Land Stations. He visited the South Pole Stations which one could never forget.

He was discharged Oct. 17, 1970, with the rank of chief master sergeant.

He and his wife, Pollyanna Long, were married on Oct. 22, 1949, and have four children and five grandchildren. Civilian employment includes AEDC, outside machinist, retiring June 30, 1960.

FREDRICK L. PUMROY, born Nov. 27, 1942, in Hammond, IN. Joined the USAF July 11, 1964, serving as a pilot. Pilot training at Del Rio, TX, 1964; flew KC-135s at K.I. Sawyer AFB, MI; flew OV-10 as FAC out of NKP Thailand, 1970; flew VC-135s, 1971-74 as command crew pilot for Adm. McCain, Generals Clay, Voight and Wilson. In 1974, was stationed at Wright Patterson AFB as a project officer for the AFLC commander-in-chief which was initial cadre for establishing the Air Force Acquisition Logistics Division. In 1976, he attended Defense Systems Management College then was assigned to the B-1A program as the chief of the Deployment Division and the Ground Launched Cruise Missile program as director of Logistics. After graduating from Command and Staff College and the Air Force Comptrollers School was assigned to Grissom AFB, IN, as operations officer for RT/EC/KC -135s. Returned to Wright Patterson AFB, OH, in 1982, as program manager for support equipment, technical data and training equipment for the B-1B program. After retiring March 31, 1985, started own business as consultant and program manager for aerospace companies.

Memorable experiences include participation in Son Tay Raid while flying OV-10s as a forward air controller. Participated in Iranian hostage raid as tanker operations commander.

Awards/ Medals: three Distinguished Flying Cross, 15 Air Medals, two Meritorious Service Medals, Joint Service Commendation Medal, Air Force Commendation Medal, plus 13 other types of medals.

He was discharged March 31, 1985, with the rank of lieutenant colonel.

He and his wife, Mary Anne, have three children, Richard, Leanne and Julia. Today he is a consultant and program manager for four aerospace companies.

ROBIN S. PURDIE, Colonel, USAF, (Ret.), born Feb. 14, 1940, in Cambridge, England. He received his commission through Air Force ROTC at San Diego State College in 1961. After attending pilot training at Williams AFB, AZ, he was assigned to the 75th Air Transport Squadron, Travis AFB, flying the C-124. In 1965, he transitioned to the C-141 performing duties as an instructor and flight examiner.

After being assigned to Altus AFB, OK, in 1969, as a C-141 Technical Training Unit Instructor, he transitioned to the C-5, also as an instructor and flight examiner.

In 1972, Col. Purdie was assigned to the Air Command and Staff College, Maxwell AFB, AL. At that time he also obtained a master's degree in political science from Auburn University.

Col. Purdie was then assigned to Nakhon Phanom RTAFB, Thailand, where he flew EC-47 aircraft. Upon completion of that tour, he returned to Maxwell AFB, to the faculty of the Air Command and Staff College.

In 1977, he was reassigned to Dover AFB, DE, where his duties included Operations Officer, 3rd Military Airlift Squadron; Commander, 9th Military Airlift Squadron; Assistant Deputy Commander and Deputy Commander, Operations, 436th Military Airlift Wing.

In 1985, Col. Purdie was assigned to HQ MAC as Director, Aircrew Standardization and Evaluation. In 1989, he was assigned to 22nd Air Force, Travis AFB, CA, as Deputy Chief of Staff, Operations.

Col. Purdie is a command pilot with over 10,000 flying hours. He flew 128 combat and combat support missions in South East and South West Asia. His decorations include the Legion of Merit with one oak leaf cluster, the Meritorious Service Medal with two oak leaf clusters, the Air Medal with two oak leaf clusters, and the Air Force Commendation Medal. He is a graduate of Squadron Officer School, a distinguished graduate of the Air Command and Staff College, and was designated an outstanding graduate of the Air War College Seminar Program.

He is married to the former Marian Abbott of La Jolla, CA, and he has three sons, Scott, Michael and Jeffrey.

Col. and Mrs. Purdie retired to Cameron Park, CA, in November 1991, after over 30 years service in the U.S. Air Force.

FRANK JAMES RAEDER, born Sept. 11, 1930, in Chicago, IL. He attended USC, UCLA, Cal-Tech Undergraduate in math, physics and engineering. He joined Jan. 21, 1948, USAF, HQ and HQ Sqdn., 62nd Troop Carrier Group, later re-designated 62nd Troop Carrier Wing, McChord AFB, Washington.

Memorable experiences include at Cold Bay, weather was so cold, it was decided to hangar the Globemaster II overnight. The next morning, the aircraft was towed to the flight line and parked, while doing preflight check, a noise like a cannon shot paralyzed everyone and the wing tips suddenly dropped about five feet. The sudden change in temperature had crushed the main wing attach bolts to fail. They had to wait over a week for Douglas to provide newly designed bolts. After the fix, this was never a problem.

Later, this same aircraft was on joint maneuvers with the Army in upstate New York when a light tank was driven aboard. The loadmaster had neglected to place a TML-strut in place and the aircraft was broken in half as the tail hit the ground. There was no safety interlock switches to prevent this from happening.

His most memorable experiences was working on a daily bases with then Col. George Brown, as a cryptographic technician. He later became chairman of the joint chiefs for two terms.

He retired April 30, 1952, with the rank of staff sergeant.

His civilian employment was North Minimum Aviation on first production F-100s while attending the University of South California, full time. Later employed as test engineer, design engineer, development engineer and program manager and consultant in numerous aerospace firms. Developed cabin pressurized components for projects Mercury and Apollo. Specialized fields of knowledge, instrumentation and control systems and components. He also worked Cal-Tech JPL Instrument Development Lab.

He and his wife, Anges T. White, have two children.

THOMAS W. RASH, "WOODY" Colonel, USAF, (Ret.), born Feb. 9, 1918, in Franklin, TN. He entered service with U.S. Army April 4, 1941. Commissioned second lieutenant from Aviation Cadet Program June 26, 1943. Began military flying career with Eastern Airlines, Atlanta, GA, then to Great Falls, Mt. Gore Field to ferry planes to Alaska for Russia, followed by C-109s over HUMP from Shamshernager, India. Left service in 1946, and flew with Tennessee National Guard until recall for Korean Conflict. Base operations officer at McGhee Tyson, TN, followed by assignments in the MATS/MAC command control system at Guam, Hickam, Andrews, Scott, Travis and Wiesbaden. Chief Airlift Operations 315th Air Division Tachikawa, Japan at beginning of Vietnam War; followed by duty at HQ MAC, 22nd Air Force, ADO 21st Air Force during Israeli War followed by DO 22nd Air Force during evacuation of Vietnam 1975. Retired Sept. 30, 1977. Flew 28 different aircraft during 36 years service with USAF and Tennessee National Guard.

Awards/Medals: Legion of Merit (three) and Air Medal (one).

RAY D. REAVES, born Aug. 6, 1935, in Jacksonville, AL. He joined the USAF March 9, 1961, serving as a command pilot. Military locations he served were Laredo, 1961-1962, pilot training; Abilene, TX, 1962-64, 1970-72; Alaska, 1964-68; Vietnam, 1968-69; Readiness Command MacDill, FL, 1972-74; CCK, Taiwan, 1975-82; Kunsan, Korea, 1983-84; Little Rock AFB, AR, 1982-83; Tinker AFB, OK, 1984-1989. In 1989 he retired retired from Tinker AFB. He participated in TET offensive in South Vietnam, 1968-69.

Memorable experiences included landing C-130 with broken nose strut; and flying ski equipped C-130s off Greenland Ice Cap, Jordanian Aircraft 1970.

Awards/Medals: Legion of Merit, Distinguished Service Medal, Meritorious Service Medal, Air Medals and JSM.

He was discharged in June 1989 with the rank of colonel.

He and his wife, Annette, have a daughter Tanya A. Walker and a son, Ronald, and three grandchildren: Ross, Hanah and Josh. Today he is manager of environmental engineering on staff of county commissioners, Oklahoma County, OK.

ROBERT A. REICHERT, born and completed high school in Napoleon, OH. He graduated from Otterbein College and received ROTC commission in 1960. He completed SOS, ACSC, ICAF and Air War College.

Reichert retired in 1990, after 30 years service, as the TRANSCOM/MAC liaison to USCENTCOM. He also served as deputy inspector general, MAC; Commander, 41st Rescue and Weather Reconnaissance Wing; Vice Commander, 39th Aerospace Rescue and Recovery Wing; Director, War and Mobilizations Plans, HQ USAF; Commander, 436th OMS; Commander 436th AMS - selected best maintenance squadron in USAF; Chief, 436th MAW Command Post; Flying Training Officer; Simulator Instructor, Flight Examiner Pilot, C-5A; Rescue Pilot, HH-3E, Jolly Green Giant; Instructor Pilot, C-124.

A command pilot with over 6,000 hours, decorations include the Legion of Merit with two oak leaf clusters, Distinguished Flying Cross and the Air Medal with six oak leaf clusters.

Bob and Gwen resides in Lakeland, FL.

KEITH T. REILING, Brigadier General. The general entered the active-duty Air Force in November 1965, through Officer Training School, Lackland AFB, TX, and earned his pilot wings at Williams AFB, AZ. He is a command pilot with more than 7,600 flying hours and a Vietnam veteran with more than 1,200 combat flying hours He left active duty in 1976 and became an air reserve technician, a full-time civil service employee and a member of the Air Force Reserve. He has been a flight examiner pilot, squadron operations officer and assistant deputy commander for operations. He was deputy commander for operations of the first Air Force Reserve wing to convert to the C-5A Galaxy strategic airlift aircraft, and was a wing commander before his present assignment as commander of the 349th Airlift Wing, an Air Force Reserve associate unit located at Travis AFB, CA. He also served as assistant test director of the Military Airlift Command's operational test and evaluation team for the installation of a triple inertial navigation system in the C-5A Galaxy.

General Reiling and his wife, Hallie, are the parents of three sons.

Education: 1965, BA degree in general science, Oregon State University; 1976, Squadron Officer School, by correspondence; 1982, Air Command and Staff College, by seminar.

Awards/Medals: Legion of Merit, Distinguished Flying Cross with oak leaf cluster, Meritorious Service Medal with oak leaf cluster, Air Medal with six oak leaf clusters, Air Force Commendation Medal with two oak leaf clusters, Presidential Unit Citation, Air Force Outstanding Unit Award with V device and two oak leaf clusters, Combat Readiness Medal with six oak leaf clusters, National Defense Service Medal with service stars, Southwest Asia Service Meal with two oak leaf clusters, Humanitarian Service Medal, Air Force Overseas Ribbon - Short with oak leaf cluster, Air Force Longevity Service Award Ribbon with five oak leaf clusters, Armed Forces Reserve Medal, Small Arms Expert Marksmanship Ribbon, Air Force Training Ribbon, Republic of Vietnam Gallantry Cross with Palm, Republic of Vietnam Campaign Medal, Kuwait Liberation Medal.

The general is a life-time member of the Reserve Officers Association, Air Force Association, Order of the Daedalians and Federal Managers' Association. He also is a member of the Oregon State Alumni Association, and St. Mary's Catholic Church, Vacaville, CA.

ROBERT N. RHODES, Lieutenant Colonel, born Sept. 6, 1949, graduated from Harrison High School, Evansville, IN, in 1967, the University of Evansville in 1972, with a BA in journalism and speech, and the University of Southern California in 1982, with a MS in systems management.

A ROTC distinguished graduate, Colonel Rhodes' first assignments after earning his wings were the 53rd Weather Reconnaissance Squadron, the Hurricane Hunters; the 3550th Air Force Recruiting Squadron; and the 21st Tactical Airlift Squadron.

In 1983, Col. Rhodes moved to Headquarters, MAC, Regional Contingency Plans Division. There he edited and published six of the nine major war plans then tasked, including rewriting the largest computer data base in existence. In 1986, he transferred to the Air National Guard Bureau, Pentagon, becoming one of the first two Long Range Planners in the Air National Guard's history. As Chief, Combat Operations Division, 322 Airlift Division he served as the Crisis Action Team (CAT) executive agent during the critical early days of Operations Desert Shield and Desert Storm. In April 1991, he became the operations officer of the 58th MAS, assuming command in January 1992. He lead the 58 Airlift Squadron (ALS) during the 1992 reorganization and transition to USAFE.

Selected for promotion to colonel in October 1992, he is a command pilot with over 2,200 hours. His military awards and decorations include the Meritorious Service Medal with three oak leaf clusters, the Air Medal, the Joint Commendation Medal, and the Air Force Commendation Medal. He has one son, Matthew.

CHARLES Z. RIDGWAY JR., born Jan. 23, 1938, in El Paso, TX. He joined the USAF in August 1960, serving with the Military Airlift Command, Air Staff and Joint Chiefs of Staff, Pentagon. Military locations he served include worldwide, Japan to Saudi Arabia.

His memorable experiences include serving as commander at squadron and wing levels.

Awards/Medals: Legion of Merit with oak leaf cluster, Meritorious Service Medal and Air Medal.

He was discharged in August 1989, with the rank of colonel.

He resides with his wife, Lynn, in Tacoma, WA. Today he is training site manager for Hughes Technical Services Company.

JACK L. RITTER, born July 10, 1933, in Van Buren, OH. He graduated from Walbridge High School, Walbridge, OH, in 1951, and attended the University of Michigan on a baseball scholarship. He received a BS in physical education and was a member of their 1953 College World Series Championship team. He signed with the Milwaukee Braves organization where he spent two seasons. He entered the U.S. Air Force in 1956, after graduation.

He was one of the first pilots to participate in the development and implementation of the C-141 Starlifter in 1962 and the C-5A Galaxy in 1969.

His career was devoted to Operational Airlift and Command positions included squadron commander, deputy command for Operations, group and vice wing commander.

He completed Armed Forces Staff College in 1971 and Army War College in 1976.

He retired as a colonel in 1986 with decorations including the Legion of Merit, Bronze Star, Meritorious Service Medal with one oak leaf cluster, and Air Force Commendation Medal with one oak leaf cluster.

He and his wife JoAnn reside in Marietta, GA, and have five children.

MYLES A. ROHRLICK, "MIKE" born Aug. 31, 1933, in Hartford, CT. He was commissioned and received his pilot wings through the Aviation Cadet Program in October 1955.

In 1964, as a C-130 instructor pilot, Mike invented the Parachute Low Altitude Delivery System and was awarded the Legion of Merit. He commanded the 21st Tactical Airlift Squadron in the mid 1970s. His last assignment was as commander of the 625th Military Airlift Support Group in Torrejon AB, Spain. Rohrlick retired from the U.S. Air Force in June 1984, at the rank of colonel. Rohrlick has been recognized in a number of books and magazine articles for his airdrop exploits at Khe Sanh AB, Vietnam. Among his awards and decorations are the Legion of Merit with two oak leaf clusters and the Bronze Star with V device.

Rohrlick was one of the founding members of the Airlift Association. He presently works for McDonnell Douglas Aerospace. He and his wife Joy reside in Oceanside, CA. They have two children, Mark and Shari.

THOMAS J. ROSS, Major, USAF, born Dec. 23, 1959, at Sheppard AFB, TX. He entered the Air Force in 1982, completed Undergraduate Pilot Training at Laughlin AFB, TX, and was assigned to the 7th MAS at Travis AFB, CA. Following a tour at Altus AFB, OK, he was assigned to the 36th MAS, redesignated the 36th ALS, 36th AS and 7th AS, McChord AFB, Washington, as a C-141B Lead Airdrop/Air Refueling Flight Examiner Aircraft Commander. He held positions as flight commander; Chief, Standardization and Evaluation; Chief, Squadron Operations Center; command post duty officer; executive officer; and pilot scheduler. He has attained over 4,770 flying hours and was McChord's Pilot of the Year for 1993. Major Ross is currently a War and Contingency Plans officer on the AMC staff. He earned associate and bachelor of applied science degrees in architectural technology from Southern Illinois University at Carbondale and master of aeronautical science in aerospace management from Embry-Riddle Aeronautical University. Decorations include the Aerial Achievement Medal with one oak leaf cluster for midwinter airdrop missions to McMurdo Station and South Pole, Antarctica; Meritorious Service Medal; Air Force Commendation Medal with one oak leaf cluster; Outstanding Unit Award; Command Readiness Medal with three oak leaf clusters; Southwest Asia Service Medal with three devices; and Kuwaiti Liberation Medal.

Memorable experience: Aircraft commander of the mission transporting the late Gen. Doolittle and family to Washington, DC, and subsequent fly-by over Arlington National Cemetery.

Tom is married to the former Sarah O'Hara of Grand Junction, CO. They have one son, Michael, and reside in O'Fallon, IL.

JOHN HARLEY RUTH, born March 25, 1922, in Herbine, AR. His education consist of Rison High School, Rison, AR, plus 62 semester hours.

He served the U.S. Army 1940-41, field artillery at Ft. Sam Houston, TX, and Ft. Sill, OK; to Army Air Corps December 1941.

Pilot training, December 1941-August 1942, Class of 42-G, staff sergeant pilot, Kelly field; first assignment, 5th Ferry Group, Hensley Field, Grand Prairie, TX, and Love Field, Dallas, TX; flight officer, January 1943, second lieutenant, April 1943; 3rd OTU, Reno, NV, December 1943, for C-46 training; over to Kalkunda, Karagpur, Paradoba and Mismari, India and Chengdu, China, April 1944-December 1944, flying support for 20th Bomb for short time in India and China then back to ATC flying the Hump for 72 round trips; returned to States December 1944, for more ferrying; separated December 1945 and re-enlisted as master sergeant January 1946; went to Albrook Field, Canal Zone as enlisted man in 1946, and returned 1949, to Ft. Campbell, KY, in the Strategic Air Command; recalled as first lieutenant in April 1953, to Hunter AFB; USAF Instrument Instructor School, Moody AFB 1953-GI; Navigation Training, James Connaly AFB 1954-HP; Wichita, KS, 1955 for B-47, transition while stationed in 96th Bomb Wing at Altus AFB, OK; 96th transferred to Dyess AFB, September 1957, where he retired as major Oct. 31, 1963, when B-47 was replaced by B-52, was in SAC 1949-1963, except for schools.

He flew all the Cessna Recips and Beechcraft, Lear Jet, Cessna Citation, Twin Otter, Aero Commander and Merlins. Last flight Dec. 9, 1992, at age of 70 and nine months; just could not hear the controllers anymore but he could still walk 36 holes on a hot day.

He was married Nov. 1, 1942, Dorothy Rae Hutchings until her death Aug. 10, 1994. They had two sons, six grandchildren and two great-grandchildren.

He was a corporate pilot and chief pilot until December 1992; still trying in 1994 with entrepreneurial business. He retired from military as major on Oct. 31, 1963, but not too tired to keep on trying.

GEORGE W. RYMER JR., born in Etowah, TN, on Sept. 20, 1932. Enlisted in the Air Force October 1949. After completing A&E Mechanics School, served with the 43rd Bomb Wing, Tucson, AZ. Discharged as a staff sergeant in October 1953. Re-enlisted in February 1954, for Aviation Cadet Training. Graduated in Class 55-P Reese AFB, TX, June 1955, and received pilots wings and second lieutenant commission. After Advanced Training in C-119, was assigned to the 50th TCS/314th TCW, Sewart AFB, TN. Transitioned into the C-130A in June 1957. In January 1960, he was reassigned to the 40th TCS/317th TCW, Evreux/Fauville AB, France. He returned to Sewart AFB, TN, July 1961, on a humanitarian reassignment. Retired in the grade of major, April 1970. While serving in Troop Carrier as a aircraft commander, lead crew, instructor pilot, flight examiner, accumulated 5,500 hours in the C-130, including 118 hours of combat time. He participated in all the major cold war crisis; Beirut, Formosa Straits, Belgian Congo, Laos, Dominican Republic, Vietnam, Pueblo capture, etc. Awards and decorations consist of Air Medal with oak leaf cluster, Air Force Commendation Medal with oak leaf cluster, Air Force Outstanding Unit Award with three oak leaf clusters, Combat Readiness Medal with oak leaf cluster, National Defense Service Medal with Bronze Service Star, Armed Forces Expeditionary Medal with five Bronze Service Stars, Vietnam Service Medal with four Bronze Service Stars and the Combat Crew Badge. In July 1970, he was hired by Lockheed Aircraft. International /AG and spent a total of 14 years on two assignments to Saudi Arabia as an instructor pilot/flight examiner with the Royal Saudi Air Force. In September 1986, he returned to the U.S. and worked for Lockheed/Georgia Company, performing production test flights, new aircraft deliveries and flight simulator training. After delivering 12 new aircraft, he spent one year in Taiwan with the ROCAF and several months with each the Korean and Chad Air Forces after receipt of their new aircraft, performing extensive ground and flight training.

He retired from Lockheed in April 1989, after 18 years with a total C-130 flight time in excess of 13,000 hours. He is a member of the China Post 1, American Legion and the Veterans of Foreign Wars. He is a Master Mason, Scottish Rite and York Rite Mason and a member of the Moslah Shrine Temple of Fort Worth, TX. He now resides in Jonesboro, GA, with his wife Sennie, of 38 years.

RONALD K. SABLE, born May 8, 1941, is a graduate of Iowa Wesleyan College (BS), Southern Illinois University (MS International Affairs) and the Harvard University Program for Senior Executives in National and International Security. He received a doctorate of humane letters degree (Hon.) from Iowa Wesleyan in 1989, following his being named Distinguished Alumni. He is an outstanding graduate of the Air War College.

Sable, currently vice president, Business Development for McDonnell Douglas Transport Aircraft was staff vice president, Legislative Affairs for McDonnell Douglas. He previously served as special assistant to President Reagan for Legislative Affairs with responsibility for Defense, Foreign Policy and Intelligence matters in the U.S. Senate. Prior to that assignment, he was special assistant to the President for National Security Affairs at the National Security Council (NSC). He retired from the USAF in February 1984. During his last Air Force assignment, he was director of Legislate Affairs for the NSC. Prior to that time, he served in assignments as the senior advanced agent for Presidential Travel, chief of Air Operations for the Secretary of the Air Force, MAC HQ staff officer and 437 MAW Wing Staff and C141 aircraft commander.

His awards and decorations include the Distinguished Flying Cross, Defense Superior Service Medal, Meritorious Service Medal, Air Medal (3) and the Valley Forge Freedom Foundation George Washington Honor Medal. A published author of professional and general interest articles and an avid sports car racer, sailor and golfer. He and his wife, Sandra, live in Alexandria, VA, and have one daughter.

RANDY D. SADLER, born December 1952, at Ft. Ord, CA. He attended Gilroy High School in Gilroy, CA, graduating in 1971. He was a distinguished Air Force ROTC graduate, receiving a bachelor of science degree in business management from San Jose State University in 1975. He completed Squadron Officer School in 1983, Air Command and Staff College in 1985, and received a master's degree in aviation management from Embry-Riddle Aeronautical University in December 1985.

After earning his pilot's wings at Williams AFB, AZ, in 1977, Sadler was assigned to the 17th Tactical Airlift Squadron, Elmendorf AFB, AK, as a C-130E pilot. He held various positions and qualifications in the squadron and upgraded to aircraft commander.

In December 1981, Sadler was assigned to the 39th Tactical Airlift Squadron, Pope AFB, NC, as C-130E adverse weather aerial delivery (AWADS) pilot. He upgraded to instructor and flight examiner, and held special qualifications in blacked out airfield operations and night vision goggle operations. His positions included chief of the 39th TAS Stan/Eval program and assistant chief of the 317th Tactical Airlift Wing's Stan/Eval division. He piloted the wing's Airlift Rodeo teams in 1985 and 1986.

Sadler moved to HQ MAC, Scott AFB, IL, in October 1986, where he served as the command's C-130 combat tactics pilot and represented the HQ MAC/DO in the command's C-17 program office.

In August 1989, Sadler started a participation with industry stint with McDonnell Douglas Corporation at Norman, OK. In June 1992, Lt. Col. Sadler assumed his present duties working for HQ AMC at Operating Location V, Charleston AFB, SC. He is helping develop the C-17 aircrew training program at AMC's initial C-17 base.

Sadler has accumulated over 4,000 flying hours, including a combat airdrop of Army Rangers at Point Salines, Grenada during Operation Urgent Fury. His decorations include the Air Medal, the Meritorious Service Medal, the Air Force Commendation Medal with oak leaf cluster, and the Armed Forces Expeditionary Medal.

He is married to the former Georgia Downing of Gilroy, CA. They have a daughter, Laura, and a son, Bryan.

RONALD T. SAMPSON, Colonel, USAF, is the commander of Civil Air Patrol-U.S. Air Force, with headquarters at Maxwell AFB, AL. In his military capacity, he directs USAF support of Civil Air Patrol and commands all Air Force personnel assigned to the headquarters, the eight regions and 52 wing liaison offices.

As the executive director of Civil Air Patrol, he serves in a civilian capacity and administers the day-to-day activities of Civil Air Patrol and is a voting member of the Civil Air Patrol National Board and National Executive Committee. Civil Air Patrol, the official auxiliary of the U.S. Air Force, is a federally chartered, benevolent, civilian corporation with more than 57,000 members.

Col. Sampson was born April 5, 1944, in Kalamazoo, MI. He attended public school in Hillsboro, OH, and received a bachelor of science degree from Ohio State University in March 1967. He was commissioned through the Air Force Reserve Officers Training Corps that same month. The colonel received his master's degree in business management from Webster College in 1978. He is a graduate of Squadron Officer School, Air Command and Staff College, and the National Security Management Course. Col. Sampson is also a graduate of the Airlift Operations School and the NATO Staff Officer's Course.

After receiving his commission, Col. Sampson entered pilot training and earned his pilot wings at Laughlin AFB, TX, in May 1968. He completed C-130E weapons systems training at Sewart AFB, TN, and was assigned to Ching Chuan Kang AB, Taiwan in November 1968, where he performed flight duty with the 345th Tactical Airlift Squadron, flying frequent missions in the Vietnam War zone. Following his combat tour, Col. Sampson was reassigned to the 778th Tactical Airlift Squadron, Pope AFB, NC, flying the C-130E.

At Pope from 1970 until August 1977, Col. Sampson served in a variety of positions including squadron pilot, duty controller and scheduler in the 317th Tactical Airlift Wing's Airlift Control Element, instructor pilot and flight commander in the 40th Tactical Airlift Squadron, evaluator pilot in the 317th TAW Standardization/Evaluation Division, and chief pilot and chief of the 40th Tactical Airlift Squadron's Standardization/Evaluation Section. In March 1976, he was assigned as the director of Concepts and Tactics, Studies and Analysis Division of the USAF Airlift Center.

In 1977, Col. Sampson was selected for the initial cadre to establish operations for the first MAC C-130 unit permanently assigned to Europe. Serving in the 435th Tactical Airlift Wing at Rhein-Main AB, Germany, from August 1977, until August 1981, he was chief of Wing Combat Tactics and Techniques and chief of Wing Aircrew Standardization/Evaluation. Upon completion of this overseas tour, he was assigned to Headquarters Military Airlift Command, Scott AFB, IL, deputy chief of staff Operations, Standardization/Evaluation Directorate, where he was the Major Command C-130 Operations Program Manager, Aircrew Evaluation Division.

In June 1984, Col. Sampson returned to overseas duty as the North Atlantic Treaty Organization Air Transport Staff Officer, Logistics and Manpower Division, Supreme Headquarters Allied Powers Europe (SHAPE), Belgium. In June 1987, he returned to Headquarters Military Airlift Command, Deputy Chief of Staff Operations, and served as a senior airlift controller in the Commander in Chief Military Airlift Command, Command Center until December 1987, when he became deputy director of Combat Techniques, Electronic Combat and Special Operations. On April 24, 1989, he became deputy commander for Operations, 375th Military Airlift Wing, Scott AFB, IL. He served in that capacity until June 22, 1990, when he was assigned to the 463rd Tactical Airlift Wing, Dyess AFB, TX, as vice commander. Col. Sampson assumed the position of vice commander of Headquarters, Civil Air Patrol-USAF, July 8, 1991, until Aug. 27, 1992, when he was selected to serve as Commander, Headquarters, Civil Air Patrol-USAF.

As captain, briefed Gen. P.K. Carlton, CINCMAC, and MAC "A" staff on required changes to airlift combat tactics. Introduced AWADS OPS to the European Theater 1977-81, while assigned to 435TAW. First C130 landing on German AutoBahn during reforcer, Gen. Bill Moore, CINCMAC on the flight. Airlift staff officer at SHAPE. Class A accident board president for Gen. H.T. Johnson, CINCMAC. Vice-commander, 463 TAW, during Desert Shield/Storm.

He is a command pilot with more than 4,900 flying hours, and has more than 670 combat flying hours in the C-130E. His military decorations and awards include the Distinguished Flying Cross, Defense Meritorious Service Medal, Meritorious Service Medal with four oak leaf clusters, Air Medal with four oak leaf clusters, Combat Readiness Medal with two oak leaf clusters, Vietnam Service Medal with six service stars, Humanitarian Service Medal, and the Republic of Vietnam Gallantry Cross with palm.

Col. Sampson's date of rank is Nov. 1, 1988.

He is married to the former Jill McGee of Hillsboro, OH. They have a daughter, Kathy.

RAYMOND P. SANCTON JR., Major, USAF, (Ret.), born in Brooklyn, NY, on Nov. 4, 1928. He attended Amityville High School on Long Island, NY. In 1950, he enlisted in the Air Force and attained the rank of sergeant prior to being accepted into the Aviation Cadet program. He received his wings and bars in May 1954.

Major Sancton's first assignment was helicopter training and then to Sewart AFB, TN, where he spent seven years flying helicopters, then C-123s and finally C-130s. Squadrons assigned were 644th, 345th heli, 345th fixed wing and the 62nd. At Dyess AFB flew C-130 D's in the 17th and 18th Squadrons. He was assigned to 7406th Support Squadron at Rhein Main AB, then to Nakhon Phanom AB in Thailand flying Jolly Greens. He is married to the former Rosemary Cowser of Center, TX, and has two daughters and one son, and four grandchildren.

KEITH N. SAWYER, born Nov. 23, 1936, Phoenix, AZ. Joined the USAF July 19, 1958, serving as command pilot. Earned pilot wings Oct. 13, 1959, at Reese AFB, TX, (Lubbock). In 1954 he graduated from Gilbert High School, Gilbert, AZ. In 1958, graduated from Arizona State University, Tempe, AZ, with a BS degree in agronomy, AFROTC Commission as second lieutenant. In 1979, received his MA degree in management from Webster University, St. Louis, MO. In 1973, Air Command and Staff College and in 1979, Industrial College of the Armed Forces. Education also includes Crisis Response Management School, Hurlburt AFB, FL, 1983, and IBM Executive Data Processing Seminar, Poughkeepsie, NY, 1985.

Awards/Medals: Legion of Merit, Distinguished Flying Cross, Air Medal with 15 oak leaf clusters.

He retired Aug. 1, 1988, with the rank of colonel.

His employment history includes: Modern Technologies Corp., principal analyst, March 1989-Present; U.S. Air Force, director, Inquiries Oversight and Enhancement, January 1987-August 1988; Chief, Airlift Management, January 1981-December 1986; Chief, Airlift Command Center and Senior Airlift Controller, July 1978-January 1981; Chief, Aircrew Standardization and Evaluation, May 1977-July 1978; Commander, Military Airlift Squadron (Heavy Transport), June 1976-May 1977; Commander, Military Airlift Support Squadron, June 1975-June 1976; Pilot and Middle Management positions, July 1958-June 1975.

His wife, Jo, is a professional musician/composer. Their children are Teri, Lisa, Kevin and Michele. Today he is principal analyst, Modern Technologies Corp., developing master airspace plan for HQ AMC use of training airspace through year 2000+.

KENNETH W. SCHAMENS, Colonel, USAF, (Ret.), graduated from Tomah High School, Tomah, WI, in 1951. He received a BME degree from Lawrence University, Appleton, WI, an MA from Webster University, St. Louis, and an MAR from Concordia Seminary, St. Louis, MO.

Upon graduating from pilot training at Goodfellow AFB, TX, he flew C-124 aircraft from 1957-65, at Donaldson AFB, SC, and Hunter AFB, GA. From 1965-69, he held various operational and management positions involving tactical airlift at HQ 315th Air Division, Tachikawa AB, Japan and HQ 7th Air Force at Tan Son Nhut AB, Vietnam. From 1969-74, Schamens was assigned to the Pentagon (DCS Programs & Resources) where he was responsible for force structure and budgeting matters for strategic airlift. He assisted in developing the C-5 program including studies to enhance the C-5 service life.

From 1974-79, he served at HQ MAC, "co-authoring" the Tactical/Strategic Airlift Consolidation Plan (October 1974) while "airborne over the Atlantic - under the watchful eye of CINCMAC!" A command pilot with 6,000 plus hours, he was assistant DCS Air Transportation at MAC when he retired in 1979.

His decorations include the Legion of Merit, Bronze Star, Meritorious Service Medal, Air Medal and Air Force Commendation Medal with one oak leaf cluster.

Ordained into the Lutheran ministry in 1982, he is the pastor of Abiding Savior Lutheran Church, Gainesville, FL, and is the Chaplain of the Airlift Association.

Schamens and his wife, Jane, have three children: David, Julie and Thomas.

ROBERT W. SCHORR, Colonel, USAFR, (Ret.), born June 20, 1938, in Jacksonville, FL, graduated from Woodward Academy in Atlanta, GA, in 1956, and received a bachelor's degree in management and ROTC commission from Auburn University in 1961. He served on active duty for eight years, and as a reservist for 25 years. As a MAC strategic, tactical, nuclear, and special operations navigator, he flew over 12,000 hours in C-124s, C-141s, C-5s, and C-130s.

He has a master's degree in political science and a Ph.D. in Soviet Studies. He completed Squadron Officer's School, Air Command and Staff College, and Air War College.

His last Reserve assignment was Individual Mobilization Augmentee at Special Operations Command (Atlantic), Ft. Bragg, NC, and U.S. Special Operations Command, MacDill AFB, FL.

Col. Schorr's awards include the Legion of Merit, Joint Service Commendation Medal, Meritorious Service Medal, and Air Medal.

An industrial consultant, he lives in Marietta, GA, with his wife, Alice and daughter Stacy. A life member of the Airlift Association. He is vice president of the Airlifter Country Chapter.

RUDY P. SCHUBERT, born Sept. 1, 1952, in Morgantown, WV. He entered the USAF in December 1974. He is a command pilot with over 5,600 hours in DC-130, HC-130H/P/N, C130E/H, C-12A/C/D, and VC-9 aircraft.

His military service is as follows: 1991-present, VC-9 instructor pilot, assistant flight commander, chief of Sq. Operations Center. VC-9 pilot transporting the Vice President, First Lady and other heads of state. From 1990-91, Chief, Senior Executive Service Mission Director 89 MAW Presidential Pilot Support (PPS); 1987-90, Chief 89 MAW C-12 Central Training Facility; 1985-87, Senior Command Post Controller, Instructor, Lapes Pilot, 317TAW, Pope AFB; 1982-85, Chief Pilot 39 TAS, Evaluator, Lapes pilot; 1979-82, Assistant Flight Commander/Evaluator/FSO 33rd Aerospace Rescue Recovery Sq. Kadena AB, Japan, credited with several saves and assist; 1976-79, Aircraft Commander DC-130E 22nd & 11th Tactical Drone Squadrons, Davis Monthan AFB, AZ; 1974-76, UPT, Craig AFB, AL.

His education includes PME: SOS/ACSC; Civilian: baccalaureate sociology, West Virginia University and master's aeronautical science technology, Embry Riddle Aeronautical University.

Awards/Medals: Meritorious Service Medal, Air Medal, Air Force Commendation Medal with two oak leaf clusters, Combat Readiness Medal with two oak leaf clusters. He is a life member of the Airlift Association and member of Daedalians.

PETER SCHWARTZ, born Oct. 25, 1918, in Salina, KS, graduated from Sacred Heart High School in 1936, and St. Joseph's College and Military Academy, Hays, KS, in 1938.

Schwartz joined the Navy, designated a Naval Aviator and released in 1970. He flew over 8,600 hours in 20 different aircraft, including the PBY, PV-1 in VPB-129, R50 in VR-5 and VR8-Berlin Airlift, AJ-1 in VAH-7 aboard straight deck carriers, and VW-2 in VW-13-15- North Atlantic Barrier. Other assignments included General Line School, first commanding officer of Naval Parachute Facility making one premeditated parachute jump and Staff Barrier Forces Atlantic Fleet.

Cdr. Schwartz's decorations include American Campaign Medal, WWII Victory Medal, Korean Service Medal, United Nations Service Medal and National Defense Service Medal 2nd Award.

After retirement earned BBS and MS, instructed in ac-

counting and became a CPA, again retiring in 1983.

He and his wife, a navy nurse retired LCDR USN Reserve, reside in Memphis, TN.

ALVIN C. SCHWEIZER II, Colonel, is Chief, Operations Plans Division, AMC. A C-130 pilot, he was commissioned in 1966 from Miami University, graduated from Reese AFB in 1968, and flew in Vietnam. He earned a Distinguished Flying Cross for flights in Blind Bat Operations. He was an RC-130A photomapping instructor pilot at Forbes AFB, a weapons controller in the Philippines, and an instructor pilot at Rhein-Main AB, Germany and Little Rock AFB. After ACSC, he worked on the JCS. At Pope AFB he command the ALCE in Grenada and Credible Cat. After MPC at Randolph AFB and Army War College at Carlisle Barracks, he was the wing ADO and 34 TATG commander at Little Rock AFB. In the Philippines, he commanded the 624 MASG. He wears the Legion of Merit, holds an MA from Ball State University, and graduated from SOS and AWC. He married Jo Holtgrewe of Lubbock, TX.

DONALD S. SCHWEITZER, Major General, born Jan. 7, 1930, in McKeesport, PA. He received his BS from the University of Maryland and his MA from Central Michigan University. He joined the USAF on May 21, 1951. He served at Lackland AFB; Tyndall AFB; Malden, MO; Reese AFB; Ashiya, Japan; Clark AFB, Philippines; Sewart AFB; Andrews AFB; Pentagon.

He married Charlotte S., and has two daughters and three grandchildren. As a civilian he was employed with American Airlines Inc., Captain. He retired Jan. 7, 1990.

DAVID W. SCOTT, Colonel, (Ret.) graduated from Michigan State University in 1959, and after completing pilot training at Vance AFB, OK, was assigned to Biggs AFB, TX, as a KB-50J tanker pilot. Following a tour as a U-10 pilot with the 1st Air Commando Squadron in Vietnam in 1964, he became a tactical airlifter with the 48 TCS (C-130B) at Forbes AFB, KS.

His career with C-130s and airlift involved a tour on the 12th AF IG Team, an RC-130A aircrew commander assignment at Rhein-Main AB Germany, and HQ USAFE/DOL staff officer position. Returning to CONUS, he served at 12AF/DOL prior to becoming executive officer to the Chief of Staff at UNC/USFK in Korea. Following attendance at the AFSC he was operations officer in the 36 TAS (C-130E) at Langley AFB, VA, programming officer at HQ MAC/XP and squadron commander of 773 TAS (C-130H) at Dyess AFB, TX. After attending the Air War College (1979) he became Chief, Airlift and Rescue Division at HQ USAF/XOOTA prior to becoming the 317 TAW/CV at Pope AFB, NC. He completed his career as Commander, 1606 ABW at Kirtland AFB and Commander, LMDC Maxwell AFB, AL.

Dave and his wife Patty currently reside in Albuquerque, NM, where he is the executive vice president of Albuquerque Economic Development, Inc.

JOEL SEIDBAND, born Sept. 29, 1964, in Plainview, NY. He joined the USAF on July 3, 1986, attended Officer's Training School Class 87-01. Military service includes Columbus AFB, MS (UPT); Charleston AFB, SC (C-141B), air drop, air refueling aircraft commander 1988-91; Maxwell AFB, AL (C-21A), evaluator pilot November 1992-present. He participated in Just Cause (Panama) and Desert Shied/Storm.

His memorable experiences includes trying to find his bags after deploying to Torajon on the first days of Desert Shield and crossing the Drop Zone in Panama and watching tracer bullets fly by.

Awards/Medals: Air Force Commendation Medal, Air Force Outstanding Unit Medal, Combat Readiness Medal, National Defense Service Medal, Armed Forces Expeditionary Medal, South West Asia Medal, Humanitarian Medal Kuwait Liberation Medal.

He is a member of the Order of Dedalians, Civil Air Patrol and Alpha Eta Rho.

He and his wife, Carolyn, have children, Scott, 4, and Steven, 2. Today he is chief of Standardization/Evaluation, 502 OPG, Maxwell AFB, AL. He is working on his master's in human resources management.

WILLIAM J. SHEEHAN, SMSGT, served three years U.S. Army and 29 years U.S. Air Force (includes two years U.S. Army Air Corps) 1942-1974.

Joined the U.S. Army at the Boston Army Base and was sent to Camp Devens, MA, for a physical examination in 1942. Put on a train and sent to Louisiana for basic training. After Basic, was promoted to corporal and sent on a cadre to help form another unit in South Carolina, then to Mississippi.

In the fall of 1943, was shipped to Camp Kilmer, NJ, where boarded a British troop ship called the *Empress of Russia,* then joined the other troop ships in a convoy escorted by U.S. Navy destroyers.

On the morning of the sixth day at sea he was awoken and handed a sheet of paper by his first sergeant. He shook his hand as he lay on his back, and told him he had been promoted to buck sergeant. Thirteen days after he left port he arrived in Scotland.

The Third Army captured the notorious concentration camp, Buchenwald. They took the town of Hershburck outside of Nuremberg and set up a Command Post in a bombed-out train station. The inmates were sent to Army units, where they were fed and clothed. He took over a three-family house, moving the families over to other houses in the town. He was assigned 12 Russian prisoners, who he put in rooms on the second and third floors.

USAF Missions flown in the following countries from 1945 to 1974: England, France, Germany, Spain, Scotland, Wales, Ireland, Italy, Norway, Sweden, Denmark, Finland, Canada, Mexico, Cuba, Austria, Greece, Portugal, Belgium, Netherlands, Turkey, Malta, Iceland, Newfoundland, Panama, Puerto Rico, Philippines, Okinawa, Hong Kong, China, Japan, Indonesia, India, Korea, Vietnam, Thailand, Greenland, Switzerland, Kenya, Morocco, Iran, Luxembourg, Lebanon, South Africa, Brazil, Dutch Guinea, Ascension Island, Bermuda, Taiwan, Gibraltar, Labrador, Libya, Martinique, Monaco, Nova Scotia, Virgin Islands, Antigua Sardinia, Wake Island, Guam, Trinidad, Azore Islands, Kwajalein Islands, Crete, Midway, Vatican City, Marshall Islands, Shetland Islands.

He flew in the following type USAF Aircraft: C-47, C-54, C-118, C-121, C-130, C-141, C-131, C-124, CG-4A, Army Gliders, HU-16 Flying Boats and C-135 Jets. His total flying time was over 15,000 hours.

Highlights of his career are as follows. Over 300 flights across the Atlantic Ocean. One flight over the North Pole from Rhein Main AB, West Germany, to Cold Lake, Canada in a C-135 type aircraft. One flight around the world from the USA to Newfoundland, West Germany (Wiesbaden), Italy, Iran, India, Thailand, Vietnam, Hong Kong, China, Formosa, Japan, Wake Island, Hawaii (Hickam AFB), California. Airlift missions in Berlin (100 missions on Berlin Airlift), Korea, Thailand, Vietnam. One flight to the Atomic Testing Center on Eniwetok Atoll, Kwajalein NAS, Marshall Islands. He observed an atomic bomb explosion from the beach and a hydrogen bomb, detonated while flying 60 miles from the "mushroom cloud." While serving with the U.S. Third Army (Gen. George S. Patton, Jr., commanding) he participated in two amphibious landings. One at Omaha Beach in Normandy and the other in southern France at Le Havre. Also served in Normandy, Ardennes offensive (Battle of the Bulge), Battle of Northern France and Rhineland Crossing as a Platoon Sergeant.

THOMAS G. SHEPKE, born Dec. 25, 1952, in Baltimore, MD. He entered the Air Force in 1971, completed Air Cargo Specialist Technical School at Sheppard AFB, and was assigned to the Aerial Delivery Section of Aerial Port at Charleston AFB, SC. During his career, he spent nearly 17 years overseas mostly with the 374 Aerial Port Squadron at Clark AB, Philippines and Yokota AB, Japan. He deployed to Israel in 1973, in support of Operation Nickel Grass, participated in the evacuation of Vietnamese and Americans from Saigon in 1975, the evacuation military personnel and their dependents from the Philippines in 1991, after the eruption of Mt. Pinatubo, and was a participant in Operation Desert Storm. MSG. Shepke retired in August 1994, where he served as NCOIC of Special Handing and Squadron Operations Superintendent at Dover AFB.

He completed both the PACAF NCO Leadership School and NCO Academy at Kadena AB, the Senior NCO Academy by correspondence, and received his AAS degree in transportation and traffic management from the Community College of the Air Force. His military decorations include the Meritorious Service Medal with one oak leaf cluster, Air Force Commendation Medal with four oak leaf clusters, Air Force Achievement Medal with one oak leaf cluster, Joint Meritorious Unit Award, and the Air Force Outstanding Unit Award with Valor Device and two oak leaf clusters. He has a daughter, Catherine and reside in Federalsburg, MD.

CHARLES J. SHORT, enlisted in the Air Force, then was commissioned in 1961. Short's first assignment after navigator training at Harlingen AFB, TX, and Mather AFB, CA, was in C-124 transport aircraft at Dover AFB, DE. He transferred to McChord AFB, WA, in January 1965, to fly C-124s and eventually C-141s. While at McChord, he became a wing training officer and classroom instructor for navigation systems. He went on to F-4 training at Davis-Monthan AFB, AZ, and followed with a DaNang AB, SVN, combat tour.

His return assignment in June 1972 was to McChord AFB for a second tour. He became the wing staff navigator and then Assistant Chief, Combat Plans and Exercises. In January 1977, he was assigned to HQ MAC, Scott AFB, IL, and became Chief, Exercise Management Branch. Short then attended Air War College, Maxwell AFB, AL, class of 1980. His next assignment was with the U.S. Readiness Command, MacDill AFB, FL, as the Programs and Policy Officer. In June 1984, he was assigned to HQ AF Reserve, Robins AFB, GA, as Chief, Contingency Plans Division.

Charlie retired June 1989 and settled in Byron, GA, where he is director of Public Works for the city. Charlie and Connie have four children: Lary, a USAF Captain; Julie, a USN nurse; Roy, working in Washington; and John, a University of Georgia medical school student.

LARY C. SHORT, graduated from Northeast High School, St. Petersburg, FL, in 1982, then entered the Air Force in December 1983, completed emergency actions controller training, and reported to Myrtle Beach AFB, SC, as a command post controller.

He was a distinguished graduate from Keesler Technical Training Center, and while at Myrtle Beach AFB in August 1985, was awarded an AFROTC scholarship under the Airman Scholarship and Commissioning Program. He won the AFROTC Vice Commandant's award in June 1987, and attended U.S. Army Airborne Training, Ft. Benning, GA, in June 1988. He received his bachelor's degree in mathematics and was a distinguished graduate and regular commission recipient from AFROTC, University of South Florida, Tampa, FL, in December 1988. He graduated from Undergraduate Pilot Training, Williams AFB, AZ, in May 1990, and was assigned to the 306th Air Refueling Squadron, Altus AFB, OK, as a KC-135R pilot and T-37 first pilot. He upgraded to KC-135R aircraft commander in March 1993, and in June 1993, became an assistant flight commander in the 11th Air Refueling Squadron, Altus AFB, OK.

His decorations include the Aerial Achievement Medal, Air Force Outstanding Unit Award with Valor and one oak leaf cluster, Combat Readiness Medal, National Defense Service Medal, Southwest Asia Service Medal with device and the Kuwait Liberation Medal. Capt. Short's father is Lt. Col. Charlie Short, retired.

JOHN F. SIEVERTSON, USAF, (Ret.), born June 26, 1936, in Glen Dale, WV. He graduated from West Virginia University with a bachelor's degree in music education in 1959. He earned a master's degree in government from Southern Illinois University in 1979, and is a graduate of Air Command and Staff College and the National War College.

Sievertson retired from the USAF on Aug. 1, 1989, with the rank of brigadier general. A varied career included an ICBM missile launch officer at Beale AFB, C-141 Aircraft Commander at Travis AFB, vice commander of 1550th Aircrew Training and Test Wing, commander of 89th Airlift Wing, and deputy for Bases and Units, HQ, USAF. During his final assignment as Vice Commander, 21st Air Force, he served as Commander, Airlift Forces for U.S. Central Command. He is currently manager of C-141 Aircrew Training Operations for Hughes Technical Service Company.

Awards and decorations include the Defense Distinguished Service Medal, Legion of Merit, Bronze Star and Air Medals.

Sievertson and his wife Anna reside in Medford, NJ. They have three children: Mallorie, Jennifer and Jay.

WILLIAM F. SIMS, born June 3, 1957, Windsor, Ontario. He enlisted June 27, 1975, basic training at Lackland. Commissioned July 13, 1984, OTS, Medina Annex, San Antonio, TX. Operations Management Officer (19XX) training at Keesler AFB, 1988. One assignment in AMC, as airlift control officer: 603 Airlift Support Group, Kadena AB, Okinawa, Japan; 1992-94. Temporary duty assignment to Mogadishu, Somalia, September - December 1993. Served as Chief, AMC Command Post, and Vice Commander, USAF Somalia. Survivor of the "Bloody Sunday" battle, Oct. 3, 1993. Repatriated the remains of over 20 Americans killed in action. Medically evacuated over 100 U.S. and United Nations wounded, including CWO Mike Durant, the conflict's only POW. Retired as captain, Oct. 1, 1994.

WILLIAM H. SISTRUNK, Major General, USAF, (Ret.), born Aug. 11, 1937, in Utica, MS. He graduated from Drew High School, Drew MS in 1955, and attended the University of Mississippi where he received a BS degree in geology and his ROTC commission in 1960. His master's degree in personnel management is from Central Michigan University. He completed National War College and all other levels of PME.

Gen. Sistrunk retired from the Air Force in February 1992, as the Chief of Staff, MAC. His command positions were: Commander 6th MAS, 436th MAW, and 322nd ALD - an Airlifter for 32 years with 7,400 flying hours.

Among his awards and decorations are: Distinguished Service Medal, Legion of Merit, Meritorious Service Medal.

Gen. Sistrunk and his wife, Sue, plan to be in their new home in Montgomery, AL, by April 1993. They have two sons, Will and Mark, two daughter-in-law, and three granddaughters.

R. SCOTT SITTON, Captain, born in San Diego, CA. Graduated from Brigham Young University in 1986, bachelor's degree, building construction management. Graduated from Embry-Riddle Aeronautical University in 1993, master's degree, aeronautical science/management. Joined the Air Force in 1986, Officer Training School. Completed pilot training at Reese AFB, 1987 (8708). Assigned to Fairchild AFB from 1987 to 1991, flying the KC-135A and KC-135R. Deployed to the Pacific Tanker Task Force during Operations Desert Shield (Iraq/Kuwait). Stan/Eval copilot, dual qualified in the KC-135A and KC-135R. Aircraft commander up-grade, 1991. Transfer to McDonnell AFB in 1991; Squadron Safety Officer, 384 ARS; instructor pilot 1994; wing safety officer, 22d ARW. TDYs include Moron AB, Spain; Ryihad, Saudi Arabia; Guam; Clark AB, Philippines; Misawa AB, Japan; Eielson AFB, AK; Santiago Chile; Mildenhall RAF; Incirlik Turkey; Mombasa Kenya, East Africa; Edwards AFB, CA; Hickam AFB, HI.

He is married to Jennifer Parker, and has five children: Michael, 8; Nathan and Caitlin, 6; Daniel, 4; and Christopher, 2.

CLARENCE BENTON SLAUGHTER, Colonel, USAF, (Ret.), born May 8, 1920, in Commerce, TX. He graduated from Commerce High School in 1939, and received a bachelor's degree in business administration from St. Edwards University in Austin, TX. He completed the Tactical Officers Squadron Course at Tyndal AFB, FL, in 1948, the Senior Marine Officers Course at Quantico Marine Corps Base, VA, in 1957, and Air War College at Maxwell AFB, AL, in 1961.

Ben Slaughter had the privilege of commanding all levels of tactical airlift units, from squadron to air division. From 1953-54, he served as squadron commander of the 782nd Troop Carrier Squadron, Evereux, France. From 1964-65, he served as group commander of the 315th Troop Carrier Group, Naha, Okinawa. From 1967-69, he served as wing commander of the 465th Troop Carrier Wing, Pope AFB, NC. From 1969-70, he served as the division commander of the 838th Air Division, Forbes AFB, KS. Ben Slaughter's final assignment in his 30 year career was that of commander of the 4500th Air Base Wing, Langley AFB, VA, from 1970 until July 1971.

Included among his many awards and decorations are the Legion of Merit with two oak leaf clusters, the Air Medal with two oak leaf clusters, both the Army and the Joint Chiefs of Staff Commendation Medals, the Presidential Unit Citation with three oak leaf clusters, and the Korean Presidential Citation, as well as, several Outstanding Unit Awards.

Slaughter and his wife, Elaine Kincaid Slaughter, make their home in Austin, TX. Frequent dinner guests include daughters Nancy and Suzanne who are also Austin residents.

BARRY H. SMITH, born Oct. 21, 1950, in Cincinnati, OH. He joined the service June 1972, graduated from the USAFA Class 1972. His military service includes USAF MAC command pilot, C-130E/AWADS, serving at Clark AB, PI, 1974-78, (776 TAS, 21 TAS, 374 TAW); USAF Academy, history instructor, 1978-82; Scott AFB, HQ MAC, 1982-84; Rhein Main AB, Germany, 1985-91, (37 TAS, 435 TAW, 435 Airlift Control SQ, Commander; Pentagon, Air Staff, 1991-92. Participated in Gulf War, Commander, 1683 Airlift Control SQ, Juball, Saudi Arabia; and Vietnam, Evacuation of Saigon, 1975, Babylift and New Life, 1975.

Awards/Medals: SWASM, Saudi Arabia Liberation of Kuwait, Air Force Outstanding Unit Award with V and seven oak leaf clusters, Bronze Star; Meritorious Service Medal with three oak leaf clusters, Air Force Commendation Medal with 10 oak leaf clusters, Air Force Achievement Medal, Combat Readiness Medal with two oak leaf clusters, National Defense Service Medal with star.

He and his wife, Debbie, have two sons, Justin and Aaron. Today he is with J-5 Regional Plans and Policy HQ EVCOM Stuttgart, Germany.

DONALD C. SMITH, born Feb. 3, 1937, in Clinton, OK. He was the valedictorian and a four sport letterman in the class of 1955, at Clinton High School. He graduated with the class of 1959, from the U.S. Naval Academy. Smith also graduated from AFIT with a master's degree in aerospace mechanical engineering. Smith served in SAC as an aircraft commander of a lead B-47 crew. He later was an 0-1 FAC in Vietnam and Korea. Smith served as a C-141 pilot, instructor, flight examiner, operations officer, and squadron commander in the 76 MAS at Charleston AFB.

His squadron was selected the best in MAC. Tours followed as wing commander of the 62 MAW and commander of American Forces Azores. Assignments included MAC headquarters and the Joint Staff. He is a graduate of the 1977 class of National War College. Smith was the assistant DO of MAC upon retirement in July 1986. He has been the president of two regional airlines, Henson and Rocky Mountain and recently retired from Continental Airlines and resides with his wife, Brenda, in Lancaster, PA.

F.P. SMITH, served as a pilot, aircrew, C-47s, C-46s, C-109, C-87s flying gas over the Hump (WWII), with ATC, MATS, MAC KC135, RC-10 etc.

GORDON A. SMITH, born April 27, 1942, in Mountain Home, AR. He entered the Air Force in 1959, completed Jet Engine Maintenance Technical School, and was assigned to Tyndall AFB, FL. Over his career, he maintained approximately 40 different types of aircraft. He retired from the Air Force in December 1989, where he was serving as the wing maintenance superintendent, Kirtland AFB, NM. He completed both the NCO Academy at Norton AFB, CA, and the Senior NCO Academy at Gunter AFS, AL.

Decorations include the Meritorious Service Medal with five oak leaf clusters, Air Force Commendation Medal with three oak leaf clusters, Air Force Achievement Medal, Presidential Unit Citation, Outstanding Unit Citation with seven oak leaf clusters, and the Air Force Longevity Service Award with seven oak leaf clusters. He is married to the former Rena Rossignol of Van Buran, Maine. They have a daughter, Brenda, and reside in Albuquerque, NM.

KATHARINE M. SMITH, born Sept. 28, 1922. He joined the U.S. Navy July 10, 1944, serving at Florida Hall, worked at Bupers, and Patuxent River NAS M2-3 MD.

Memorable experiences include returning from trip to Frankfurt, Germany, supplying Airlift, lost an engine at point of no return between England and Azores over Atlantic Ocean. Cargo Flights perhaps 15-20 passengers, one a very pregnant lady. Second engine developed oil leak and had to be feathered. It would be restarted to gain loss altitude and then again feathered as they lost altitude. They limped along this way till arrival at Azores, MAT-VR-3. Many approaches in GCA when weather was socked in. Once they couldn't get landing gear down and had to do it manually obviously they made it or he couldn't write about it. He feels honored to have served his country.

He lost his oldest brother in the battle for Rime. He is buried at Gettysburg.

Previous to MATS they were MATS flying Hatsbrat and U-Route across U.S.

Worked at a few airlines at PAX River and Cleveland.

DAVID R. SORUM, born on Feb. 27, 1942, in Bowman, ND. After graduating from high school in Faulkton, SD, he attended South Dakota Tech in Rapid City, where he graduated with a bachelors degree in electrical engineering in 1965. He entered the Air Force via OTS in 1966 and completed Undergraduate Pilot Training at Reese AFB, TX, in 1967. His initial assignment was to Dover AFB, DE, where he flew C-141 aircraft. Except for a tour as an 0-2A pilot in Vietnam in 1969-70, Sorum served with MAC throughout his career. Other assignments included McGuire AFB, Charleston AFB, Altus AFB and Pope AFB.

He concluded his Air Force career flying C-5 aircraft at Travis AFB, CA. His awards include Distinguished Flying Cross, Air Medal with eight oak leaf clusters, Distinguished Service Medal with two oak leaf clusters, and Air Force Commendation Medal. He retired as a lieutenant colonel in 1987, and currently resides in San Jose, CA, where he is employed by Loral Aerospace. He has two children - Robert of Spearfish, SD, and Leslie Ann of Seattle, WA.

COVERT A. SOULE' JR., "CO" was born in New Orleans; he graduated from Freyhan High School; obtained a BS degree from Louisiana State University, and an MBA from Webster University in 1976.

Soule' retired from the Air Force in 1986, with the rank of colonel. During his last assignment he served as the director of Manpower and Organization, HQ Military Airlift Command. He holds the aeronautical rating of master navigator having logged over 6,800 flying hours including 209 Vietnam combat missions in the 0-2 aircraft.

Among his many awards and decorations are: Legion of Merit with one oak leaf cluster, Distinguished Flying Cross, Air Medal with nine oak leaf clusters, Meritorious Service Medal and Air Force Commendation Medal.

Soule' and his wife, Iris, reside in O'Fallon, IL. They have three children: Chip, Cathy and Carolyn. Soule' is currently the St. Louis district manager for TRW.

JAMES ROGER SPATAFORA, was born May 19, 1929, New York City. He graduated from Manhattan High School and enlisted in the USAAF on June 10, 1947. Military stations/locations include: SAACC Field (now Lackland AFB) for basic training; Keesler Field, aircraft mechanic and P-51 specialist courses; Kearney AF, NE, hydraulic mechanic.

He volunteered for "Berlin Airlift" in June 1948 and was sent to Rhein/Main AB, Germany as C-54 hydraulic mechanic and assistant crew chief. He studied German, Russian, photography and taught photography. When lift ended, he transferred to C-47 used in support of "air attache" to Prague. East Kelly AFB, taught depot level hydraulics to students headed for "Depot" at Chateauroux AS in France; North Kelly, served as senior hyd. tech. selected as cadre to start new C-124C Sqdn. at Hill AFB; served as sen. hyd. tech., hyd. shop chief, asst. hangar chief, asst. insp., dock chief, NCOIC of all OJT and flying and ground safety programs, specialist chief of all other a/c specialist, and selected for Air Force OCS in December 1954.

At Lackland AFB, he was commissioned on June 17, 1955; Chanute AFB, attended aircraft maintenance officer course; Edwards AFB, went from 2nd lieutenant to captain; Brookley AFB, AL, F-105 Military Support Unit, as team chief took depot specialists to Bitburg and Spangdahlem ABs. Was assigned to the 36th and 49th Fighter Wings and worked as senior controller in the Depot Control Center. Volunteered for Vietnam in 1964 and attended indoctrination for counter-insurgency and "Jungle Bunny" courses at Hurlburt Field, FL. Sent to Maxwell AFB, AL; Clark Field, P.I.; Bien Hoa AB, RVN; Chateaurous AS, France; RAF Mildenhall, promoted to major and attended C-141 Familiarization and C-141 Flight Simulator School at Charleston AFB.

At Scott AFB, he was assigned to Standardization and Evaluation Unit as team chief. Was required to attend FAA air carriers maintenance inspectors course at FAA Academy, Oklahoma City. Was senior aircraft maintenance advisor to the Royal Saudi AF at Riyadh, Saudi Arabia and was promoted to lieutenant colonel. At Davis-Monthan AFB, he was chief industrial directorate and provided services to all U.S. Armed Forces, NASA, FAA, foreign governments with approval of DOD and U.S. State Dept. Sent to Sembach, Germany 17th AF, PCS; transferred to Zweibrucken AB due to sudden loss of maintenance officer; and 10 months later transferred back to Sembach AB as assistant deputy director of Logistics for 17th AF.

Spatafora retired April 30, 1976. His awards and decorations include the Bronze Star, Meritorious Service with two OLCs, Army Good Conduct with two clasps, Army of Occupation with Germany Clasp and Berlin Airlift Device, Medal for Humane Actions, National Defense Service Medal with one OLC, Vietnam Service Medal with three Campaign Stars, Armed Forces Reserve, Air Force Outstanding Unit Award, with two OLCs, Air Force Commendation, Joint Service Commendation, RVN Campaign, RVN Gallantry Cross with Palm and Small Arms Expert Ribbon.

Married Dec. 24, 1949 to Elfriede Kruger of East Prussia. They have daughter, Jacqueline, and sons, Ronni and Jeffrey.

DONALD W. SPEAREL, born April 2, 1929, in Lowell, MA. Enlisting in the AAF in 1946, he became an aviation cadet in 1947. After receiving his wings in June 1948, he served as a troop carrier pilot flying C-47s, C-82s and CG-15 gliders at Lawson AFB, GA. He then flew C-47s and C-46s for 15 months in Korea. Spearel flew with the 7167th SAM Squadron out of Rhein Main for 30 months 1954-56, before reassignment to CINCUSAFE's flight crew.

Operational and staff assignments followed at Bolling, MacDill, Andrews and Wiesbaden prior to a 1969 combat assignment at NKP, Thailand. He served as Wing DO at McChord C-141s in the early 1970s, and retired with over 15,000 hours on Oct. 1, 1978, while director of Training, 22AF, Travis AFB.

After five years as chief pilot of a commuter airline, he returned to college receiving a business management degree in 1988.

His many awards include the Legion of Merit, Distinguished Flying Cross with 10 oak leaf clusters, Meritorious Service Medal and Air Medal with 11 oak leaf clusters.

He resides in Clearwater, FL, with his wife, Ellen. Sons Warren and Bruce survive their sister Leslie Ann, who died in a car accident in 1979, and Capt. Don Jr., who was killed in an F-4S accident while serving an exchange tour with the USMC in 1983.

Operation Haylift, 1949. The accompanying photos were taken at Kearney AFB, Nebraska, on Jan. 27, 1949. The aircraft and crews in these photos were assigned to the 4408th Air Base Squadron, Lawson AFB, Ft. Benning, GA, although other units, including the 314th Troop Carrier Wing from Sewart AFB, TN, were involved in flying the 10 C-82 and 10 C-47 aircraft utilized. The mission, officially 9AF Mission 2444 and nicknamed Operation Haylift or Operation Hayride was conceived as a result of the unusually heavy accumulations of snow in the western states that winter. The concept was to drop (free fall) bails of hay from the rear doors to the starving cattle from an altitude of 50 to 100 feet. Although initially frightened, it didn't take long for these critters to figure out that the big, noisy bird roaring overhead meant chow-time. Airlift of equipment such as snow cats throughout the West was also accomplished during the approximate two-week period of Operation Haylift.

ROBERT D. SPRINGER, Lieutenant General, USAF, (Ret.), prior to his retirement on Oct. 1, 1988, Springer was Vice Commander in Chief, Military Airlift Command, with Headquarters at Scott Air Force Base, Illinois. In that position he had responsibilities for more than 92,000 personnel... 1,000 aircraft ... and a 7 billion dollar budget.

Springer was born Jan. 7, 1933 in Millheim PA, and graduated from the Patton Masonic School, PA in 1950. He earned a Bachelor of Arts Degree in Social Sciences in 1964 and a Master's Degree in International Affairs in 1969, both from George Washington University. He completed Squadron Officer School in 1959, Industrial College of the Armed Forces in 1966 and was a distinguished graduate of both the Air Command and Staff College in 1964 and the Air War College in 1972.

He entered the U.S. Air Force in October 1952 and received his commission and navigator wings through the aviation cadet program in August 1953. He entered B-29 combat crew training school at Randolph Air Force Base, Texas, and in January 1954 was assigned to the 344th Bombardment Squadron, Yokota Air Base, Japan. In March 1957 General Springer entered pilot training with the 3307th Pilot training Group at Marana Air Base, Arizona and in June 1957 with the 3301st Pilot Training Group at Moore Air Base, Texas. He was awarded his pilot wings at Reese Air Force Base, Texas, in March 1958. General Springer then returned to Randolph where he attended KC-97 co-pilot training and later became the 4397th Air Refueling Wing intelligence officer.

In May 1960 General Springer became an instructor at the Squadron Officer School and the Warfare Systems School, both at Maxwell Air Force Base, Alabama. In January 1964 he entered the Air Command and Staff College. From July 1964 to September 1965 he served as a C-130 pilot at McGuire Air Force Base, New Jersey.

The general then was assigned to the Republic of Vietnam, first at Bien Hoa Air Base as an intelligence officer with the 6251st Tactical Fighter Wing and later as Chief of Intelligence with the 1st Air Commando Squadron at Pleiku Air Base. Upon his return to the United States in September 1966, he was again assigned as an instructor at the Squadron Officer School. In August 1969 he transferred to the Directorate of Doctrine, Concepts and Objectives, headquarters U.S. Air Force, Washington D.C., as a planning and programming officer. After completing Air War College in May 1972, he was assigned to Air Force headquarters as special assistant to the Air Force Vice Chief of Staff.

In May 1974 Springer was assigned to Pope Air Force Base, NC, where he served successively as Assistant Deputy Commander for Operations of the 317th Tactical Airlift Wing. In June 1978 he took command of the 435th Tactical Airlift Wing at Rhein-Main Air Base, West Germany. He then transferred to Ramstein Air Base, West Germany, as commander of the 322nd Airlift Division. Springer became Deputy Chief of Staff for Personnel at Military Airlift Command Headquarters in August 1981.

The general returned to Randolph Air Force Base in July 1982 as Assistant Deputy Chief of Staff, Manpower and Personnel for Military Personnel, headquarters, U.S. Air Force, and Commander, Air Force Manpower and Personnel Center. The general became commander of the 21st Air Force, McGuire Air Force Base, in October 1984. In August 1985 he again was assigned to Air Force headquarters as Air Force Inspector General. He serves as Vice Commander in Chief Military Airlift Command from July 1987 until his retirement in October 1988.

The general is a command pilot with more than 6,000 flying hours, and 72 combat missions in Southeast Asia. His military decorations and awards include: Defense Distinguished Service Medal, the Air Force Distinguished Service Medal, Legion of Merit, Distinguished Flying Cross, Bronze Star Medal, Meritorious Service Medal with three oak leaf clusters, Air Medal with one oak leaf cluster, Presidential Unit Citation, Air Force Outstanding Unit Award with three oak leaf clusters, National Defense Service Medal with one service star, Korean Service Medal, Vietnam Service Medal with one service star, Korean Service Medal, Vietnam Service Medal with three service stars, Humanitarian Service Medal, Air Force Overseas Ribbon–Long, Air Force Overseas – Short, Air Force Longevity Service Award Ribbon with six oak leaf clusters, Armed Forces Reserve Medal, Small Arms Expert Marksmanship Ribbon, United Nations Service Medal and Republic of Vietnam Campaign Medal.

He was promoted to Lieutenant General August 2, 1985, with the same date of rank.

Springer is married to the former Bonnie Brubaker of Elizabethtown, PA. They have five children: Robert Jr., Debra Miller, Curtis, Michele, and Tania. General and Mrs. Springer retired to Southern Pines, NC, where is a consultant ... lecturer ... and the Executive Director for the Air Force ROTC's Arnold Air Society. He is also the Executive Director, Air Force Memorial Foundation and the Senior Vice President and life member of the Airlift/Tanker Association.

BENJAMIN FRED STARR, born Dec. 14, 1924, in Atlanta, GA. He joined the USAF in September 1943, serving as a command pilot and navigator in MAC: Scott, McChord, Norton, Andrews; in SAC: D-M Biggs, Smokyhill March. He served in Vietnam 1968-69, and Dir. Transportation, HQ USAF 1975-77.

Awards/Medals: DJM, Distinguished Flying Cross, Legion of Merit with one oak leaf cluster, Bronze Star, Air Medal with eight oak leaf cluster, Air Force Commendation Medal, Air Force Presidential Unit Citation with one oak leaf cluster, Air Force Outstanding Unit Award.

He was discharged Feb. 29, 1980, with the rank of major general.

He and his wife, Virginia White Starr, have two boys, Rick (deceased) and Mark and a daughter, Cindy.

DUANE B. STEINKE, born Nov. 5, 1932, in Hope, ND. He joined the USAF Feb. 24, 1954, serving at Lackland AFB, Mission AB, Laredo AFB, Ashiya AB and Tachikawa AB, Japan, Sewart AFB, CCK AB, Langley AFB and APO San Francisco 96307.

Awards and decorations include Bronze Star Medal, Meritorious Service Medal, Air Medal, Joint Service Commendation Medal, Air Force Commendation Medal with one oak leaf cluster, Air Force Outstanding Unit Award, Armed Forces Expeditionary Medal, Vietnam Service Medal.
He was retired March 1, 1974, with the rank of lieutenant colonel and 20 years and seven days of service.

He and his wife Kathlyn J., have three children and eight grandchildren.

SUSAN STEWART, born Nov. 19, 1961, in Ebensburg, PA. She enlisted in the USAF in 1979, as a Romanian Linguist. After a tour at the National Security Agency, she retrained as a Russian linguist. She was then stationed at Misawa, Japan, as the noncommissioned officer in charge, Voice Processing. Receiving her BA in management from the University of Maryland and her MS in education from USC, she entered OTS in 1988.

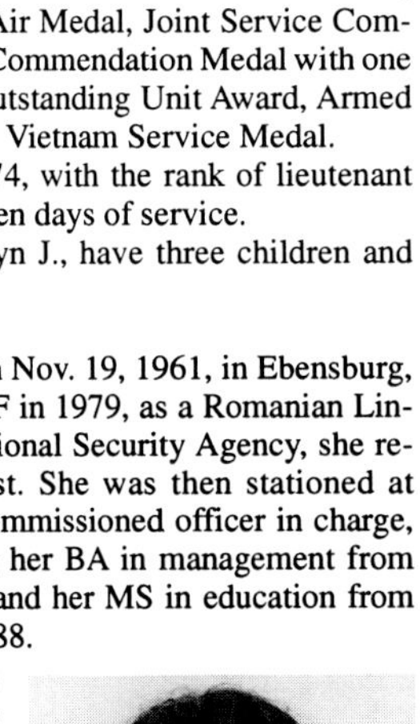

Since her commission in 1989, she has served as the squadron section commander of the 19th Organizational Maintenance Squadron and the 99th Air Refueling Squadron, Robins AFB, GA. She has recently been selected to attend the Air Force Institute of Technology in May 1993, to earn a master's in management information systems.

Among her many awards and decorations are: Meritorious Service Medal and Air Force Commendation Medal, as well as nominated for the 12 Outstanding Airman Award for the Electronic Security Command in 1987, the Robins AFB Company Grade Officer of the Year for 1990, and the winner of Strategic Air Command's Essay Contest for the Year of Reviewing Fundamentals in July 1989.

Stewart and her husband, Eugene, have three sons: Scott, Keelan and Niklaus.

JOHN LARRY STRUBE, Colonel, was commissioned through the ROTC at Auburn University in 1974. Following navigator training at Mather AFB, CA, he was assigned to the 41 MAS at Charleston AFB, SC, flying C-141s. He held a variety of positions in the squadron including flight examiner navigator. In 1979, he became the 437 MAW tactics navigator and was selected for the initial crew which developed procedures for blacked-out landings using night vision goggles.

In 1981, he was assigned to Headquarters MAC, Scott AFB, IL, working first in the Airlift Management Division as a C-130 airlift director and later as C-141 command navigator in the Aircrew Standardization/Evaluation Division. While assigned to Stan/Eval, he was the project for Volant Rodeo, the command's international airlift competition.

In 1985, Col. Strube attended the Armed Forces Staff College and upon graduation was assigned to Headquarters USAF, the Pentagon, as an action officer in the International Negotiations Division. He was later the executive officer for the director of Regional Plans and Policy.

Returning to flying in 1990, he was assigned to McChord AFB, WA, where he served as operations officer, 8 MAS and then as commander, 4 ALS. He has the distinction of being the first navigator selected to command a C-141 flying squadron. Col. Strube is currently a student at the Army War College, Carlisle Barracks, PA.

Col. Strube is a master navigator with over 4,600 flying hours. He has flown on a variety of missions including the 1993 Antarctica midwinter airdrop to McMurdo Station and the South Pole. He and his wife, Patty, currently reside in Carlisle with children Lauren, John Jr. and Steven.

DAVID O. SWAIN, Senior Vice President of Transport Aircraft, McDonnell Douglas Aerospace, born in 1942 is responsible for the business, technical, and production operations of all government business which focuses on airlift and tanker aircraft. Major programs include the C-17 and the KC-10.

In October 1992, Dave was assigned as senior vice president. Since April 1991, Dave had been assigned as the executive vice president-government segment, Douglas Aircraft Company. Prior to this, he was vice president-general manager of the C-17 Program since October 1990.

Dave began his career with MDC in 1964, when he joined McDonnell Douglas Astronautics Company to become an engineer on the Gemini Project.

Dave is a native of Indiana, graduating from Purdue University in 1964, with a bachelor's degree in aeronautical engineering. He is the recipient of the 1993 Distinguished Engineering Alumnus Award from Purdue University.

Dave is a member of the American Institute of Aeronautics and Astronautics (AIAA), the MDC representative on the National Defense Transportation Association (NDTA) Military Airlift Committee and also is a member of the C-17 Chapter of the Airlift Association.

JAMES V. SWEENEY, Chief Master Sergeant, (Ret.), married with three children and four grandchildren. Enlisted December 1956. After completing aircraft mechanic school, assigned to the 9 Bomb Wing, Mountain Home AFB, ID, as a B-47 aircraft mechanic. In June 1960, assigned to the 4321st Strategic Missile Wing at Offutt AFB, NE as an aircraft mechanic on the EC-135C SAC Airborne Command Post (Looking Glass). Volunteered for C-130 flight mechanic school at Sewart AFB, TN, in 1966, with follow-on assignment to Ching Chung Kang (CCK) AB, Taiwan with the 50th TCS/314 TCW. Transferred to Hamilton AFB, CA, in 1968, with the 41st ARRS, returning overseas to Clark AB, R.P with the 772 TAS/773 TAS/463 TAW from September 1968 to July 1971 as a C-130B instructor/flight examiner flight engineer. Assigned to the 48 TAS, Forbes AFB, KS, from July 1971 to May 1972, returning to CCK AB and Clark AB as a flight examiner flight engineer with the 774 TAS from June 1972 through July 1973. Assigned to the 6594 Test Group from July 1973 through July 1978, as a JC-130B/H/P/N instructor/flight examiner flight engineer. Returned from overseas to the 36th TAS, McChord AFB, WA, in July 1978, as chief flight engineer. Assigned to HQ MAC/DOT as command chief flight engineer for Training from mid-1983, until retirement in February 1987. Awards and decorations include Air Force Distinguished Flying Cross with two oak leaf clusters, Meritorious Service Medal with one oak leaf cluster, Air Medal with eight oak leaf clusters, Air Force Commendation Medal with two oak leaf clusters, Army and Air Force Good Conduct Medals with oak leaf clusters, Vietnam Service/Campaign Medals. Retired with over 10,000 hours C-130 flying time. Employed by Flight Safety Service Corporation in Altus, OK, where he resides with his wife JoAnne.

LAUREL LOUISE TAFT, born Dec. 17, 1959, in Phoenix, AZ. She joined the USAF on May 27, 1978, in Phoenix, AZ. Military service included: 1978-83, Norton AFB, CA, 63 MAW, vehicle operator; 1983-89, Norton AFB, CA, 53 MAS, loadmaster; 1989-93, Altus AFB, OK, 57 ALS, instructor loadmaster; July 1993, moved to Travis AFB, CA, to finish out career as a loadmaster. Presently she holds the rank of E-5 staff sergeant. Will be discharged May 27, 1998.

Memorable experiences include 1977, graduating from Phoenix West High School; February 1990, becoming the first female airdrop qualified in the C141B.

Awards/Medals: Air Force Commendation Medal, Air Force Achievement Medal, Air Force Outstanding Unit Award, Combat Readiness Medal, Air Force Good Conduct Medal, National Defense Service Medal, Southwest Asia Service Medal, Air Force Longevity Service Award Ribbon, NCO Professional Military Education Graduate Ribbon, and Air Force Training Ribbon.

She is married to SSGT Bruce Ivan Taft of San Jose, CA. He is a flight engineer on the C141B. They presently have a son, Michael with a desire for two more and upon retirement (1998) plan to live in Colorado.

JOHN L. TAYLOR JR., born July 16, 1928, in Cumberland, MD. He graduated high school and attended one year of college.

He joined the USAF on Feb. 14, 1949. Military locations and stations include: Lackland, Chanute, Sewart, Komaki, Ashiya AFBs. He served with the following units: 37th and 50th Troop Carrier Sqdn. and 314th Combat Cargo.

Taylor's memorable experiences include being a crew member of the C-82 Packet during Operation Portrex in the Caribbean, crossing the Pacific by Air at start of the Korean War, and receiving a commendation letter from FEAF Commanding General Tunner.

He achieved the rank of staff sergeant and was discharged on Oct. 7, 1952. Civilian employment includes being an aircraft inspector for Fairchild Aircraft Company in Hagerstown, MD. He retired Sept. 25, 1987.

SCOTT E. TAYLOR, graduated from Redford Union High School in Detroit, MI, in June 1968. Entered the Air Force four months later, completed Aircraft Loadmaster School, and reported to Lockborne AFB, OH (presently Rickenbacker AFB) to begin his flying career on Lockheed C-130A's. He was qualified in minimal time due a rotation to Germany that the squadron had been obligated. After nine months in Ohio, he returned to Clark AB, RPI, flying C-130B's with the Red Foxes of the 772nd Tactical Airlift Squadron. For the next 18 months he flew over 368 combat missions shuttling cargo and people throughout Southeast Asia from Cam Rahn Bay, RVN.

His next assignment took him to Headquarters TAC, Langley AFB, VA, where he flew C-130E's with the 36th TAS. After five years of rotations to England, Panama and several humanitarian and disaster relief missions he left for McChord AFB, WA, to deliver the first C-130, after MAC assumed control of all airlift aircraft.

His background as an instructor loadmaster, joint airdrop inspector, combat tactics and evaluator loadmaster made him a logical choice to join the Flight Test Center at Edwards AFB, CA. While stationed there he participated in numerous experimental flight test programs including the Space Shuttle Crew Escape System, KE-3A cargo loading, Low Altitude Retro Rocket System, MC-130H Combat Talon II, AC-130U Gunship and C-17 Globemaster III.

He earned an associate degree in transportation management from Community College of the Air Force in 1985. He has been selected as NCO of the Quarter and SNCO of the year for his organizations several times.

His decorations include the Distinguished Flying Cross, Meritorious Service Medal, Air Medal (5), Aerial Achievement Medal, Joint Service Commendation Medal, Air Force Commendation Medal, Air Force Achievement Medal (3), Combat Readiness Medal (2), Air Force Good Conduct Medal (5), National Defense Service Medal (2), Air Forces Expeditionary Medal, Vietnam Service Medal (4), Republic of Vietnam Gallantry Cross with Device, and Republic of Vietnam Campaign Medal. He is married to his wife Judi and has one son.

F. KEITH TEDROW, Brigadier General, Commander, Chanute Technical Training Center, Chanute AFB, IL.

Gen. Tedrow was born Sept. 4, 1942, in Kansas City, KS. He earned a bachelor of science degree in business in 1964, and a master's degree in education in 1965, both from Kansas State Teachers College. The general completed Squadron Officer School in 1972, Air Command and Staff College in 1979, and Air War College in 1985.

He completed basic military training in May 1966, and was commissioned in October 1966, as a distinguished graduate of Officer Training School, Lackland AFB, TX. He remained at Lackland as a personnel officer, first with the 3700th Air Base Wing, then with the 3504th USAF Recruiting Group.

From December 1968 to December 1969, Gen. Tedrow was a personnel officer with the 366th Tactical Fighter Wing, Da Nang AB, South Vietnam. He entered pilot training at Moody AFB, GA, and received wings in December 1970. His first operational assignment was flying the EC-121 with the 964th Airborne Early Warning and Control Squadron, McClellan AFB, CA.

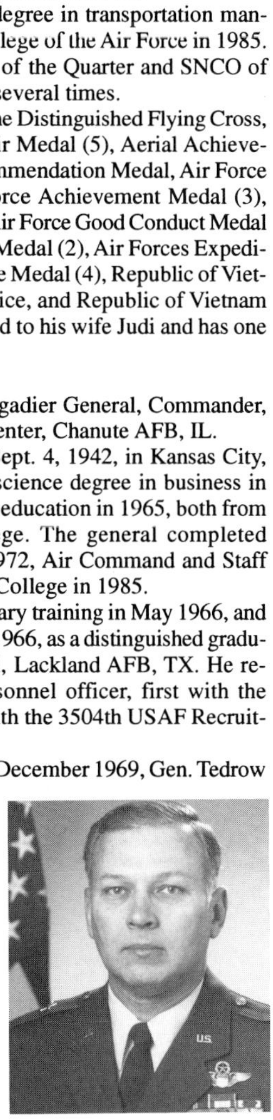

In 1973, the general participated in the Air Staff Training (ASTRA) program as a plans officer, deputy chief of staff for personnel, Headquarters U.S. Air Force, Washington, DC. Upon completion of the ASTRA tour of duty in 1974, Gen. Tedrow was assigned to Military Airlift Command, Travis AFB, CA, and flew the C-5A with the 22nd Military Airlift Squadron.

After graduation from the Air Command and Staff College, Maxwell AFB, AL, in 1979, he returned to the 22nd MAS as assistant operations officer. In September 1981, he assumed command of the 75th Military Airlift Squadron, Travis AFB.

Gen. Tedrow was assigned to Headquarters Military Airlift Command, Deputy Chief of Staff, Operations, Scott AFB, IL, in July 1983, as a senior airlift controller and chief of the MAC Command Center. He departed Scott to attend the Air War College in July 1984.

He returned to the pentagon in 1985, as director of military support and chief of the Airlift and Rescue Division, Deputy Chief of Staff, Plans and Operations, Air Force Headquarters. In July 1987, he became vice commander of the 436th Military Airlift Wing, Dover AFB, DE, and in February 1988, assumed command of the wing. In September 1989, he returned to Scott AFB, as executive officer to the commander in chief, U.S. Transportation Command/Military Airlift Command. In April 1990, Gen. Tedrow became inspector general for the Military Airlift Command. In February 1991, he was assigned as deputy chief of staff for Technical Training, Headquarters Air Training Command, Randolph AFB, TX. He assumed his present position in November 1991.

The general is a command pilot with more than 4,000 hours, including 320 combat hours. His military awards and decorations include the Legion of Merit, Bronze Star Medal, Defense Meritorious Service Medal, Meritorious Service Medal with two oak leaf clusters, Air Medal and Air Force Commendation Medal with two oak leaf clusters.

He was promoted to brigadier general Jan. 1, 1991, with same date of rank.

Gen. Tedrow is married to the former Anne Alexander of Sacramento, CA. They have a son, Jason.

EDWIN E. TENOSO, Major General, is director of Operations and Logistics (J3/J4), U.S. Transportation Command (USTRANSCOM), Scott AFB, IL. A native of California, he joined the Air Force as an aviation cadet, earning his commission and navigator wings. He later attended pilot school, then flew 552 combat missions in 0-1 Bird-dogs as a forward air controller. After Vietnam, he became an airlifter at Travis AFB, CA, flying C-141s. He later commanded the 18th Military Airlift Squadron, McGuire AFB, NJ, and the 62nd Military Airlift Wing, McChord AFB, WA. During Desert Shield/Storm, he was the commander of Airlift Forces in Southwest Asia with 149 C-130s assigned. The general is a navigator and command pilot with over 7,800 flying hours. He is a distinguished graduate of Squadron Officer School and Air Command and Staff College and attended the Industrial College of the Armed Forces. He has a BA degree from Florida Southern College, a MS degree from Troy State University, and a MBA from Marymount College of Virginia. Gen. Tenoso is married to the former Kathleen Crandall of Orlando, FL. They have a son, Eric.

JAMES T. THOMAS, graduated from Nanticoke High School in Nanticoke, PA, in 1957. He entered the Air Force in October of that same year, completed Medical Administrative Specialist Training at Gunter AFB, AL, and reported to the USAF Hospital, Sheppard AFB, TX, in May 1958.

At Mather AFB, CA, he was selected as USAF Hospital Mather Senior NCO of the Year, Base Senior NCO of the Year, and First Sergeant of the Year for 1984. Completed the ADC NCO Academy and USAF Senior NCO Academy in residence.

His decorations include the Meritorious Service Medal with two oak leaf clusters, Air Force Commendation Medal with three oak leaf clusters, and Air Force Outstanding Unit Award with V device and two oak leaf clusters.

CMSgt. Thomas retired in 1987 last serving as DGMC Senior Enlisted Advisor at Travis AFB, CA. He has two sons: James T., II and James Edward.

STEVEN F. TOMHAVE, Colonel, born Jan. 11, 1946, in Austin, MN. He graduated from the University of Missouri in 1968, with a bachelor's degree in political science and in 1976, received a master's degree in operations management from the University of Arkansas. He is a graduate of Squadron Officer School, Air Command and Staff College, the U.S. Army Command and General Staff College, the Air War College, and the Army War College.

Commissioned through the Reserve Officer Training

Corps in 1968, he completed pilot training at Vance AFB, OK, and flew the C-130 at Ching Chuan Kang AB, Taiwan. Other C-130 assignments include Forbes AFB, KS; Little Rock AFB, AR; and then as squadron commander at Dyess AFB, TX. He was a readiness initiatives analyst at Headquarters Military Airlift Command, Scott AFB, IL, and an exchange officer with the Canadian Forces in Trenton, Ontario. In Washington, DC, he served as the JCS Senior Military Representative to the Department of State Operations Center and Chief of the Command Center Integration Branch of the Command and Control Division, Operations Directorate, Joint Staff.

His decorations include the Defense Superior Service Medal, Meritorious Service Medal with two oak leaf clusters, Air Medal with one oak leaf cluster, and the Air Force Commendation Medal. He is a command pilot with over 4,500 flying hours.

In August 1991, Col. Tomhave was named vice commander of the 314th Airlift Wing, Little Rock AFB, AR. He and his wife, the former Jetty Ann McDonald of Austin, TX, have a daughter, Joy Christine, and a son, Scott David.

PHILLIP B. THOMPSON, graduated from Pulaski High School, June 1952, attended the University of Wisconsin (1952-53), Milwaukee Institute of Technology (1954) and California Polytechnic (1960-61) majoring in M.E. and Aero.

He was employed at Lockheed, Burbank, CA, 1956-62, where he worked on the design of the F-104 Starfighter. He worked at McDonnell Aircraft, St. Louis, Space General Corp. and Lockheed Georgia on various military design projects in the 1960s. Employed at Douglas Aircraft in Long Beach, CA, working on the design of commercial airliners, 1964-85, he was re-assigned in early 1986, to the C-17A Globemaster III program as a group leader in the Mission Systems Accommodations Team overseeing the design of the crew seats, troop seats, ditching system, crew galley, crew lavatory, potable water system, insulation blankets and sidewall liners. The design years on the C-17A were particularly rewarding leading up to the 1st flight of the C-17A on Sept. 15, 1991.

Mr. Thompson has been employed in aviation over 40 years and continues his career looking forward to the successful deployment of the C-17A Squadrons. He also is looking forward to being involved in the design and development of the future trend setting aircraft like the MD-12, MD-XX and the supersonic civil transport which represent the future of McDonnell Douglas Aircraft. He is a member of the See Seventeen Chapter.

Mr. Thompson and his wife Sharon have four grown children.

CHARLES G. TRESER, Chief Master Sergeant, born in Indianapolis, IN, on March 26, 1939. His formative years were spent as a farm boy in central Indiana. Following entry into the Air Force in February 1958, he received his basic training at Lackland AFB, TX, and formal technical training at Chanute AFB, IL, in the Aircrew Protection and Environmental Life Support Systems career field.

He held assignments as a life support specialist and technician at Dyess AFB, TX; Dover AFB, DE; Tan San Nhut AB, Vietnam; Naha AB, Okinawa; Perrin AFB, TX; and McClellan AFB, CA. With the assignment to Tan San Nhut as an A1C (E-4) in May 1963, he became the first Life Support person to be assigned PCS to Southeast Asia. With the initial Advisory Program, he taught Aircrew Life Support Systems to the Vietnamese Air Force. As a technical sergeant at McClellan, he was the NCOIC of one of the largest pools of Life Support equipment in the Air Force. His unit supported 90 EC-121 aircraft with 26 crew positions per aircraft. This mission was conducted on a global scale from several worldwide operating locations.

In 1972, Chief Treser entered training in the weather observer career field. Upon completion of formal technical school, he was assigned to Carswell AFB, TX, where he underwent training as a detachment chief observer.

The chief was the chief observer at Carswell AFB, TX; Shemya AFB, AK; and Dyess AFB, TX. In 1977, he was assigned as the Squadron Weather Operations Superintendent, 2nd Weather Squadron, Andrews AFB, MD, supporting Air Force Systems Command (AFSC). His duties included monitoring weather observing activities at all AFSC bases and assistant to the manager of the Air Force Environmental Rocket Sounding System (AFERSS). Annual visits were made to each Rocketsonde site at such remote locations as Shemya Island in the Aleutians, and Thule, Greenland. Chief Treser completed Weather Forecasting Training in 1979, and was assigned to Detachment 7, Air Force Global Weather Central, Carswell AFB, TX, in March 1980, as the Detachment Weather Superintendent; he was responsible for the management and training of 36 weather personnel operating as the hub of the Automated Weather Network that provided meteorological data worldwide to Air Weather Service customers. In October 1982, Chief Treser reported to Hickam AFB, HI, as the senior enlisted advisor to the Commander, 1st Weather Wing (MAC), supporting PACAF, with 28 units spread over a geographic area of more than 30 million square miles. Frequent visits to each unit required extended periods of travel that resulted in his being TDY about 40 percent of the time.

Chief Treser was actively involved in the 165 member USAF Chiefs' Group of Hawaii as chairman of the Ways and Means Committee, 1983-84, and president in 1985. He served as membership committee chairman for the rapidly growing Hawaii Chapter of the NCO Academy Graduate Association. As chairman of the "Hawaii MAC Top 4" Association Awards Committee, he coordinated the selection of quarterly Airman, NCO, and SNCO Outstanding Performer awards.

The Chief was assigned to Scott AFB, IL, in November 1985, as the senior enlisted advisor to the Commander, 7th Weather Wing. He frequently visited the 30 weather units throughout the CONUS and the Azores at MAC, AFLC and ANG bases.

In December 1987, he was selected to be the senior enlisted advisor to the commander of the 375th Aeromedical Airlift Wing (later redesignated as 375th Military Airlift Wing). He spent a great amount of time visiting the wing's 28 worldwide geographically-separated units.

Additional education accomplishments of Chief Treser are: NCO Preparatory School, 1962; MAC NCO Academy, 1974; AF Rocketsonde Manager Course, 1975; Senior NCO Academy correspondence course, 1975; Squadron Officer School correspondence course, 1975; AF Rocketsonde Indoctrination course, 1978; and Senior NCO Academy in residence, 1979. Chief Treser has completed three years of college, working toward a degree in business management. He was promoted to his current rank on Jan. 1, 1982. Chief Treser's decorations include the Meritorious Service Medal with one Silver Oak Leaf Cluster, the Air Force Commendation Medal, the Air Force Achievement Medal, the Air Force Good Conduct Medal, the National Defense Service Medal, the Armed Forces Expeditionary Medal, the Vietnam Service Medal, Vietnam Gallantry Cross with palm and the Vietnam Campaign Medal.

Chief Treser is married to the former Sophia Withrow of Abilene, TX. They have six children and 11 grandchildren. During a few months prior to his retirement in February 1991, at Scott AFB, he was the active duty enlisted person with the most service time in the USAF. Charles and Sophia live in Belleville, IL, where he manages a large vending operation.

ROBERT L. TRIMPL, Lieutenant Colonel, born April 26, 1938, in Peoria, IL. He graduated ROTC June 5, 1960, Coe College, IA, second lieutenant effective June 18, 1960. He served the USAF as a master navigator and parachutist. His final assignment was as Chief Policy and Doctrine Division, Headquarters MAC, Scott AFB, IL. Serving at Sewart AFB, TN, AFIT Kent State University, OH; Clark AB, PI; Langley AFB; Scott AFB.

Memorable experiences include planning Clove Hitch 1, Commandant, Airlift Operations School.

He received the Vietnam Campaign Medal, Distinguished Flying Cross, Defense Meritorious Service Medal, Meritorious Service Medal, Air Medal with five oak leaf clusters. He was discharged June 30, 1982, with the rank of lieutenant colonel.

He married Alice T. Kelly, on Oct. 19, 1963, at Sewart AFB. They have three children and two grandchildren. Today he is vice president, Education Loan Servicing Center, USA Grove, national board member, TROA, 1992-93, Who's Who in the Midwest.

PAUL D. TUCK, started his career in transportation while still in high school. He drove with and for various truckers, loading things like coal and railroad cross ties. He was assigned to run the largest Air Terminal in England supporting the Normandy Invasion, The Breakout and as the allied advanced across western Europe.

In 1945, as executive officer to Col. Bernt Balchen (Arctic expert and polar explorer with Adm. Byrd) he participated in a classified special airlift operation in Neutral Sweden which airdropped and airlanded cargo and troops into Northern Norway. This project was controlled by the Air Transport Command and the office of Strategic Services (forerunner of the CIA).

Following the above, Paul had several important assignments running air terminals for the European Air Transport Service in France and Germany. Following these he served with the Military Air Transport Service.

His introduction into the Pentagon came with a 1955 to 1958, tour in the Directorate of Transportation. His MATS service culminated in the position of deputy director for Transportation from 1963 to 1964.

After retiring from active duty in 1964, Paul was the Transportation Planning Officer for the Army deputy chief of staff for Logistics and a traffic management specialist for Military Traffic Management and Terminal Service. In 1967, he assumed duties as Transportation Analyst, in the office of assistant chief of staff for studies and analyses, routinely directing major study efforts. He served here until 1984 and became an independent consultant.

He completed a BA degree from the University of Massachusetts and MA from the University of Maryland.

Awarded the Exceptional Civilian Service Award. Listed in Who's Who in Aviation and Aerospace.

WILBERT TURK, joined the Air Corps March 1942. Rated pilot and commissioned in January 1943, he was assigned to the 65th TCS. For nearly two years with the 5th Air Force in WWII, he flew air lift missions throughout the Southwest Pacific. Significant operations included the 1943, airborne assault in the Markham Valley, New Guinea and the highly successful rescue of Allied POW's from the Japanese prison camp, Los Banos, Philippines in 1945.

Later operational assignments included: Commander, 61st TAS when it pioneered ski-equipped C-130 operations on the Greenland ice cap and in Antarctica; Commander 483 TAW in RVN and Commander, 322 TAW in Germany. Other assignments included staff duty at Headquarters, USAF, Joint Staff (OJCS), and NATO.

Awards and decorations: Legion of Merit with two oak leaf cluster, Distinguished Flying Cross with two oak leaf clusters, Air Medal with eight oak leaf clusters, Joint Staff, Air Force and Navy Commendation Medals, and numerous campaign and service awards.

He is a graduate of the Air Tactical School, Air Command and Staff School, Armed Forces Staff College, and Air War College. Retired in 1971. He and his wife Katherine live in Montgomery, AL. They have two children and three grandchildren.

MICHAEL W. TURNER, is a native of Georgetown, SC. He attended high school in Morristown, NJ. After graduating from the University of North Carolina in 1967, he received his commission through OTS. He completed pilot training at Laredo AFB, in 1968, and flew C-141s at McChord AFB. He was then assigned to C-130s at Ching Chuan Kang AB, Taiwan, where he flew 113 combat missions in Vietnam. In 1972, he became a special operations staff officer at Udorn RTAFB, Thailand. From 1973-78, he was a C-130 instructor and flight examiner at Little Rock AFB. He then served as an initial cadre academic instructor at the HQ MAC Airlift Operations School. He was selected as CINCMAC's speechwriter in 1983. From 1984-86, he was commander of the 435th Transportation Squadron, Rhein-Main AB, Germany. In 1985, the squadron's maintenance and operations branches were selected Best in MAC. He became Chief of Public Affairs at Wright-Patterson AFB in 1986. He retired in the rank of colonel on Sept. 1, 1991.

Decorations: Defense Meritorious Service Medal, Meritorious Service Medal with four oak leaf clusters, Bronze Star Medal, Air Medal, Humanitarian Service Medal, etc. Alive in '65 and can shoot straight.

THOMAS M. TURNER, Colonel, Commander and Professor of Aerospace Studies at the University of South Carolina, Columbia, South Carolina.

Col. Turner was born Jan. 21, 1943, in Duncan, SC, where he also graduated from James F. Byrnes High School. He was commissioned in the Air Force through the Reserve Officer Training Corps in 1966, and then attended pilot training at Craig AFB, AL. In 1968, he was assigned to the 606th Special Operations Squadron, Nakhon Phanom Royal Thai AB, Thailand, as a UC-123 Forward Air Controller, followed by assignment to the 18th Military Airlift Squadron, McGuire AFB, NJ, as a C-141 pilot. In 1971, he was assigned to the Headquarters Military Airlift Command Briefing Team at Scott AFB, IL. Between 1973 and 1977, Col. Turner served as a career advisor, resource manager and personnel programs officer at the Air Force Military Personnel Center, Randolph AFB, TX. He was then assigned for two years in Jakarta, Indonesia, as operations and training officer with the U.S. Defense Liaison Group. In 1979, he became the assistant operations officer for the 20th Aeromedical Airlift Squadron and in 1980 moved to the wing executive officer position, 374th Tactical Airlift Wing, Clark AB, Philippines. Upon returning to the states and Scott AFB in 1981, he was assigned to Headquarters Military Airlift Command, DCS Personnel as Chief, Officer Assignments Division, and subsequently as deputy director of Assignments. In May 1984, he became commander of the 1401st Military Airlift Squadron at Scott, and then moved to the position of Deputy Commander, 375th Air Base Group, in December 1985. After completing the National War College in June 1987, Col. Turner was assigned to the Pentagon as Chief, Office of Inspection for The Inspector General, Office of the Secretary of the Air Force. In June 1989, he was selected to be the military assistant to the administrative assistant to the secretary of the Air Force. Col. Turner assumed his present duty in August 1991.

Col. Turner has a bachelor of science degree in business management from the University of South Carolina and a master's degree in personnel management from Webster University. He completed The National War College, Air War College, Industrial College of the Armed Forces, Air Command and Staff College, and Squadron Officer School.

Col. Turner is a command pilot with over 4,000 flying hours in the C-21, C-141, C-9, C-12, T-39, C-131, T-29, and UC-123, including over 600 combat hours in Southeast Asia. Military awards and decorations include: the Legion of Merit, the Distinguished Flying Cross, Purple Heart, Defense Meritorious Service Medal, Meritorious Service Medal with three oak leaf clusters, Air Medal with 10 oak leaf clusters, and Air Force Commendation Medal with one oak leaf cluster.

Col. Turner is married to the former Mary Ann Bennett of Duncan, SC. They have two children, Tommy, age 21, and Jennifer, age 19.

DARRELL VAUGHN, born on Jan. 25, 1942, in Maysville, KY. He entered the Air Force in 1962, he was assigned to SAC. He cross trained into the loadmaster career field and qualified in the C130B and E model aircraft. He completed a tour in Southeast Asia in 1969, taking part in the Tet Offensive and evacuation of Kham Duc during his tour. He left active duty in 1970, and entered the Air Force Reserve, he continues to serve as a C141B loadmaster examiner. He currently holds the rank of senior master sergeant. Darrell's decorations include the Distinguished Flying Cross, Air Medal, Aerial Achievement Medal, Air Force Outstanding Unit Award with V device, Vietnam Service Medal, Southwest Asia Service Medal and Kuwait Liberation Medal. He has accumulated 5,500 hours in the C130, C141B and C123 type aircraft.

GERARD G. VAUGHAN, Lieutenant Colonel, (Ret.) of O'Fallon, IL. Born in Boston, MA. Received commission through ROTC at Trinity College, Hartford, CT, in 1958. Following Air Force pilot training at Malden AB, MO, and Craig AFB, AL, his first assignment was to the Strategic Air Command, 2nd BW at Hunter AFB, GA, flying B-47 bombers. In 1963, he transferred to the 100th BW, Pease AFB, NH, up-graded to aircraft commander and continued flying B-47s.

In January 1966, he was assigned to the Military Airlift Command, 437th MAW, Charleston AFB, SC, to fly C-141s. In 1970 he was in the initial cadre from Charleston to transition to the newly acquired C-5. From 1970 through 1972, he flew the C-5 and worked in Airlift Operations. In 1973, he was an electronic reconnaissance pilot with the 361st Tactical Electronic Warfare Squadron, Nakhon Phanom (NKP) Thailand. While at NKP he was chief of Wing Standardization for all C and EC-47 operations.

In 1974, he was assigned to Headquarters, MAC, Scott AFB, IL, as senior controller in the MAC Command Center. Follow-on positions at MAC HQ were as Chief, Operational Aerial Refueling in 1975 ; Chief, MAC Special Assignment Airlift Missions in 1978; and Chief, MAC Command Center in 1980.

In 1982, he was assigned to Hill AFB, UT, as Chief, Operations Plans for the Ogden Air Logistics Center and Director, Command and Control for the 388th TAC Fighter Wing (F-16s). Jerry was a flight examiner and command pilot with over 7,000 hours of flying. His decorations include the Distinguished Flying Cross, three Meritorious Service Medals, four Air Medals, two Commendation Medals and the Combat Readiness Medal.

Following his retirement in 1985, Jerry accepted the position of national operations manager and director of retail, supply and distribution for a petroleum company in Irving, TX. He is married to Sandra Sivier of Pembroke, Bermuda and they have three sons, Michael of O'Fallon, IL, Gerard Jr. (Jay) of Ogden, Utah and Richard a college student in Ft. Worth, TX.

THEODORE R. VENTURINI, born May 28, 1932, in Luxembourg, Europe and immigrated to the United States via Canada in 1957. He entered the Air Force in 1958. His career spanned 22 years as an Air Force loadmaster and he is still today the McDonnell Aerospace-Transport Aircraft's chief loadmaster for the C-17. He claims the title as the oldest loadmaster still actively flying with the USAF on the C-17 Test Program.

Military service: 1965-68, loadmaster for Satellite and Space Vehicle Mid Air Recovery on JC-130s. From 1969-74, project loadmaster for the C-5A Test Program at Edwards AFB, the First Minute Man I Loading on the C-5A for Airdrop Testing; First F-111A Loading on the C-5A for Air Transport; First F-15A Loading on the C-5 for Air Transport; First F-16A Loading on the C-5 for Air Transport. From 1974-78, project loadmaster for the YC-14 and YC-15 AMST, Advanced Medium Shoot Takeoff and Landing Test Program from inception to the Flight Test Program at Edwards AFB. Member of the YC-14 "First Flight" crew. From 1979-93, chief project loadmaster for the CX and later the C-17 program from inception to first fight and the present test program at Edwards AFB. Member of the First Flight Crew. Loaded the First X24B on the C-5. Participated in a multitude of firsts too numerous to mention. He also included an article he wrote on C-17. He was discharged Aug. 31, 1979, with the rank of master sergeant.

He and his wife, Marcelle Venturini, have one daughter, Lydia Jowyk. Today he is chief loadmaster C-17 flight test McDonnell Douglas Corp.

CARL A. VOIGTSBERGER, born Feb. 6, 1939, in Philadelphia. He graduated from Roman Catholic High School in Philadelphia in 1957, and attended McMurry College in Abilene, TX. He received a bachelor of science in mechanical engineering from the University of Wyoming, Laramie, through the Airman Education and Commissioning Program in June 1964.

Voigtsberger retired from the U.S. Air Force in October 1977. During his last assignment, he served as Commander, 90th Civil Engineering Squadron at Warren AFB, WY. Previous assignments include: missile engineering officer (Minuteman III) at Warren AFB, 1972-77; senior navigator, C-130E at Langley AFB, 1968-72; and C-130B at Clark AB, 1966-68; and bomb/nav enlisted technician, B-47E at Dyess AFB, 1958-62.

Voigtsberger spent the years at Clark AB in the 29 TAS, F Troop, shuttling in and out of Tan Son Nhut AB, Vietnam and at Langley AFB in the 36th and 37th TAS, C Flight in support of Air Force STRIKE Command at MacDill AFB, FL.

Voigtsberger was married to Leora Joan, who passed away in 1973, and he now resides in Cheyenne, WY. They are the parents of four children: Gary, Joan, Cathy and Carl.

EARL VON KAENEL, was born May 17, 1918 in Forest Park, IL. He received a bachelors degree.

He joined the USAF on Feb. 11, 1942 and was located at the following stations: Love Field, Dallas, TX, North Africa, Hamilton Field, Fairfield/Suisun Army Base (now Travis AFB, CA), Hickam Field, HI, Fassberg, Germany (Berlin Airlift), Kelly AFB, West Palm Beach, Command and Staff (Alabama), Malstrom AFB, Oslo Norway, and Travis AFB.

He served with the following units: ATC, MATS, SAC, Spec Activities Branch and MAC.

Memorable experiences include pioneering airways in North Africa, the Berlin Airlift, Hawaii, Oslo, Norway, also participated in "Deep Freeze" - the Antartic supply mission.

He achieved the rank of Lt. Colonel and was discharged on Sept. 1, 1969. He is married to the former Mitzi Kappel, they have three children and one grandchild. Civilian employment includes being a R. E. Broker. He retired in 1977.

MICHAEL M. WELCH, Chief Master Sergeant, (Ret.), born in Newport, RI, on Nov. 11, 1945. He graduated from Fletcher High School, Jacksonville Beach, FL, in 1964,

Southern Illinois University in 1979, with a BS degree in education and Central Michigan University in 1982, with a MA degree in management and supervision. He entered the Air Force on Feb. 7, 1964. His assignments included McCoy AFB, FL; Hickam AFB, HI; Moody AFB, GA; Sheppard AFB, TX; Pope AFB, NC; Nakhon Phanom, RTAFB, Thailand; Charleston AFB, SC; Altus AFB, OK; and Wright-Patterson AFB, OH. He is the recipient of the Distinguished Flying Cross, Meritorious Service Medal with one oak leaf cluster, Air Medal with nine oak leaf clusters, Air Force Commendation Medal with two oak leaf clusters, Air Force Achievement Medal, Presidential Unit Citation, Air Force Outstanding Unit Award for Valor with five oak leaf clusters, Air Force Organizational Excellence Award with three oak leaf clusters, Combat Readiness Medal with five oak leaf clusters, and other numerous foreign and unit decorations. He accumulated more than 5,000 flying hours in C-130E, C-141A/B, KC-10A, C-23A and 151 combat missions in C-123K aircraft as a loadmaster. He retired May 1, 1988, and currently is the business unit manager, commercial product support, technical publications for McDonnell Douglas Corporation in Long Beach, CA.

WILLIAM WELSER III, Colonel, Commander, 436th Airlift Wing, Dover AFB, DE.

Col. Welser was born Aug. 2, 1949, in Mineola, NY. He attended public school in Baldwinsville, NY, and received a bachelor's degree in biology from the University of Buffalo in 1971. He earned a master's degree in personnel management from Webster University in 1981. His military education includes Squadron Officers School, Armed Forces Staff College, and National War College.

Col. Welser received his commission through the Air Force Reserve Officer Training Corps program at the University of Buffalo in 1971. He entered the aircraft Maintenance Officers Course at Chanute AFB, IL, in September 1971, and following graduation, was assigned to the 631st Organizational Maintenance Squadron, Norton AFB, CA, where he served as flightline maintenance supervisor. In 1973, he was assigned to the 8th Tactical Fighter Wing, Ubon Royal Thai AFB, Thailand, as an F-4 branch chief and job control duty officer.

Upon his return to the U.S., Col. Welser completed initial flight training at Hondo AFB, TX, and was awarded his wings at Columbus AFB, MS, in November 1975. Following graduation, he was assigned to the 20th Military Airlift Squadron, Charleston AFB, SC, as a C-141 pilot.

In 1979, Col. Welser was selected as initial cadre for the U.S. Air Force Officer Retention Group, Air Force Manpower and Personnel Center, Randolph AFB, TX. Following attendance at the Armed Forces Staff College, Norfolk, VA, in 1982 he was assigned to the 438th Military Airlift Wing, McGuire AFB, NJ. He first served as a C-141 instructor pilot with the 6th Military Airlift Squadron, then as chief of the Wing Current Operation Division, and finally as the operations officer and commander of the 6th Military Airlift Squadron.

Following graduation from the National War College in 1988, Col. Welser served as special assistant for General/Flag Officer Matters, Officer of the Director, Joint Staff, Washington, DC. In 1990, Col. Welser became vice commander of the 443rd Airlift Wing, Altus AFB, OK. He assumed his present position in August 1992.

Col. Welser is a command pilot with more than 3,000 flying hours in the C-141 and C-5. His military decorations include the Legion of Merit, Defense Superior Service Medal, Meritorious Service Medal with two oak leaf clusters, Combat Readiness Medal with one oak leaf cluster, Humanitarian Service Medal, and the Southwest Asia Campaign Medal. He was promoted to colonel on Oct. 1, 1989.

Col. Welser is marred to the former Susan Marie Rea of Syracuse, NY. They have four children Amie, William IV, Michael and Sarah.

JOHN V. WERNER, graduated from Thomas B. Doherty High School in Colorado Springs CO, in 1977. He entered the Air Force in 1979, completed weather specialist training and reported to Hunter AAF, GA, in support of a mechanized infantry division.

He was a distinguished graduate from the weather technician course at Chanute AFB in 1984. While serving at Malmstrom AFB, MT, since 1984, he has been selected base NCO of the Quarter on five occasions, chosen as the 15th Air Force Forecaster of the Year, nominated as one of the Twelve Outstanding Airmen of the Year, and awarded a laudatory finding on a no-notice Air Force Inspector General inspection. He is a 1991 winner of the Airlift Association Young Leadership Award.

He earned an associate degree in Applied Science from the Community College of the Air Force in May 1989, completed his bachelor's degree in mathematics at the College of Great Falls in May 1990.

His decorations include the Air Force Commendation Medal, Air Force Achievement Medal, Air Force Outstanding Unit Award with two oak leaf clusters, and the Air Force Good Conduct Medal with three oak leaf clusters. He and his wife, Cathy, have three children.

D. SCOTT WHITE, Lieutenant Colonel, USAF retired, was born Sept. 28, 1948, and graduated from N. Attleboro, MA, High School in 1966. He received a BA in political science from Brown University in 1970. He completed pilot training at Webb AFB, TX, in 1971. He received a MBA from Southern Illinois University in 1974. He completed Air Command and Staff College in 1984, and received an award for his paper on the "History of Airdrop."

He spent 18 years at Charleston and Altus accumulating over 9,200 C-141 flying hours. In 13 years in the 41 MAS at Charleston he worked for 13 different commanders. He flew missions in Vietnam, Nickel Grass, Just Cause, and Desert Storm. One of his most memorable experiences occurred in 1982. He planned, briefed, and led Reforger 82, a 20 ship flow of C-141s from Charleston to an airdrop in Germany recovering at Charleston. The mission included three air refuelings and logged 21.5 hours, the longest C-141 flight at that time. He retired with 22 years service in 1992 as Chief, Standardization/Evaluation in the 41 MAS with over 1,000 airdrops. He is currently employed by McDonnell as a C-17 simulator pilot instructor.

Scotty and his wife, Betty, reside in Summerville, SC. They have two daughters: Tara and Heather.

JAMES E. WHITE, Colonel, (Ret.), of Lyman, SC, was commissioned through the U.S. Naval Academy in 1951. In 1952, following pilot training at Spence Airfield/Reese AFB,

TX, he was assigned to Donaldson AFB, SC, flying C-119s.

In 1954, he transferred to Ashiya AB, Japan, again flying C-119s. He was then reassigned to Donaldson AFB, SC, in 1955, transitioning to the C-124 and participating in both the Congo and Lebanon Crises.

In 1960, he attended the Radar Maintenance Officers Course, Keesler AFB, MS. After graduation, he was assigned to Air Weather Service, Scott AFB, IL. He transferred to Tin City AFS, AK, in 1963, as a radar maintenance officer in an Airborne Early Warning Radar Site. Completing this remote tour, he move to Charleston AFB, SC, flying C-124s, later transitioning into C-141s in 1965.

In 1968, Jim transferred to Wright Patterson AFB, OH, serving as MAC C-5 liaison officer to the C-5 SPO. In 1971, he moved to Scott AFB, IL, as Chief, Airlift Division, Operational Requirements, DCS Operation and later DCS Plans. He later served as Deputy Director, Operational Requirements, DCS Plans until his retirement in July 1979.

Jim was a command pilot with over 7,900 flying hours. His decorations include the Legion of Merit, Air Medal, Air Force Outstanding Unit Award, Armed Forces Expeditionary Medal, Korean Service Medal and Vietnam Service Medal.

In 1980, he went to work for the Boeing Aircraft Company in Wichita, KS, until his retirement from there in 1991. Jim and his wife Loretta now reside in Greer, SC.

BRUCE NAIRN WHITMAN, born Aug. 4, 1933. He was commissioned in 1955, USAF, serving at Lackland, Bainbridge, Goodfellow, Homestead, Andrews. He was discharged in 1962, with the rank of captain. He and his wife have three sons. Today he is executive vice president of Flightsafety International.

ROY P. WHITTON, Colonel, (Ret.), born Dunreith, IN, Dec. 28, 1922. Enlisted Regular Army Air Corps Jan. 6, 1941. Completed Airplane Mechanics Course Chanute Field, IL, arrived at Albrook Field, Panama Canal Zone, November 1941. Assigned Aircraft Instrument Shop, 19th Material Squadron. June 1943 to June 1944, Instrument Shop Chief, 538th ABG, Galapagos Islands. Returned ZI July 4, 1944, and assigned Heavy Bomber (B-29/B-32) Instrument Shop Chief Wright Field, OH. On way to Okinawa with 41st Air Depot when WWII ended Assigned ATC, Romulus Field, MI. Discharged as staff sergeant November 1945.

Signed up AAF Enlisted Reserve Corps; June 1948, received direct commission second lieutenant. Volunteered active duty for Korean War, called to active duty Jan. 16, 1951. served as public affairs officer, Williams AFB, AZ, released from active duty September 1954.

Was Mobilization Augumentee public affairs officer HQ 9th AF TAC Shaw AFB, SC, 1954-58; HQ MATS/MAC 1958-68, Scott AFB, IL; SAFPA Pentagon, 1968-78.

Graduate Armed Forces Information School, Ft. Slocum, NY, 1951-53; Defense Information School, Ft. Harrison, IN, 1975; BS and MA Ball State University, Muncie, IN. Completed SOS, ACSC, AFSC, AWC, ICAF.

In civilian life was director of Civil Defense Education in Indiana, 1959-61; executive secretary Indiana State Highway Commission 1961-65; executive director of Indiana Health Care Association 1977-80; radio station manager; public relations agency account executive; industrial developer, Taylor County, FL, 1980-83.

Active in Air Force Association, Florida Highlands Chapter president; Florida State president and AFA National AFA vice-president. Named top AFA communicator for 1991. Retired and resides in Lake Placid, FL, with wife Kay.

THOMAS P. WILLIAMS, was raised in Waynesburg, OH. He attended Mount Union College and enlisted in the USAF, April 28, 1951, and was commissioned from aviation cadet status, Class 52-G, Oct. 25, 1952. His overseas tours include Korea (B-26), Saudi Arabia (C-47/C-54), Turkey (C-130 Airlift Advisor), and CCK AB, Taiwan (C-130). His airlift assignments include Dyess AFB, TX, where he flew the C-130 A/D/E and Forbes AFB, KS, as Commander, 48 TAS.

In these assignments, he completed many overseas rotational deployments and contingency operations. He was an airlift planner at HQ 19 AF and associated with aircrew standardization and evaluation at HQ TAC and, after airlift consolidation, HQ MAC. He was Deputy Commander, Resource Management, 314 TAW, Little Rock AFB, AR, and retired in 1982. He and wife Carol live in North Little Rock, AR, they have two daughters, Ann and Nancy. He is a graduate of the University of Maryland, George Washington University, SOS, ACSC and AWC. He is treasurer, Airlift Tanker Association, and past president, Air Force Association Arkansas and the Retired Officers Association Central Arkansas Chapter and State of Arkansas. His decorations include: Legion of Merit, Air Medal, Joint Service Commendation Medal, Air Force Meritorious Service Medal, and Air Force Commendation Medal.

DAVID D. WILSON, Colonel, (Ret.) from Ft. Worth, TX, was commissioned through the ROTC program on graduation in 1960, from TCU, and after pilot training at Bainbridge and Craig in 1961, he was assigned as a C-130 pilot in the 345 TCS at Sewart AFB, TN. Dave served as a pilot, maintenance officer and/or safety officer in C-130 and AC-130 unit at Naha, Lockbourne, Ubon, Pope and Rhein Main.

Dave served on the Air Staff from 1975-79, and on the MAC staff in Ops Requirements from 1981-82. He was honored to be the commander of four units during his career: 322 OMS 73-75; 772 TAS 80-81; 316 TAG 83-85; 34 TATG 85-87.

Dave is a command pilot with over 6,000 flying hours in the C-130. His decorations include the Legion of Merit with two oak leaf clusters, the Distinguished Flying Cross with one oak leaf cluster, the Air Medal with 14 oak leaf clusters, and the Purple Heart. Dave is married to he former Gayle Howe; they currently live in Abilene, TX, where he is a site and program manager for CAE-Link Corp.

SAMUEL A. WILSON JR., is the vice president for the Air Force/Command, Control, Communications, and Intelligence (AF/CI) business center of HFSI, a worldwide systems integrator based in McLean, VA. He is responsible for marketing and supporting computer solutions and services to the C3I and Air Force communities, evaluating and managing proposals, implementing programs, and developing new business initiatives. In this role, he is accountable for maintaining a balanced profit and growth perspective for one of HFSI's largest business entities, which includes a worldwide field organization serving the entire federal community as well as commercial accounts in the state of Hawaii.

Since 1963, Wilson has served in marketing and management positions with the information systems division of General Electric Company and later with HFSI, a former Honeywell subsidiary, where he distinguished himself as a program director for numerous federal procurements. From 1974 to 1983, he was instrumental in capturing more than $800 million in competitive federal contracts, including the Navy's vast shipboard non-tactical ADP program, SNAP.

The head of C3I operation for more than a decade, Wilson is largely responsible for HFSI's position as a top information systems vendor within the command and control community. He directed the company's integrated workstation campaign for the Defense Department's Worldwide Military Command and Control System, clearly establishing HFSI in the industry as a systems integrator. In 1990, he assumed control of Air Force operations, where he has promoted and improved HFSI's services business among key customer accounts. Most recently, he directed the company's successful bid for the $150 million Air Force integrated database machines contract.

Mr. Wilson holds a BS in civil engineering from Drexel University and an MS in industrial management from Purdue University. He is affiliated with numerous professional associations and is a member of the Armed Forces Communications and Electronics Association and the National Military Intelligence Association.

WARREN E. WITMER, born May 9, 1921, in Middletown, PA. He received his education at Annville High School , electronics fundamentals I & II. He joined the USAAFATC Aug. 29, 1942, serving with Technical Training Command and Air Transport Command. He served at Gulfport, MS; Yipsalanti, MI; Nashville, TN; and Sahmshernager, India.

He was discharged Nov. 14, 1945, with the rank of staff sergeant.

He and his wife, Betty J., have three children and eight grandchildren.

As a civilian he was employed with Air Division Center, as a electronics technician, retiring July 28, 1975.

CHARLES H. WITTROCK, "COKE", Colonel, born April 11, 1945, in Uvalde, TX; graduated Desert High School, Edwards AFB, CA, in 1963; earned a BS in aerospace engineering from Texas A&M in 1968, as a Distinguished Military Graduate; received an MS in counseling from Mississippi State. He is a Distinguished Graduate of Squadron Officer School and a graduate of ACSC and the National War College.

Col. Wittrock's service includes: UPT at Williams; flying C-130s in 47 TAS/313 TAW at Forbes; 776 TAS/374 TAW at Ching Chang Kang, Taiwan and Vietnam (over 360 combat hours); T-37s in 37 FTS/14 FTW at Columbus where he was SOF, flight commander, and chief of Wing Standardization/Evaluation; squadron commander at Keesler of 3401 and 3403 TTS; returning to C-130s in 1981 at Clark AB, PI in 21 TAS/374 TAW as instructor, tactics chief, chief pilot, assistant operations officer, and chief of Combat Operations; Little Rock as assistant commandant, instructor school, squadron commander, 16 TATS, and DO, 34 TATG. After National War College he was assigned to the AF secretariat in Acquisition as chief of Airlift and Training Division (C-17, AF-1, JPATS, Snowstorm). He returned again to C-130s in 1991 as DO, 317 TAW, Operations Group Commander, and Commander, 624th Airlift Support Group.

A command pilot with over 4,500 hours, his awards include the Legion of Merit, Air Medals and Meritorious Service Medal among others. He is married to the former Rosemary Essenwine and has one daughter, Kristin Dianne.

DOUGLASS G. WOOD, born Dec. 18, 1938, in Ft. Worth, TX. He graduated from Texas Military Institute in San Antonio, TX, in 1956, and graduated with a bachelor's degree in international relations from Lehigh University in 1961. He graduated from Squadron Officers School in 1966, Armed Forces Staff College in 1975, and the Air War College in 1979.

Wood retired from the Air Force in 1984. During his last assignment he served as Chief, Special Air Missions, Officer of the Vice Chief of Staff. Prior to that he was Commander, 14th MAS, Norton AFB, CA. Wood 20 years in MATS and MAC, five of which were in the 89th MAW, where he flew C-140 aircraft in support of the President and Vice President. With over 6,000 hours flying time, Wood flew C-124s, C-141s and C-140s. He is currently vice president, military marketing, Gulfstream Aerospace Corporation.

Wood and his wife, Glenn, reside in Alexandria, VA. They have two children, Doug Jr. and Chris.

JOHN D. WOODRUFF, Colonel, born in Bonham, TX, on Feb. 12, 1944. He graduated from Bonham High School in 1962. He received a BS degree from East Texas State University and an MS degree from the University of Southern California. He attended the Air Command and Staff College and U.S. Army War College in residence.

Col. Woodruff entered the Air Force through AFROTC in 1966, as a distinguished graduate. In 1967, after completing the Air Intelligence Course, he served in the intelligence field at 15 Air Force, March AFB, CA; Task Force Alpha, Nakhon Phanom, Thailand; and HQ USAFE, Wiesbaden AB, GE.

In 1971, after completing UPT, he was assigned to the 41 TAS, 317 TAW at Pope AFB, NC, where he advanced to aircraft commander, IP, and worked in Current Operations. In 1977, he was assigned to AFISC as a C-130 Safety Action Officer. At Dyess AFB, TX, from 1982-85, Col. Woodruff served as 772 TAS Operations Officer, 773 TAS Commander, then Assistant DCM and DCM, 463 TAW.

At the Pentagon, from 1985-87, Col. Woodruff served as Chief, Office of Inspection, Office of the Inspector General, HQ USAF and OSAF. From 1987 to 1993, he served as Vice Commander, 41 RWRW at McClellan AFB, CA; Vice Commander, Air Rescue Service; then Commander, Air Rescue Service. In 1993, Col. Woodruff assumed duties as the Vice Commander, 314 AW, Little Rock AFB, AR.

Col. Woodruff's Military decorations include the Legion of Merit and the Meritorious Service Medal with three oak leaf clusters. He is a command pilot with over 3,900 flying hours in six different aircraft.

Col. Woodruff is married to the former Carol Thompson of Bonham, TX. They have three children and one grandchild.

MIKE WOOLEY, Colonel, Michael W. Wooley was born Aug. 29, 1950, in Natchitoches, LA. He graduated from Ouachita High School in 1968, and holds a BBA from Northeast Louisiana University and MA from Webster University. Professional military education includes Squadron Officer School, Air Command and Staff College, Air War College, and The Industrial College of the Armed Forces.

He received his commission from Officer Training School in 1972. Colonel Wooley was a distinguished graduate from UPT and assigned to fly C-141s. He was an instructor pilot in the 20th MAS, Operations Officer in the 41st MAS, and reactivated and commanded the 17th MAS. He has held staff positions at HQ Military Airlift Command, the Pentagon, and U.S. Forces Korea.

His decorations include the Defense Meritorious Service Medal, Meritorious Service Medal with two oak leaf clusters, and the Air Medal.

Col. Wooley and his wife Kathy have two daughters, Amber and Kristy.

SCOTT W. WUESTHOFF, Lieutenant Colonel, is a native of San Francisco, CA. He entered the Air Force through the AFROTC program in 1980 and went directly to pilot training at Williams AFB. After flying KC-135s at Fairchild AFB and KC-10s at Barksdale AFB, he was reassigned to the Pentagon as an ASTRA officer in SAF/PA. Following his next tour at March AFB in KC-10s, which included time on the squadron, 15 AF, and Gulf War planning staff, he PCS'd to Maxwell AFB for both ACSC and SAAS. Eager to return to the "real Air Force," he ended up on the AMC staff at Scott AFB as Assistant Chief, Commander's Staff Group, and then his present job as executive officer to the Vice Commander, AMC. LTC Wuesthoff is married to the former Jodi Percell of Denver, CO.

MORLEY E. YAGEN, born July 16, 1922, at home in Erie, ND. He was the son of the village barber and a local farmer's daughter. He was graduated from Erie Public High

School in May 1940. He enlisted in the U.S. Army in July 1940 and was immediately assigned as battery clerk for Battery B, 81st Field Artillery, Ft. Lewis, WA. Here he frequently saw Gen. David Eisenhower (lieutenant colonel then) and Gen. Mark Clark (captain then). The 4th Infantry Regiment and the 81st Field Artillery Battalion were first combat team formed for WWII. They were ordered to Ft. Richardson, AK, June 26, 1940, under the command of Gen. Simon Bolivar Buckner, III (colonel then).

On June 23, 1942, following the bombing of Dutch Harbor, AK, Gen. Buckner commandeered nine Northwest Airline planes and ordered the 4th Infantry and 81st Field Artillery personnel flown to Nome, AK. This was the first combat air movement of the U.S. Army. The mission was only slightly edited and later formed the basis for creation of the Military Airlift Command (MAC). In December 1942, Yagen was appointed warrant officer at 20 years old and assigned to the post engineer's office on Amchitka Island in the Aleutian Chain. His concluding assignment was as administrative officer in the 1412th Base Engineer Depot, Saipan Island, Mariannas Group in October 1945. In January 1946, he enrolled in a liberal arts course at the University of Iowa. Dis-satisfied with civilian life, Yagen requested enlistment in the Army Air Force in October 1946. He was accepted for enlistment in grade of technical sergeant and assigned to the Air Provost Marshal General's office at the Pentagon. In October 1948, he volunteered for duty with the Berlin Airlift and was assigned to Celle, Germany in the British Zone of Occupation. Following the end of the airlift Yagen was called to active duty and served as non-appropriated welfare fund officer at Furstenfeldbruck, Germany under command of Gen. Robert Scott. Following this active duty Yagen reenlisted as master sergeant and was assigned to U.S. National Military Representative Office, SHAPE, Paris France. He retired as first sergeant, Field Maintenance Squadron, 4039th Strategic Wing, (re-numbered 461st Bomb Wing) on Oct. 31, 1961.

Yagen and his wife, Christa, have two sons, Arnold and Walter. Walter is a senior at the U.S. Merchant Marine Academy, Kings Point, NY.

JEROME YARCHEVER, born March 11, 1934, in Pittsburgh, PA. He graduated high school in Brooklyn, NY. He completed Air Command and Staff College in 1972.

After enlisting in the Air Force in 1952, he was commissioned through the Aviation Cadet Program in 1954. In 1969, he entered the Air Reserve Technician Program and in 1976, was selected as one of the first navigators to command a C-141 flying squadron.

Upon promotion to full colonel, he accepted assignment as Reserve Forces Advisor to the 21st Air Force Commander.

In 1984, he retired while serving as vice commander of the 315th MAW Charleston AFB, SC.

His decorations include the Legion of Merit, Air Force Outstanding Unit Award, Air Force Good Conduct Medal, and Air Force Reserve Longevity Award.

Col. Yarchever and his wife, Diane, reside in Myrtle Beach, SC. They have three children, Denise, Randy and Dana and one granddaughter, Taylor.

ELMER R. ANDREWS, CMSgt., USAF (Ret.) was born April 1, 1925 in Lima, OH. On Feb. 5, 1943 he enlisted in the Regular Navy at age 17. In May that year he graduated from Lima South High while on Navy leave. He flew as a top turret gunner in TBM "Avenger" torpedo bombers in the Pacific during WW II. After the war he was a Plane Captain at NAS Anacostia, DC and the carrier USS *Leyte* until his discharge as an Aviation Machinists Mate 2nd Class on April 20, 1948. Enlisted in the USAF on May 5, 1948 as a Staff Sergeant at Bolling AFB, where he became a flying crew chief on C-45s.

In 1953 he completed Flight Engineer training at Chanute AFB and was assigned as a C-124 Flight Engineer in the 8th TCS at Larson AFB, 15th ATS at Dover AFB and 48th ATS at Hickam AFB. Promoted to CMSgt. on March 1, 1961 he was assigned as Chief Flight Engineer of the 44th and 307th Air Refueling Squadrons at Selfridge AFB, flying KC-97G tankers. He was organizer and charter president of Chapter 768 at Selfridge in 1963. At the 1st AFSA Convention in 1964, he was elected First Trustee on the International Council, and he served as International President from 1965 to 1968. He was AFSA's first Senior Advisor and he spent four years as Chairman of the board until 1974. He has been on the AFSA Past President's Council since its inception. He was the Coordinator and President pro-tem of Division Five in 1977 and managed the first 15 Division Conventions. He received his bachelors degree from the University of Omaha in 1962, From 1966 to 1973 he served at Dover AFB, DE as Chief Flight Engineer in the 9th and 20th MAS, flying C-141A Starlifters, then as CMSgt. of the 436th MAW. He retired on June 1, 1973 with 30 years service. His wife, Audrey, was the first three term International Auxiliary President. Son, Jonathan Andrews, served as AFSA Chapter President at Clark AB and Dover AFB and a term as International Active Duty Trustee. Daughters Sherren and Cathy are Life Members of the Auxiliary. He was instrumental in organizing Chapter 564 at Port St. Lucie, FL in 1977 where he and Audrey still reside.

THOMAS M. GLASSBURNER, SR. was born on Jan. 19, 1917 in Conway, PA. He attended the University of Maryland for two years. He joined the USAF on May 13, 1942. Military locations and stations include: SA, MAC, Training Command. He flew most planes and has 25,000 flying hours. He was the Commander of the 20th Military Airlift Squadron in 1966.

A memorable experience includes one dark and stormy night when he grabbed a hold of the Ben Gay and thought he had lubricant.

Awards/decorations include: Air Medal, 13 clusters. He achieved the rank of colonel and was discharged on July 1, 1971.

He is married to Flora Margaret Glassburner, they have two children and four grandchildren. He is a real estate broker.

SAM McGOWAN, is a veteran airlifter and aviation writer currently living in Houston, TX. McGowan served 12 years in the Air Force, with 11 of those years as an aircraft loadmaster on C-130s, C-141s and C-5s. He began his airlift career at Pope AFB, North Carolina as a flight line mechanic before cross-training into the loadmaster field in 1964 at age 18. McGowan served with the 779th TCS at Pope, the 35th TCS as Naha, Okinawa, the 58th MAS at Robins AFB, Georgia, the 29th and 773rd TAS and the 3rd MAS at Charleston and Dover.

McGowan's airlift career included special airlift operations on flare missions aover the Ho Chi Minh Trail and North Vietnam and leaflet missions over North Vietnam and North Korea. He flew more than 50 bombing missions in 1969-1970 in Operation Commando Vault. McGowan holds the Distinguished Flying Cross and the Air Medal with 12 Oak Leaf Clusters. Today he is a corporate pilot based in Houston and a free lance aviation writer and author.

MARK A. SMITH, was born Nov. 17, 1957 in Hickory, NC. He entered the Air Force as an Aircraft Loadmaster in 1977 flying C-141A aircraft at Charleston AFB, SC. In 1980, he was selected as one of six initial cadre loadmasters to begin flying C-141B aircraft in covert operations. (Special Operations Low Level), in preparation for the American rescue mission in Iran. In 1983, SSgt. Smith was selected as one of the "12 Outstanding Airmen in the United States Air Force." In 1985, he transferred to Wright-Patterson AFB, OH where he served as the only C-17 System Program Office Loadmaster. MSgt. Smith was responsible for design, development, and acquisition of cargo compartment systems and mission equipment for the $37 billion C-17 weapon system.

SMSgt. Smith deployed as a C-141B loadmaster during Desert Shield/Storm and upon return proposed to his girlfriend, Cynthia of West Milton, OH. In 1991 Chief Master Sergeant Smith transferred back to Charleston AFB, SC to begin flying the C-17 as an initial cadre loadmaster. Chief Smith is presently the Wing Chief C-17 Loadmaster at Charleston. Mark and Cynthia have a one year old son, James Anthony.

INDEX

Symbols

A

B

C

I

J

K

L

M

N

O

P

Q

R

S

T

U

V

C-124/Thor over SanFrancisco, CA.

C-141B STARLIFTER
BACKBONE OF U.S. STRATEGIC AIRLIFT

GENERAL CHARACTERISTICS

Engines (4) Pratt & Whitney	TF33-P-7
Engine Thrust	21,000 pounds
Length	168.3 feet
Height	39.3 feet
Wing Span	160.0 feet
Wing Sweep	25.0 degrees
Range With Maximum Payload (94,508 pounds)	2,501 nautical miles
Ferry Range	5,551 nautical miles
Takeoff Distance Over 50 Ft. (Gross Weight)	6,800 feet
Landing Distance From 50 Ft. (Design Landing Weight)	3,750 feet
Maximum Cruise Speed	491 knots
Long Range Cruise Speed	430 knots
Rate of Climb (Maximum Gross Weight)	2,600 feet per minute

WEIGHTS SUMMARY

Maximum Ramp Weight (2.25g)	344,900 pounds
Design Landing Weight (10 ft/sec sink rate)	257,500 pounds
Fuel Capacity	23,592 gallons
Operating Weight Empty	144,492 pounds
Maximum Payload (2.25g)	94,508 pounds

CARGO COMPARTMENT

Length (including ramp)	104.4 feet
Length (excluding ramp)	93.3 feet
Width	10.3 feet
Height	9.1 feet
Total Cargo Volume (including ramp)	11,399 cubic feet

Lockheed Aeronautical Systems Company

86 South Cobb Drive, Marietta, Georgia 30063

C-5 GALAXY GLOBAL REACH, HEAVY AIRLIFT CAPABILITY

C-5 GALAXY GLOBAL REACH, HEAVY AIRLIFT CAPABILITY

GENERAL CHARACTERISTICS

Length	247.8 ft
Height	65.1 ft
Wing Span	222.8 ft
High Speed Cruise Speed	0.79 Mach
Long Range Cruise Speed	0.77 Mach
Fuel Capacity	332,500 lb
Engines (4) General Electric	TF-39-1C Turbofans
Engine Thrust	41,100 pounds of thrust

WEIGHTS SUMMARY

	C-5A lb	C-5B lb
Operating Weight	370,000	374,000
Max. Design Gross Takeoff Wt. (2.5g)	769,000	769,000
Max. Design Gross Takeoff Wt. (2.25g)	837,000	837,000
Maximum Payload (2.5g)	220,000	216,000
Maximum Payload (2.25g)	265,000	261,000
Max. Design Landing Wt. (9fps sink rate)	635,850	635,850
Max. Design Landing Wt. (6fps sink rate)	920,000	920,000

CARGO COMPARTMENT

Length (including ramp)	144.6 ft
Length (excluding ramp)	121.1 ft
Width	19.0 ft
Height	13.5 ft
Total Volume (including ramp)	34,795 cu ft

Lockheed Aeronautical Systems Company

86 South Cobb Drive, Marietta, Georgia 30063

U.S. AIR FORCE C-130H HERCULES

The C-130H ~~is the latest model~~ Hercules. Many new design changes have been incorporated into this Hercules model which improve the maintainability, reliability, and performance of the airplane.

More than 40 different versions of the C-130 series airplane have been built and over 2000 have been sold to the U.S. Government and 63 countries throughout the world.

Weights	
Operating Weight	81,546 lb
Maximum Payload, 2.5g	40,174 lb
Maximum Takeoff Weight (MTW) , 2.5g	155,000 lb
Design Landing Weight (DLW)	
Normal - 9 fps sink	130,000 lb
Maximum - 5 fps sink	155,000 lb

Overall Dimensions	
Length	97.8 ft
Wing Span	132.6 ft
Height	38.1 ft

Power Plants

Engines - Four Allison T56-A-15, 4508 eshp

Propellers - Four Blade, 13.5 ft diameter, Hamilton Standard

Auxiliary Power Unit - Supplies air for engine starting and ground air conditioning; also drives 40 KVA generator and is operable in-flight

Fuel (Total usable with reticulated foam installed)	
Internal Wing Tanks	6,820 U.S. gal
External Pylon Tanks	2,800 U.S. gal
Total Capacity	9,620 U.S. gal

Personnel Capacity	
Crew	5
Troops	92
Paratroops	64
Litters (plus attendants)	74+2

Cargo Compartment	
Length	40.4 ft
Width	10.0 ft
Height	9.0 ft
Ramp	10.0 ft (W), 10.25 ft (L)
Total Volume (incl. ramp)	4,500 cu ft

Performance (Std. Day Conditions)	
Range with 38,536 lb payload	2,000 nm
Ferry Range	4,400 nm
Takeoff Distance (MTW)*	5,142 ft
Landing Distance (DLW)*	2,405 ft
Maximum Speed	325 knots
Cruise Speed	300 knots

*Takeoff and landing distance over a 50 ft obstacle.

FROM REGIONAL CRISIS TO PEACEKEEPING AND HUMANITARIAN RELIEF THE C-17, AIR MOBILITY FOR A CHANGING WORLD

The C-17, our nation's newest airlifter, is undergoing very successful flight testing at Edwards Air Force Base in the California desert. This quality aircraft, designed to be the cornerstone of America's global reach well into the 21st century, is capable of delivering outsize cargo over long distances into small austere airfields.

The C-17 combines the best attributes of today's airlift fleet with proven state-of-the-art technology to support our air mobility needs in a changing world. The C-17 will fly cargo and troops over intercontinental distances for airland or airdrop delivery. It will also provide needed humanitarian relief and aeromedical airlift, augment special operations, and move United Nation's peacekeeping forces to troubled spots throughout the globe.

Following extensive development and operational testing, the first C-17 squadron will begin initial operations at Charleston Air Force Base, South Carolina, in 1993.

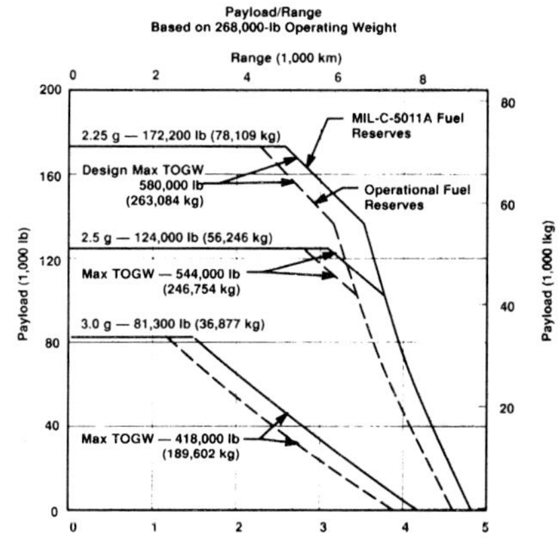

C-17

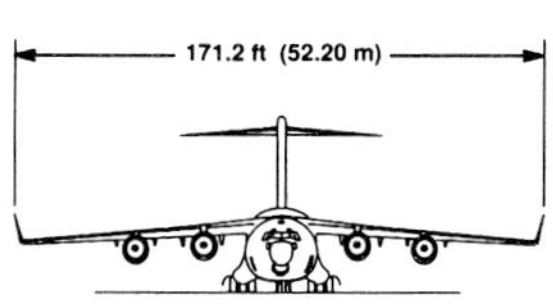

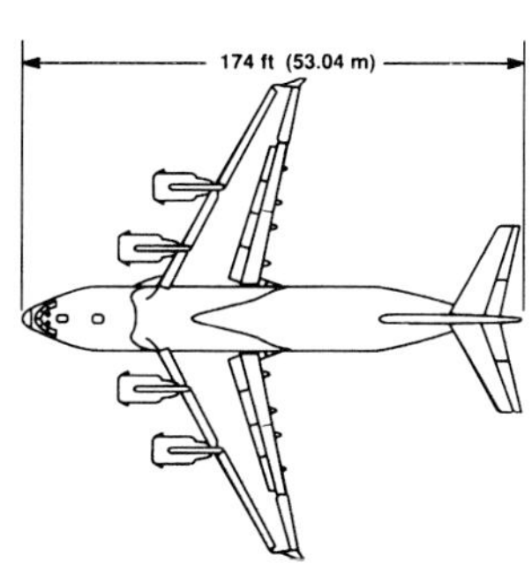

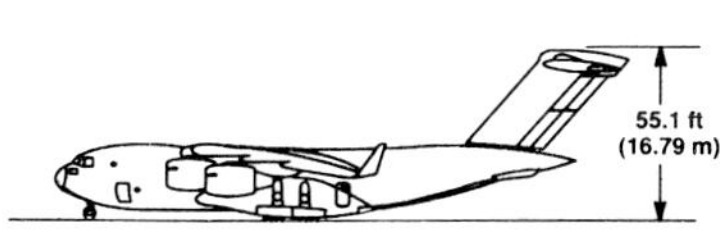

Max Takeoff Gross Weight	580,000 lb	(263,084 kg)
Max Payload	172,200 lb	(78,109 kg)
Max Fuel Capacity	176,200 lb*	(79,923 kg)
Service Ceiling	45,000 ft	(1,566 m)
Cruse Speed	0.77 Mach	
Wing Sweep	25 degrees	
Wing Area	3,800 ft^2	(353 m^2)
Engines	F117-PW-100	
Thrust Rating	40,700 lb	(181 kN)
Cargo Compartment Length	88 ft	(26.8 m)
Cargo Compartment Width	216 in.	(5.48 m)
Cargo Compartment Height	148 in.	(3.76 m)

*27,100 gallons at 6.5 lb/gal

USAF/MCDONNELL DOUGLAS C-17

LIFELINE TO THE FRONT LINE

KC-10A EXTENDER — THE USAF SUPERTANKER

The KC-10 Extender is an advanced tanker/cargo aircraft designed to provide rapid, worldwide force projection for the armed services. In addition to providing a major increase in air refueling capability, the KC-10 is a versatile personnel and cargo carrier. It deploys fighter aircraft, including support equipment and personnel, nonstop to any point on the globe. It increases the payload/range of cargo airlifters with air refueling. And it augments the airlift fleet by carrying cargo and personnel.

A military derivative of the commercial DC-10 Series 30 convertible freighter, the KC-10 is powered by three General Electric CF6-50C2 turbofan engines. All non-essential systems and equipment were removed from the DC-10 and those necessary for the KC-10 mission added. These include additional fuel tanks in the lower fuselage compartments, advanced air refueling boom, independent hose reel/drogue system, air refueling operator station, an air refueling receptacle, and military avionics.

By employing its refueling boom or hose/drogue system, the KC-10 can refuel a wide variety of U.S. Air Force, Navy, and Marine Corps aircraft as well as aircraft from allied nations. It can refuel both receptacle-equipped and probe-equipped aircraft on the same flight.

Unexcelled Reliability and Maintainability

	Goal (%)	Achieved (%)
Fully Mission Capable Rate	70	95
Mission Completion Success Rate	96	99.95
Contractor Logistics Support	Rated "OUTSTANDING" by USAF	

Engine Thrust (Each)	52,500 Lb
Maximum Takeoff Gross Weight	590,000 Lb
Maximum Fuel Capacity	356,065 Lb
Operating Empty Weight	241,027 Lb
Maximum Cargo Payload	169,409 Lb

1 Air Refueling Receptacle

- Allows the KC-10 to be refueled by other tankers

2 Cargo/Passenger Versatility

- Entire main deck available for various cargo and/or passenger configurations
- Carries up to 27 463L cargo pallets in the all-cargo mode
- Or 20 passengers and 23 cargo pallets
- Or 75 passengers and 17 cargo pallets

3 Fuel Tanks

- Fuselage bladder fuel tanks for maximum fuel payload
- Isolation capability for two fuel types

4 Independent Hose Reel/Drogue

- 600 gallons per minute maximum flow rate
- Automatic tensioning
- Surge protection

5 Advanced Air Refueling Boom

- Increased length and larger maneuvering envelope
- 1,500 gallons per minute maximum flow rate
- Active "fly-by-wire" control
- Independent disconnect
- Automatic load alleviation

TANKER
Fuel Off-load (1,000 Lb)

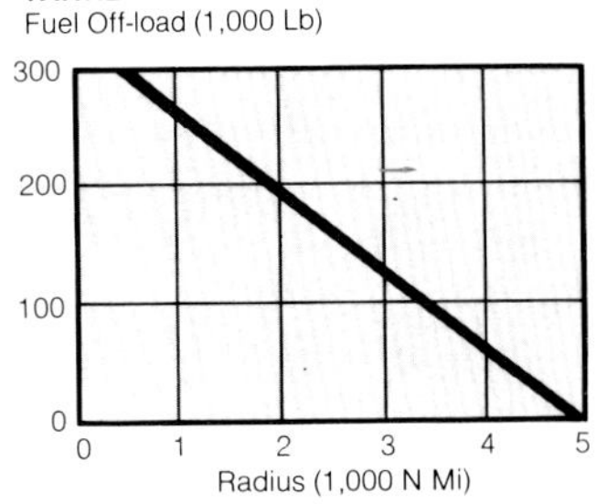

6 Air Refueling Operator Station

- Comfortable seated positions for boom operator, instructor, and observer
- Large viewing window
- State-of-the-art controls, instruments, and displays

CARGO
Payload (1,000 Lb)

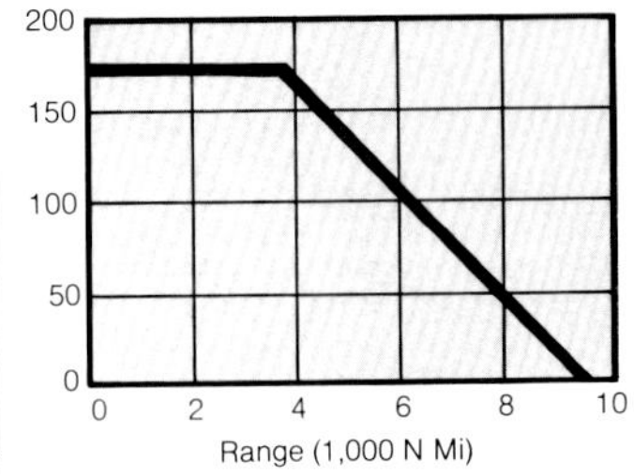

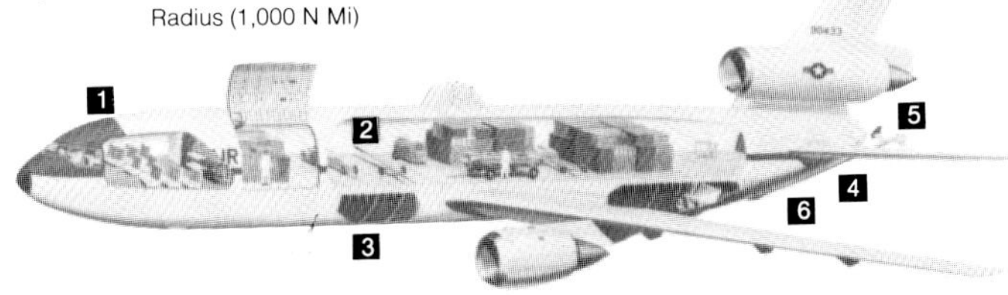

Printed in U.S.A. B87-1722 11-87

U.S. AIR FORCE